Tourism in China

Routledge Advances in Tourism

Tourism in China

Destination, Cultures and Communities

Edited by Chris Ryan
and Gu Huimin

Routledge
Taylor & Francis Group
New York   London

First published 2009
by Routledge
270 Madison Ave, New York, NY 10016

Simultaneously published in the UK
by Routledge
2 Park Square, Milton Park, Abingdon, Oxon OX14 4RN

Routledge is an imprint of the Taylor & Francis Group, an informa business

© 2009 Taylor & Francis

Typeset in Sabon by IBT Global.

Library of Congress Cataloging in Publication Data

Tourism in China : destination, cultures, and communities / edited by Chris Ryan & Gu Huimin.
p. cm. -- (Routledge advances in tourism ; 14)
Includes bibliographical references and index.
1. Tourism--China.   I. Ryan, Chris, 1945–   II. Huimin, Gu.

G155.C6T68 2009
338.4'79151--dc22   2008025972

ISBN10: 0-415-99189-7 (hbk)
ISBN10: 0-203-88636-4 (ebk)

ISBN13: 978-0-415-99189-6 (hbk)
ISBN13: 978-0-203-88636-6 (ebk)

Contents

For
Dragon and Phoenix

List of Tables

List of Figures

Preface

This book was motivated by a wish to introduce not only issues of tourism development and destination management to western scholars, students and those interested in China and its tourism, but where possible to also introduce the work of Mainland Chinese colleagues who rarely obtain a chance for their work to be published in English. This is not to deny the significant and at times insightful work produced by many colleagues of Chinese ethnicity working in western universities, and they have acted as a bridge by which many western academics have been able to visit and learn about China. However, it is also true that many of these overseas Chinese scholars have not worked in the Chinese tourism industry or taught for any length of time in Chinese Universities. This is not stated as a criticism, but as a simple fact, but, as is ever the case, to be an outsider is not to have experienced full immersion in the daily realities of the situations of others. For that reason we wished to include work by our Mainland colleagues to better capture the nuances and insights that they alone can possess. To be honest we have not been wholly successful in this aim, but nonetheless we feel that we have at least caught a flavour of the work being undertaken by our Mainland colleagues, and the book is probably richer for including the work of western scholars like Geoff Wall, Trevor Sofield, and Wolfgang Arlt, to name but a few, who have worked in China for at least a decade.

Indeed, these liaisons between Mainland academics and their Chinese colleagues based in western universities, or work undertaken with western colleagues are important to Mainland scholars for a number of reasons. One major reason is the lack of resource that is available to so many of our PRC colleagues. Access to state-of-the-art computing facilities is not always present, and while this can be a frustration, of more importance perhaps is the lack of easy access to the databases of journals, statistics and books that western colleagues tend to take for granted. Consequently these exchanges enrich both groups of scholars. Equally our Chinese colleagues can access materials not easily available to us in the west.

As always though, cross-cultural work is prone to the mistaken phrase or lost nuance, and both of us have been aware of this as we have sought to translate or in some instances work from 'Chinesei' transcripts. Even though our contributors have all been shown and approved the chapters we recognize that in some cases issues of language lay in their reading of our attempts to catch the sense of their work, and thus if we have mistakenly misconstrued a sentence or a meaning, we accept that responsibility, seek understanding that we did try to get it right, and perhaps take some pride that our work can now be corrected and in that sense future contributions to knowledge and understanding can occur.

A second motivation for the book was an appreciation of the fact that within China tourism is heavily contextualized within broader economic and social policies, and thus any text that concentrates on tourism to the exclusion of the wider picture would therefore be incomplete. We have therefore tried to provide that wider framework and have sought to locate the contributions within wider policies and to make more explicit the relationships between tourism and the drive to lift millions from poverty, and the context of socialist market economies with a Chinese perspective and the nuances of Confucianism that help make China the society that it is.

Another issue of concern to the editors has been the choice of illustrations and photographs. Why this concern? Many readers will note that they are, for example, of indigenous peoples within China, or of aspects of China's cultural heritage. A problem with this, as with all tourism, is that it can be interpreted as signalling a form of neocolonialism where that perceived as being of interest to the western gaze is decontextualized, framed and shown for reasons of curiosity without reference to the norms and values of those being illustrated. The photographs have been selected on the premise of illustrating points within the text, and are to be framed within the text which does attempt, however imperfectly, make explicit such framing. It is recognized that different interpretations can occur, and the editors and authors can only state that the process has been reflexive, and seek tolerance from those who may have different opinions.

Finally, any book owes much to many people. We would wish to thank Professor Stephen Page who believed in this project and introduced us to our publishers. They too we thank, for to state the obvious, without Benjamin Holtzman and Jennifer Morrow of Taylor and Francis, there would be no book. We obviously owe a tremendous debt to our different contributors, and thank them for their patience in the front of our perhaps at times silly questions as we chased a reference or misread a script. We hope that they believe we have created a framework within which their work sits comfortably and perhaps is enhanced by the juxtaposition of the different chapters and themes we have sought to identify. We owe debts to our own colleagues who by their tolerance give us the space within which to find time to write this book, and of course, we would wish to thank our own families and friends for their understanding of our own need to complete this work.

There are also some formal acknowledgements that need to be made. Four of the chapters appeared in the journal *Tourism Management* and we thank the editor for permission to reproduce these and to amend them slightly so as to better fit within the book's format. The chapters concerned carry an acknowledgement at the end of their text. Many of the other chapters initially appeared as conference papers at different Asia Pacific Tourism Conferences and that organised by Professor Wu Bihu of Pekin University in 2006. In most cases they have been substantially amended in editing.

It has been a privilege to work with so many in completing this book and we hope that it will provide a source of information and ideas for those interested in the development of China's tourism in the early years of the twenty-first century.

Chris Ryan
Gu Huimin
Beijing, China
Dubai, UAE and Hamilton,
New Zealand.
February 2008

1 Introduction

The Growth and Context of Tourism in China

Chris Ryan and Gu Huimin

INTRODUCTION

This book is about tourism in China and the context within which Chinese tourism develops—a context that is formed by a wish to lift many millions out of poverty, a need for economic growth, a political system that is in transition with a dismantling of statism to form a socialist market system with a Chinese perspective, and a culture still imbued with Confucian values yet influenced by Buddhism, Taoism and Maoism. To concentrate upon the purely touristic without reference to these wider frameworks would lead to an incomplete understanding of tourism development in China today. Such developments are seen 'on the ground', and thus this book concentrates upon destination developments and policies with reference to tourism.

The Chinese data relating to the growth of tourism are generally well known in terms of the rates being exceptional with the outcome that within a few decades China will not only be the major tourist generating country in the world but also the major tourism receiving destination. However, as noted in the following, much of the current tourism flows are intraregional, and thus some caveats have to be noted with reference to such predictions. A brief review of those data and the accompanying growth in hotel accommodation is provided in the following, but impressive though the figures are, the wider context needs to be established if one is to understand the motives that lie behind these developments, and their implications. For all of the high-rise architecture of Shanghai and Shenzhen, and the impressive buildings associated with the Beijing Olympics, China remains, at least for the moment, a developing country. While the populations of its major cities number in total about 76 million and thus exceed the populations of many countries, and its total urban population was estimated at 562 million in 2007, thus far exceeding that of the United States (approximately 305 million in 2007), in the rural areas almost a further 767 million people live, and for many of them the promise of economic development has yet to arrive. In 2005 the average annual disposable rural per capita income was 3,255 RMB, that is approximately US$400 a year. Not that all living in the urban areas enjoy much higher standards of living, for again in 2005 the mean average per capita urban disposable income was 10,493RMB (US$1,281). Yet any visitor to Shanghai or Beijing will see as many expensive cars as in London, Paris or any other major global city while the luxury car manufacturers see China as a market with an exciting future. The nature and speed of development in China, and the stresses it involves can be traced in the movement of population from the rural areas to the urban. In 2001, 64 per cent of the population lived in rural areas, by the end of 2006 that figure was 56 per cent (data derived from State Statistical Bureau of China—http://www.stats. gov.cn/english/). However, such changes are not solely due to rural emigration. There is also the physical enlargement of the cities themselves as in their growth they encroach upon once surrounding countryside.

THE CONTEXT OF TOURISM POLICIES IN CHINA

Poverty alleviation is thus a driving motive behind much of China's economic development and strategies. In 1996, at a speech at the Central Work Conference on Poverty Eradication on 23 September, the then premier, Li Peng, declared that the country would see an end to poverty in its rural areas by the end of the century. This objective meant the lifting of more than 65 million people out of poverty (Gustafsson and Zhong 2000). Gustafsson and Zhong proceed to note that poverty is very much associated with location. Poverty is very much associated with rural areas, with mountainous areas and being in the west. Wealth is associated with the cities of the Yangzhi River delta, Pearl River delta and eastern coastal areas.

The new affluence is thus unequally distributed, as are the opportunities to acquire such wealth. The reforms introduced after the Fall of the Gang of Four and the ushering in of the 'Open Door' policy has meant radical changes in the fortunes of many, but it is marked by unequal spatial, generational and educational opportunities. Younger people tend to gain advantage, and access to higher education becomes a key to affluence, thereby explaining the annual tension of university entrance examinations in China. Yet arguably it is the economic revival of the rural areas that helped China initiate the economic boom that is seen in the early decade of the twenty-first century. An important mechanism ushered in by the reforms enacted between 1978 and 1983 was the end of collective farming and the restoration of private ownership of land (Peng 1999), although land can still remain in village collective ownership effectively controlled by local political structures. As described by Gu and Ryan in their chapter on Hongcun and Xidi, and Zhou and Ma in their chapter on Baiyang Lake, this collective village entrepreneurship and Township and Village Enterprises (TVEs) (or *xiangzhen qiye*) came to be important in the development of rural-based tourism initiatives. Indeed, more generally, as described in Chapter 15, the TVEs were an important contributor to GDP until the latter part of the 1990s when banking practices began to change. However, Peng (1999) argues that the land reforms tended to produce a one-off economic impetus because although rural production rose, food prices tended to remain relatively flat for long periods with exceptions as in 1987–1989, and again in 2006–2007 when pork production was affected by swine fever and inflationary pressures began to be exerted in the Chinese economy. Given these slow increases in rural incomes, it is not surprising that small-scale tourism rural enterprises emerged based primarily on farm stays and bed and breakfast accommodation that had easy access to major urban populations. Important keys that account for the success in the removal of so many people from poverty include migration to the cities and expatriation of monies back to rural family members as in the population rural–urban distribution noted earlier, and second, especially prior to 1995, the encouragement and emergence of rural enterprises as discussed in Chapter 15. Today China's domestic private sector or non-state sector is now used to include the following entities: (a) TVEs; (b) small-scale household enterprises or family businesses with fewer than eight employees (*getihu*); (c) private businesses with eight or more employees (*siren qiye*); (d) state-owned enterprises (SOEs) that have been privatized; and (e) publicly listed joint stock companies. Excluded from this definition are firms completely owned by investors from Hong Kong, Macau and Taiwan, and joint ventures formed by these investors with their partners in China (Li 2005; Tsui *et al.* 2006). Hence another contextual theme within which tourism development needs to be located is the dismantling of statism, as is discussed in Chapter 8. This is evident at many different levels. For example, at the time of writing the 11th Chinese People's Political Consultative Conference (CPPCC) was taking place (March 2008). In the government work reports by Premier Wen Jiabao and Jia Qinlin, the chair of the 10th CPPCC, several references are made to the development of a socialist market economy with Chinese perspectives, and the need for socialist responsibility on the part of private sector entrepreneurs.

With reference to the evolution of a socialist market economy with 'Chinese characteristics' a debate exists as to the real reasons for Chinese economic growth. Woo (1999) entitles competing schools of thought as the experimentalist and convergent, and comments past research has tended to find that economic growth almost always accompanies the dismantling of any centralized state. Woo's analysis is based in part on the performance of TVEs not categorized as being in the private sector, but one problem that is noted in the processes of classification is that categorization was and is not, in itself, always an accurate guide to the *de facto* situation of 'on the ground' ownership, especially in the period prior to about 1995. Chen (2000) draws upon the distinctions and interactions between *de jure* and *de facto* ownership of TVEs in his analysis of China's market liberalization. Be this as it may, given the lack of importance attributed to tourism prior to 1978 and the commencement of economic reform, the tourism sector was able to fully take advantage of the trend towards freedom from state control, but did so within the 'birdcage' of centralized planning. Tourism has thus gained from

Chinese entrepreneurialism within the Chinese socialist market system, and as will be discussed within the book, one key emergent theme has been a role assumed by rural communities, albeit one primarily motivated by a need for economic growth.

A further context is the recognition of minority peoples and religious minorities within China. It is part of the concept of a harmonious Chinese socialist state that these peoples are recognized and brought into a harmonious relationship with the Chinese state. Thus, as illustrated in this book, various initiatives exist to endorse tourism as a means of economic growth, but a key issue remains, namely that of recognition. The recognition of minorities such as the Bai or Sani peoples, or of the Buddhist and Taoist faiths, for tourism purposes possesses advantages for destination promotion and economic revival, but the aspirations of such peoples may not be constrained within a touristic–economic dimension alone. Any review of the literature of indigenous-based tourism in other countries reveals a linkage between tourism patronage and the use of tourism as a legitimization of cultural, economic, social and political aspirations (e.g. see Ryan and Aicken 2005). The Chinese government thus seeks to develop well-being for its various peoples as a whole, and to a western eye there may appear longer term difficulties as minorities may seek further rights beyond the purely economic, as became evident in Greater Tibet in March 2008. To argue from this that potential problems lie in the future becomes an evident pattern of potential discourse, but it needs to be remembered that Buddhism, Taoism and the different minority peoples share cultures that tend to the collective, and share a similar understanding of harmonious relationships. Given this, the official statements of harmonious development within a socialist market system becomes possibly less paradoxical, yet more nuanced, than would be understood in western societies, and also perhaps generates ways forward for peaceful development.

Another context that must be considered if one is to understand tourism in China are the very levels of economic growth that have been achieved. As Weymes (2007) describes, in 1983 a visitor to China would have found Beijing airport in desperate need of maintenance, a poor road connecting the airport to Beijing, and in Beijing itself, little traffic. In 2008 the city hosted the Olympic Games, state-of-the-art buildings existed and traffic congestion was a problem for the Beijing Olympic Games Organizing Committee. In 1978 there were only 137 star classified hotels in the country. Today more than 18,000 exist (CNTA 2007). For each of the last 30 years China has achieved GDP growth of over 9 per cent per annum. Foreign trade has averaged a 15 per cent annual growth rate over the same period, and the economy is now one of the largest in the world. Commentators speak of the twenty-first century as the Chinese millennium (Naisbitt 1995). China is, after all, the Middle Kingdom from the Chinese perspective, and thus will stand as the central point in a global network of business interests. One consequence of this is the growth of domestic and Chinese outbound tourism.

With reference to inbound tourism, in 1996 51.3 million visitors were hosted, and these visitors spent 10.2 billion RMB. By 2007 total tourism income exceed 1,000 billion RMB for the first time with an average increase of 22.6 per cent. In the same year the inbound visitors numbered 131.87 million and the first estimates of visitor expenditure indicated total spending of about 35 billion RMB (CNTA 2007). The main markets, (other than Hong Kong, Macau and Taiwan which continue to dominate the statistics), by numbers of visitors in 2007 were Japan (3.7 million), South Korea (3.9 million), Russia (2.4 million) and the United States (1.7 million) (CNTA 2007). Domestic tourism numbered 639 million traveller arrivals in 1996, and an estimated 1,610 million in 2007 involving an expenditure of 777.1 billion RMB. The pattern of growth has almost been unbroken, other than in 2003 with the crisis of SARS, but in 2004 growth rebounded quickly. With reference to outbound tourism, in 1995 4.5 million Chinese tourists travelled overseas, and a decade later that figure had increased to 31 million. In 2007 the figure had climbed to 40.95 million with an average increase rate of 18.6 per cent which made China the number one tourism generating country in Asia (CNTA 2007). Impressive though these figures are, predictions that China will be the leading tourism generating country by a given date need to be tempered by the actual patterns of travel. Much of this travel is accounted for by travel to Hong Kong (which in many instances are day trips), Macau and other bordering countries such as Russia. The longer stay visitors tend to travel on package tours, and these account for about one-quarter of all private trips (G. Zhang 2006). Indeed, Hong Kong and Macau together account for about 70 per cent of all outbound Chinese tourism, and other Asian nations account for about 20 per cent of that tourism. Only about 5 per cent of outbound Chinese tourists visit Europe, and approximately the same amount visit North America, and of these many are on official business. Another sector of importance here is a VFR market involved in visiting relatives who are studying overseas (see Table 1.1).

To summarize, economic growth has helped to create a marketplace for tourism. A newly emergent, younger, professional, often western-educated and urban-market segment has emerged that wishes to emulate the consumer spending of their western counterparts. This is, however, in spite of appearances, more than a simple replication of the purchasing of consumer goods, as the ties of family responsibilities remain high, and, especially for those who gain access to urban wealth from rural areas through the portals of higher education, an awareness of the sacrifices made by parents must weigh heavily on their minds. Confucian mindsets of respect and familial duty are still present in Chinese society, and the current middle-aged beneficiaries of the reform policies exist as a generation in transition that seeks the best for its children while recognizing the role of their parents. As stated in different chapters in this book, studies of Chinese visitors, both domestically and internationally, show a socio-economic bias in the samples. They tend to be under the age of 45 years, tertiary educated and coming from the

Table 1.1 Tourism Statistics in China

Year	Inbound Tourist Arrivals (10,000)	Inbound Tourist Receipts (US$100 million)	Domestic Tourist Arrivals (100 million)	Domestic Tourist Income (Yuan 100 million)	Outbound Tourists (10,000)
1978	180.09	2.63		18.40	
1979	420.39	4.49			
1980	570.03	6.17			
1981	776.71	7.85			
1982	792.43	8.43			
1983	947.70	9.41			
1984	1285.22	11.31	2.00		
1985	1783.31	12.50	2.40	80.00	
1986	2281.95	15.31	2.70	106.00	
1987	2690.23	18.62	2.90	140.00	
1988	3169.48	22.47	3.00	187.00	
1989	2450.00	18.10	2.40	150.00	
1990	2746.00	22.20	2.80	170.00	
1991	3334.98	35.12	2.97	187.50	
1992	3811.50	39.47	3.30	250.00	
1993	4152.69	46.83	4.10	864.00	374.00
1994	4368.45	73.23	5.24	1023.51	373.36
1995	4638.65	87.33	6.29	1375.70	452.05
1996	5112.75	102.00	6.39	1638.38	506.07
1997	5758.79	120.74	6.44	2112.70	532.39
1998	6208.91	124.21	6.93	2310.85	834.61
1999	7279.56	140.99	7.19	2831.92	923.24
2000	7830.98	162.24	7.44	3175.54	1047.26
2001	8901.00	177.92	7.84	3522.37	1213.31
2002	9791.00	203.90	8.78	3878.36	1660.23
2003	9166.21	174.06	8.70	3442.50	2022.19
2004	10900.00	257.39	8.80	4710.71	2885.30
2005	12029.00	292.96	12.12	5286.00	3102.63
2006	12494.00	339.49	13.94	6230.00	3452.36
2007	13200.00	419.00	16.10	7771.00	4095.00

Source: Adapted from CNTA, Yearly Tourism Statistic Book, 1979–2007.

upper-income groups in China. The social changes are, however, significant and commentators like Weymes (2007) observe distinct generational differences beginning to emerge.

TOURISM AS A CREATOR OF ECONOMIC GROWTH.

It has been noted that tourism in China has emerged as a result of economic growth. However, over the last decade the growth in domestic and inbound tourism has tended to be faster than the overall rate of economic growth with, for example, CNTO (2007) data on international arrivals providing evidence of fast growth rates, such as in 2004 with a growth rate of 47.87 per cent in the rebound from the SARS-affected tourism of the previous years. Indeed, for the period since 1978, international arrivals have exceeded 15 per cent per annum in 16 of the 28 years to 2005. Given this it can be asked if tourism itself has become a contributor to economic growth. This is an especially pertinent question if one of the motives for tourism policies is the relief of poverty and low incomes.

Many studies have found strong contributions to economic growth that emanate from tourism development. In a study of OECD and non-OECD countries Lee and Chang (2008) developed a model of economic and tourism growth based on cointegration tests. They concluded that 'unidirectional causality relationships exist from tourism growth to economic development in OECD countries, but bidirectional causality relationships are . . . found in non-OECD countries' (Lee and Chang 2008: 191). They also found that the relationships are weaker in the case of Asian than non-Asian countries. The work of Kim, Chen and Jang (2006) in Taiwan independently supports these observations, for they found a mutual bidirectional relationship between tourism and more general economic growth in the case of Taiwan. In a Spanish context Parilla, Font and Nadal (2008) examine the issue of perceived lower productivity rates in tourism and the degree to which these lower productivity rates might inhibit longer term economic growth. They find that specialism in tourism in areas of Spain like the Balearics and Canary Islands has been a driving force for development, but longer term economic growth is dependent upon linkages between the tourism industry and other economic sectors, and the nature of diffusion effects and technology transfers. They also note that the public sector, mainly education, research and infrastructures, is also a key to sustained increases in productivity. They conclude that 'the problem does not lie in the abundance of natural resources nor in specializing in tourism *per se*, but rather in the failure of economic agents to attend to the determinants of long-term growth. Thus, public policies must be aimed at palliating the lower growth in productivity that stems from the productive specialization itself and correcting the low propensity of economic tourism to address innovation and education' (721). There is reason to believe that as the Chinese economy matures the same issues will arise. Namely that tourism is a generator of economic growth, but its longer term capabilities and productivities rest on its incorporation into wider economic systems including those provided by the public sector.

INTRODUCTION TO THE BOOK

Arising from these contextual issues this book will consider problems and management practices in China with reference to destinations and their planning. The book is divided into three main themes, namely 'Destination Change and Planning', 'Destinations and Cultural Representations' and 'Community Participation and Perspectives'. Almost inevitably there is some overlap between these three themes, but the various contributions tend to possess different emphases that justify these distinctions. Each section is introduced by a separate chapter that picks up the contexts outlined here and which then serves to establish a context within which the following chapters provide case studies, illustrations and conceptual materials. A number of these chapters also provide further information about structural processes current in China.

The first section, 'Destination Change and Planning' comprises, after the introduction, two specific studies on China's National Parks. One relates to issues of efficient use of resources, and the role of proximity as creating external economies of scale, while the second utilizes theories of destination development and spatial patterns to illustrate how economic growth emanating from tourism growth can have potential negative environmental and social impacts. Other chapters pick up similar themes. Thus the changing retail patterns of Luzhi are described, while Chinese Overseas Town provides a very western-oriented example of property development that incorporates both residential, leisure and recreational facilities.

The second section on 'Destinations and Cultural Representations' contains a chapter on Huangshan that discusses the importance of cultural interpretation of parks from a Chinese perspective, and the issues that creates when seeking World Heritage Site status. The uses of culture as a 'pull' factor in attracting visitors to a destination, and the ways in which culture is manipulated form the subject matter of chapters on Qufu, Ganzi, and Wutaishan, while another chapter considers the messages enunciated by tourist guides.

The third section of the book considers the role of the community in tourism. Examples are provided from Hongcun, Xidi, Baiyang Lake, and villages within wider Beijing as well as from a Beijing hutong.

As noted, throughout the text, cross-references can be found, and those references also refer to structural aspects of Chinese tourism including, for example, the role of TVEs and the modes of site classification adopted by the Chinese tourism authorities.

Part I

Destination Change and Planning

2 Destination Planning in China

Chris Ryan, Gu Huimin and Fang Meng

INTRODUCTION

'Tourism planning' as a term is often found in the academic tourism literature, yet in practice as Zahra and Ryan (2005) point out, in western societies it is more articulated in word than actually practiced. Indeed Stokes (2007: 261) has questioned the degree to which planning in tourism is truly strategic, concluding that 'Commercial realities and time-frames involved in event acquisition were top-of-the-mind more than the opportunities afforded by grassroots event development or expansion'. In the western democracies a tension exists between the public and private sectors. The former possesses, ideally, concerns of social well-being and a wish to not simply minimize the worst excesses of tourism on natural and social environments, but wherever possible to use tourism to the betterment of those environments. Tourism has thus aided the recovery of previously degraded waterfront areas such as the Sydney Harbour Rocks. It has generated income and employment among peoples previously marginalized by the economic and political mainstreams, and in many instances has generated economic benefits by reinvigorating interest in minority cultures for all of the criticisms that are often voiced about commodification of those cultures. On the negative side of the accounts, tourism has been blamed for (a) overcrowding that degrades natural environments, (b) threatening the habitats of wild life, (c) emasculating cultures to be but performances devoid of meaning, and (d) negatively impacting on patterns of community life. Both sides of the equation are used by proponents of planning to justify the role of the public sector as a means of addressing what are perceived as examples of market failure. On the other hand, the western democracies tend to equally constrain the role of the state, arguing that a number of functions are best left to the marketplace to ensure an efficient distribution of resources through the forces of competition, the marketplace and demand and supply. Hotels, theme parks, resort complexes, attractions, festivals and events—all are often managed in the private sector, albeit subject to limitations imposed by the public sector. Thus hotels may require planning permission to build in certain locales, may be subject to requirements as to disposal of waste water, provision of car parking facilities, or amounts of green space required around a hotel, but once built its management, operations and levels of service are left to private enterprise that seeks a rate of return to ensure that such operations are profitable, thereby permitting future refurbishment and development. Planning is therefore in a continuous state of tension between competing rights—the rights of communities to have their environment protected as far as is possible, and the right to have access to the highest levels of service and product that can be profitably developed. The private entrepreneur has rights to use resources in manners consistent with public policy to develop products and profits and so meet the needs of the community within a market system of demand, supply and prices. Therefore, in western societies, destination planning is never total, is always constrained by the rights of the private sector and is often subjected to short-term time-frames induced by a political framework of elections that are essential to the notion of democracy.

The one exception where the public sector, in western societies, may hold total sway in being both provider of services and protector of the environment lies in the provision of conservation zones or National Parks. Selected for reasons of natural or human heritage values, the marketplace is largely excluded in attempts to sustain and maintain an environment thought better left free from development.

From these somewhat generalized descriptions it becomes immediately evident that (a) exceptions can be found to these scenarios within the western democracies and (b) they do not wholly lend themselves to the Chinese environment. Consequently the remainder of this chapter will first concern itself with a closer examination of what is understood by the principles of tourism planning, and then examine how these might apply within the Chinese context before outlining the contributions made by the following chapters in describing the application of destination and tourism planning in the Chinese environment.

DEFINING TOURISM PLANNING.

Wang and Wall (2009) provide a review of a series of definitions of planning derived from earlier publications including those of Gunn (1988), Inskeep (1991) and Hall (2008). Thus for example, Evans (2000: 308) defined planning as 'a process considering social, economic and environmental issues in a spatial context in terms of development, conservation and land use'. Timothy (1999: 371) argued tourism planning 'is viewed as a way of maximising the benefits of tourism to an area and mitigating problems that might occur as a result of development'. An influential approach towards tourism planning was developed by Inskeep (1991)—influential not only in influencing writers but also, to at least some extent, policymakers. Thus, for example, in 1997 the Organization of American States, in its policy document, *Sustaining Tourism by Managing its Natural and Heritage Resources*, borrowed heavily from Inskeep's principles, while Augustyn (1998) argues that the same principles are implicit in Polish planning of rural tourism. Citing the former document, underlying principles of tourism planning include:

1. The first requisite is to identify, list, assess and develop as many visitor attractions as possible that have roots in the local community or complement local activities. Local culture and heritage, if properly developed, can improve the overall ambiance of the area and also add to the quality of community life. At the same time, local pride and cleanliness of the area may evolve as tourism moves throughout the community.
2. Development within each local community should strive to keep the uniqueness of the environment preserved. If there are period historical buildings, special natural resources, or sensitive culture traits available, capitalize on these and let further development or restoration take these aspects into account. This approach keeps the authenticity of the area intact which enriches its value to the visitors and local people alike.
3. It is not possible to develop realistic guidelines for sustainable tourism development without community involvement. Not only is it good public relations to encompass the local population in the tourism development process, but it will also result in the ultimate success of the endeavour. The community then becomes an effective force in assisting in the implementation of the program.
4. A local community should seek to measure tourism development in light of environmental and social costs and benefits to the community. Sustainable tourism should be viewed in terms of both short term and long term value to the community. Intangible values such as 'quality of life' should be added or included in the overall quantification of sustainable tourism development. (Organization of American States 1997)

Augustyn (1998) thus comments in the light of Polish experience that implicit in the application of these principles is the existence of environmental standards and regulations, that there are institutions that permit representation, that educational and awareness programmes exist, that research is sanctioned and sponsored, that developmental models at the appropriate level are understood and applied, that systems by which stakeholders can be held to account exist and that genuine processes of involvement take place. However, it can also be argued that empirical evidence does not always support the assumptions that indulge these principles. Local communities are often neither small nor homogenous. Not all will care for tourism and may prefer alternative sources of income, and there are dangers of over-dependency on tourism in small communities that may have long-term negative impacts.

For their part Edgell Sr and others (2008) and Hall (2008) both trace changing patterns of tourism development over the post-war period. For example Edgell Sr and others (2008: 52–62) reproduce in full the US *National Tourism Policy Act*, 1981, as an example of forward-looking legislation of its time, but note the elimination of the then existing United States Travel and Tourism Administration in 1996 and suggest that the 1981 legislation would now require significant updating if it were to be applied today. Among the new factors of the contemporary world, it is argued, are greater networking through transport and information technology networks, and arising from these developments the needs for more collaborative approaches and adherence to ethical guidelines. Potts and Harill (1998) propose an alternative mode that reinforces a point made by Wang and Wall (2009), namely, that these processes are inherently political. Potts and Harill (1998: 5) propose a system based on what they term *political ecology*, the inquiry 'into the causes of environmental change, with the goal of facilitating sustainable development through the reconstruction of social and political institutions'. Within this system lie principles of discovery, mutuality, locality, historicity, potentiality and enhancement.

DISSEMINATION OF PLANNING PRACTICES WITH REFERENCE TO CHINA

One factor that has also played a role has been the internationalization of these approaches through their adoption and dissemination by international bodies. Within the field of tourism two of the foremost are the United Nations World Tourism Organization (UNWTO) and the United Nations Environment Programme (UNEP). One much cited document of the former is the Global Code of Ethics for Tourism, adopted by resolution A/RES/406(XIII) at the thirteenth WTO General Assembly held in Santiago during 27 September to 1 October 1999, and confirmed by the United Nations General Assembly in 2001. Comprising 10 articles the declaration confirms a belief that tourism leads to a mutual understanding of people based on respect, is a means of personal development, should be planned in ways conducive to the sustainability of environments and cultures, should be beneficial for host communities, requires the adoption of responsible courses of action from all stakeholders, even while all have a right to tourism and liberty of movement, while workers and entrepreneurs in the industry equally have rights subject to constraints imposed by the need to respect cultures, peoples and environments, and all should seek to behave in ethical ways. To support these ideals the UNWTO have sought examples of good practice, commissioned reports on examples of good planning practices, sponsored initiatives in dealing with poverty reduction through local tourism initiatives, sought to encourage ecotourism and, in short, through meetings, conferences and debate has attempted to encourage and articulate an agenda of good tourism planning. To that end, as an appendix to this chapter, a sample of these reports has been added. On occasion the UNWTO has been asked to play a role in planning, but if so it has usually been in conjunction with other organizations.

One example in China is that its advice was sought in a UNEP-funded project on Leshan in order to better protect a World Heritage Site (Zhang 2003). This location meets the criteria in IUCN Category V, having previously been initially approved as a scenic area in 1982 by the State Council of the People's Republic of China. In 1988 the Mount Emei Administration was established to administer the area, and plans were drafted by the provincial government (Proposal for Approval of the General Plan of the Mount Emei Scenic Area; see Chuanfufa 1994) and approved by the State Council in 1995, before being declared a mixed natural/cultural World Heritage Site in 1996 (UNEP 1996). Its main cultural feature is the Giant Buddha of Leshan, carved out of rock in the eighth century, and being 71 metres in height. Mount Emei reaches a height of 3,099 metres and thus has five vegetation belts, and some 3,200 plant species have been defined, as have 2,300 animal species. Yet, the UNEP (1996) notes that 'pressure from tourism is evident at the Golden Summit'. Observations would indicate that this still remains the case. However, it can also be pointed out that just as great a threat has been the environmental degradation caused by industrialization and the toxic waste from nearby smoke-stacks. By the end of the 1990s the Buddha's nose was turning black, fragments were falling from his hair due to acid rain, and it can be argued that it is only the economic value generated by tourism that gave rise to a debate about site protection and means by which the process could be reversed and whether cleaning was a permissible option. Wang (2003) lists the plans for the period 1999 to 2005, and scanning those lists shows the focus of the plans relate to land usage, tourist flow planning, information provision, market research, environmental planning and monitoring and site facility management such as the provision and location of car parking and toilets. Wang's own work related to a wider range of stakeholders that include not only the Park officials but also residents and tourists. The samples are, however, small, but tourists seemed to want yet more development in terms of new attractions related to urban environments, a travel agent wanted yet more tourists, complaints were made about the manners of local people, and 'several respondents refer [*sic*] that . . . regional protection hurts legitimate competition' (Wang 2003: 13), while an implied view is that environmental protectionism has value because it attracts yet more tourists. Wang (2003: 13) writes 'All stakeholders applaud on [*sic*] the ongoing protection proposal of the World Bank for the Giant Buddha, which not only draws world concerns on the protection but also touts more tourists to come to the city.'

While it is dangerous to argue from a single example, it might be said that such an example is not atypical if the review offered by Sofield and Li (1998) and other contributions in this book are taken as a whole. There is a tension existing between, on the one hand, an increasing awareness and understanding of the principles of planning as espoused by official bodies like the World Bank, ENEP, UNWTO and, as described in the following, the Chinese State Bodies, and, on the other hand, implementation of those plans. To better understand these tensions requires consideration of at least three variables: (a) the means by which wider planning principles have been adopted into Chinese planning, (b) the development of the Chinese tourism planning system and (c) the factors that inhibit the proper implementation of principles of planning. In writing this the authors would wish to emphasize that the problems being described are not unique to China. Indeed, almost any society can point to plans that have been commissioned, evidence that has been collected, details and principles that have been enunciated—yet the documents line shelves gathering dust, and over time, the plans being disregarded, the whole cycle might commence yet again as problems remain unsolved.

THE ROLE OF ACADEMICS IN CHINESE PLANNING REGIMES

In light of the review of the tourism planning literature, it might be said the underlying philosophies and principles, even if expressed differently by different authors, are widely understood. Equally widely understood are the steps required to achieve these goals, even if they are again expressed differently by various authors. Murphy's (1985), Gunn's (1988) and Inskeep's (1991) works from the mid-1980s to early 1990s are replete with functional and logical steps in planning, while Hall's later work (2008) continues this tradition with lists of steps to be taken, criteria to be adhered to and means of measuring performance. China's planning processes have not been oblivious to these processes for a number of reasons, one of which has been the role played in the last decade by academics in the development of regional plans. This phenomenon has been noted by a number of commentators. Lai, Li and Feng (2006) and Wang and Wall (2009), for example, both separately comment on this, the former commenting 'Tourism planners in China are mostly university academics capable of conducting master planning on a theoretical basis, yet far from being capable of managing detailed (site) planning practically' (Lai, Li and Feng 2006: 1175). Consequently, many of the planners are familiar with the western texts, and thus have taken these into a Chinese context, but as Wang and Wall (2009) note 'Few planning cases in China follow the Western style of planning.' This gives rise to the obvious question, where and why do the differences arise? To begin to answer this question a historical context may be of help.

A BRIEF HISTORY OF TOURISM PLANNING IN CHINA

The Period Prior to Economic Reform (circa 1978)

Prior to the commencement of the 'Open Door' policy in the late 1970s, tourism was not accorded a high priority within Chinese governmental policies. Tourism planning was not generally undertaken, what travel existed was subordinated to the political objectives of promotion of the State's policies, and tourism assets were directly managed by central government. The main agency for this was the China Bureau of Travel and Tourism (CBTT) that oversaw the operations of the government owned and operated travel agencies—China International Travel Service (CITS). This existed to handle non-Chinese travellers, and was part of the Ministry of Foreign Affairs. The other main agency was China Travel Service (CTS), which sought to serve Chinese outside of Mainland China. As noted, destination planning was noted by its absence, and overall the infrastructure was poor in terms of assets, management, staff training and, for the most part, in attitude towards the general public in that it served the needs of its administrators above any perceived client demand.

1978–1985

As Fan and Hu (2006) describe, in July 1979 Deng Xiaoping visited Anhui Province and Huangshan, a location popular with several heads of state like Jiang Zemin as seen in Figure 2.1. While in Anhui in 1979, Deng stated the Province should work on a tourism plan—the first time a head of state had made such a suggestion. In September 1979 the first National Tourism Conference was held at the seaside resort of Beiddaihe, and an outline plan was drafted concentrating on overseas arrivals and foreign exchange earnings. Indeed, Wang and Wall (2008) comment that in this period, policy was primarily dictated by a perceived need to earn foreign exchange. However, such policies were piecemeal and embryonic, although given the encouragement of Anhui, not all such policies were at a Central State level. Local tourism bureaux were being established, in part motivated by the new opportunities being presented by overseas hotels now being able to operate in China with local partnerships. One example of urban planning was provided by Beijing, when, in 1985, a team of 100 was gathered to commence the planning of Beijing's tourism (Fan and Hu 2006). Tourism was becoming established as a separate function of State, instead of being tied to, for example, foreign policy. The China National Tourism Administration (CNTA) was established and the first steps in national planning commenced in 1983. However, such local and regional initiatives were unco-ordinated and bore relatively little relationship with wider economic planning other than the promotion of tourism was seen as providing economic benefits.

Figure 2.1 Display board commemorating President Jiang Zemin's visit to Anhui at Xidi. Photograph taken in 2006—courtesy of Chris Ryan.

1986–1991

As 1985 turned to 1986, the Standing Committee of the State Council began to consider tourism as part of a national development plan, and in early 1986 the State Council adopted the *Development Planning of Tourism Causes from 1986 to 2000* sponsored by CNTA. Premier Zhao Ziyang, at the second National Tourism Conference hosted in Beijing, developed the theme of tourism as part of integrated economic planning and by 1988 meetings and conferences were being held to consider how best tourism might be incorporated into the next state Fifth Year Development Plan. Again Wang and Wall (2008) characterize this period as being dominated by the economic role of tourism, where planning had little theoretical basis, was primarily based on the then existing planning models but with little understanding of the potential impacts of tourism. The emphasis was on growing visitor numbers, and to some extent this became easily achievable with little consideration of the longer term impacts. What made this possible was the ending of state monopolies and the emergence of profit-led organizations allied to an emergent marketing effort aimed simply at the promotion of various locations and destinations. The initial destinations that received attention were key scenic resources that included Hangzhou, Mount Emei and the Giant Buddha noted earlier, Mount Taishan, Huangshan and Qian. These attractions all combined scenic beauty with cultural and heritage resources as well as supporting literary heritages. In 1982 the State Council had authorized 44 national scenic areas to be designated as the first group of National Parks and on 7 June 1985, *The Temporary Rules of Administration of Scenic Areas* were promulgated. This possessed significant importance as, for the first time, Codes for Scenic Area Planning (GB50 298 ~ 1999) specified practices with the concept of standardization to be applied at three levels, the local city or county area, the Provincial level and the National level.

However, while regulation was being strengthened and tourism was beginning to grow, the industry and the planning of both the industry and destinations were still in relatively inexperienced hands. Wang and Wall (2008) argue that there was a lack of expertise, tourism-related education was still relatively embryonic, and certainly there was little conceptualization of stakeholder consultation with most planning being 'top down'.

1992–2001

The period 1992 to 2001 becomes a period when the national government increasingly began to redefine its role from one of central control to the encouragement of decentralization and the encouragement of the private sector, but within the constraints imposed by regulations to which the provincial, regional, city governments and private sector had to adhere. Coupled with this, inbound investment and foreign companies were being encouraged through not only the opportunities created by the growing Chinese economy but also through the implementation of tax concessions. At the same time, sensitive to criticisms of, and aware of the dangers implicit in, too large a surplus on the trade balance, China began to accede to a growing demand by the emergent Chinese middle class for overseas travel with the relaxation of regulations previously imposed on outbound Chinese tourism.

One outcome of this noted by Wang and Wall (2008) was the emphasis provided to resort complex development, especially at beach locations. Aimed at meeting the needs of an emergent middle class seeking hitherto difficult to access levels of luxury, wanting to attract overseas visitors, and consistent with seeking to establish propositions attractive to overseas management companies and the domestic construction industry, China commenced on a significant development of resort development. This process still continues at the time of writing. Hickman (2007) describes developments at Hainan thus:

> Hainan Island, which lies off the southern coast of the Chinese mainland, is often called 'China's Hawaii' due to its silky sand, palm-fringed beaches and year-round tropical climate. That it has attracted this nickname says something about the type of tourism on offer. The main resort's hub, Sanya, with 18 golf courses (and 10 more planned), a thick swathe of beach front and high-rise hotels means it has the look of

pretty much any identikit tropical beach resort in the world—clearly inspired by western hotels.

But it's China's domestic tourists who are really fuelling Sanya's building boom. Yalong Bay, a special 'tourism national reserve' about 12 miles east of Sanya, is where most of the luxury hotels are now located. International hotels chains that have located there include Sheraton, Crowne Plaza and Marriott. Further along the coast still is Haitang Bay where there are plans for a $200m 'seven-star deluxe hotel project' to be constructed by 2011. (Hickman 2007)

Hickman also reports the comments of Tourism Concern's director, who is reported as saying:

What the tourist will never see, though, is who might have been forcibly moved from their homes to make way for the new resorts and golf courses, how much biodiversity has been lost in their development, how local needs might be sacrificed—such as water—and the extent of the corruption. The tsunami in 2004 unexpectedly and conveniently cleared land that had been longed for by the developers. It's particularly frightening because the enormity of the resorts is often obscured by their luxury. Nowadays, these developments aren't necessarily high-rise and in your face. They can be more subtle than that and are cleverly sculpted over huge acreages of land. (Hickman 2007)

These are not new sentiments. Wang and Wall (2008: xxxx) reiterate the point, stating that such developments in China have meant 'Large scale development further plundered the opportunity for normal people equitably sharing the benefits'. Overall, they argue that during this period there was an under-estimation of the contribution that could be made by local communities, while at the same time the desire for what was perceived as a modern approach to tourism planning effectively permitted large overseas developers and construction companies to articulate a voice in the planning procedures that overruled the concerns of local grassroots communities. Another aspect identified by Zhang, Pine and Lam (2005) with the advent of such complexes was the added use of outsourcing by the Chinese hotel industry—which practice also added to representations by business interests who could justify such developments on the economic grounds of income and employment creation.

One factor that contributed to the development of such resorts was that they easily reinforced 'think big' planning approaches based on major investments. Fan and Hu (2006: 81) note a series of five-year plans that emanate from this period and describe the plans created for Hainan as 'systematic and penetrating, epoch making'. From 1983 to 2003 they note the involvement of many academics, government officials and the UNWTO in a series of plans and note that this period formed a five-year peak in regional tourism planning.

2002–Present

Almost inevitably a countervailing perspective began to emerge as the ramifications of the 'think big' approach to planning emerged—while paradoxically the success of previous policies in terms of generating greater employment and income helped establish higher levels of concerns about the costs of such progress. Public movements and opinions began to seek reassurance about protecting environments—both natural and human—and this discourse increasingly has been articulated in terms of 'people-centred' development while, in tourism, adopting UNWTO agendas on tourism as a mean of poverty reduction through local action. Evidence for a changing environmental dialogue continues to be evident in many different spheres— indeed even while the authors are penning these words in February 2008 opposition to a chemical plant in Xiaman, initiated by Zhao Yufen, (a professor with the Chinese Academy of Sciences who wrote an open environmental assessment of the impact of the plant) are celebrating the news of its relocation to the Gulei Peninsula in Zhangzhou, Fujian, although Zhangzhou's population of more than 4.5 million may not be so celebratory. Again, citizens in Shanghai are protesting (in the same month) against the extension of the Maglev rail system. In short, new agendas, discourse and means of articulation of concerns are being increasingly evidenced in China today.

These present new challenges for tourism planning. As noted elsewhere within this book, community involvement in tourism planning is being increasingly mooted as an appropriate path forward, but Wang and Wall (2008) still note two important caveats. First, planning is still perceived as an issue for the 'experts' and hence for communities to be involved requires assertive action on their part, thereby leading to uneven developments across the country in terms of who becomes involved, how they become involved, and the stages at which different stakeholders become involved. The processes of planning are thus weakened by a failure to incorporate important local stakeholders from an early stage as of right now. Again, it should be noted that this is not restricted to tourism alone. Tan, Wong and Lau (2007: 60) highlight this when they write of an all too prevalent view that there is 'an assumption that the public still lacks the knowledge and capacity to participate in policy formulation and decision making'—a view that reinforces the 'expert-cult' phenomenon (*zhuanjia changbai*) in Chinese planning. They provide evidence of the problems of this approach in the case of the Guangzhou Development District and the relocation of villagers, in that the issue was seen as one to be 'managed' rather than, initially at least, one of open consultation with a series of genuine alternatives existing. The second caveat noted by Wang and Wall (2008) is that in some senses the 'experts' are failing due their adopting particularistic and not holistic perspectives. Tourism planning is still perceived as planning of the physical resources and assets at a given destination—it is still perceived as an issue of infrastructure to move people from one location to another, to accommodate them, provide activities and in the process create favourable economic impacts. This perspective leads to a failure to note implications for not only cultural, community and environmental issues, but, strangely for a country previously given to state planning, for a failure to link tourism with other industrial development. How else is one to explain examples where tourism is being promoted and planned for within areas negatively affected by industrial pollution generated by other planned developments in close proximity to tourist zones, such as in Leshan as noted earlier, or in Xiamen where a coastal residential area found itself the neighbour of a US$1.4 billion chemical plant project built by Tenglong Aromatic PX (Xiamen) Co. Ltd leading to Zhao Yufen's actions and its coverage by the paper *China Business* on 18 March 2007.

THE GAP BETWEEN PRINCIPLE AND IMPLEMENTATION

To argue that there has been a gap between the principles of tourism planning and their implementation raises the question as to what factors inhibit successful planning. A second related question is whether participatory action, as espoused by many writers in the West with reference to tourism planning, is appropriate in the context of China. Indeed, it should be noted that within the West, voices have questioned just how successful is the participatory process in achieving better outcomes. For example, in a series of papers pertaining to tourism planning in the late 1980s and early 1990s in Banff Valley, Canada, Getz and Jamal have noted the time taken in consultation—time that can militate against company needs for certainty before undertaking investment, the financial and emotional costs involved in participation in consultative processes and then finally, have the outcomes been any better than those arrived at through a more 'top-down' approach? (For a list of Getz's and Jamal's work please refer to the Bibliography.) Again, as Ryan (2003), among others, has questioned, are communities homogenous, is it realistic to expect consensus to emerge, and might not consultative process intensify differences and simply reinforce or create new political hegemonies that will then dictate practice for at least a period of time? For his part Pearce (2000) notes the lack of follow-up research that examines to what degree plans have been implemented, and in what ways have they been implemented. Within a Chinese context Yang (2003) notes a failure of the Village Self-Management Programme, with a lack of education being a problem while Chen and Chun (2004) indicate the prevalence of bribery.

Lai, Li and Feng (2006) suggest a number of factors that inhibit successful planning, citing the 2001 China Tourism Planning Summit Workshop and the Conference on Tourism Planning and Management in Developing Conference of the same year as providing evidence of *zhishang huahua, quiangshang guagua* (drawn on paper, then hung on the wall)—a Chinese equivalent of the commissioning of reports that simply collect dust as paper exercises. Based on research undertaken at Guniujiang National Nature Reserve that involved interviews conducted with park staff, La, Li and Feng (2006) found that 9 out of 19 recommendations of the 2001 Park plan had not been implemented, while the remaining 10 had only been partially undertaken, or completed in ways different to that envisaged by the plan. Seven factors were identified as impeding the implementation of the plan, namely:

1. Lack of detailed measures—plan objectives might be couched in general terms without sufficiently detailed measures of success, contrary to the principles required of tourism plans by the 2003 CNTA *General Rule of Tourism Planning*.
2. Insufficient background research by the original planners—the original planners had been found to be deficient in some of the original research, thus leading to poorly defined problems and/or solutions and a poorly defined context within which the plans were to be supposedly implemented.
3. Inaccurate forecasts of future trends. In this instance forecast numbers of arrivals had been badly overestimated. Lai, Li and Feng (2006) also suggest that problems can emerge because of unexpected exogenous effects, although Gu (2008) argues that crisis management should be an on-going management process of anticipation and not simple reaction.
4. Inability of the Parks/destinations to employ and retain suitably qualified staff. One problem being experienced because of the expanding nature of the Chinese economy is that qualified planners and personnel responsible for implementing plans are able to quickly obtain better and higher paying jobs, leading to high labour turnover and problems with both the process of planning and subsequent management of plan implementation (e.g. see Tang, Wong and Lau 2008).
5. The lack of practical experience on the part of the planners—as previously noted there has been a heavy reliance on academics who may lack marketing, park management and other experiences and skills.
6. A divergence of views between the planners on the one hand, and park management and operational staff on the other. These differences of views matter, in that being those who implement the plans, park management themselves effectively act as planners through their partial implementation. This highlights the lack of monitoring processes in the plans.
7. The problems generated by the role of private investment. As discussed elsewhere in this book, unlike their western counterparts, particularly in the United States, Chinese National Parks have significant investment and infrastructure provided by the private sector. Park personnel

reported that, in their view, the private sector entrepreneurs are not only motivated by profit, but seek returns within quite short periods of time. Wider economic and environmental impacts are thus secondary in the eyes of the private sector, which may have little long term commitment to an area, seeing it simply as a means of revenue generation for further investment elsewhere.

CATEGORIZATION IN CHINA'S TOURISM PLANNING

As noted earlier, China's tourism development is generally directed and supported by all levels of government. With regard to the development of a variety of tourist cities, attractions, resorts, scenic areas and parks, the government has been both decisive and fully involved in playing an essential role in guiding China's tourism planning at regional and national levels. Major categories have been created for use in China's tourism planning and have been established and applied with different regulations and policies as described in the following. These categories are very heavily promoted, and one cannot progress far in any understanding of China's tourism assets and the planning for those resources unless some familiarity exists with these classifications. Consequently some of the major categorizations are listed here.

China Tourist Attractions: Categories and Rating Standards

Tourist attractions are the essential core components of China's tourism industry. China enjoys a rich variety of tourist attractions, which include scenic attractions, National and State Parks, museums, zoos, gardens, theme parks, ski resorts, forests, mountains, temples, historical and/or modern buildings and so on. The premier evaluation standards of tourist attractions were originally established and implemented by the CNTA in 1999. The tourist attraction rating system set the standards of overall quality of destinations and identified four categories: A, AA, AAA and AAAA tourist attractions, in which AAAA represented the category of highest quality. The rating system was modified in October 2004 to represent a more detailed and rigorous set of evaluation standards. The new "Categories and Rating Standard of Tourist Attractions" issued by CNTA became effective 1 January 2005 (CNTA 2004b).

A new, more upscale AAAAA category of tourist attractions was added in the rating system, and the evaluation criteria of each original category (A to AAAA) were further specified. To be eligible for AAAAA tourist attraction, the original AAAA tourist attractions were required to meet higher quality standards in terms of cultural features, uniqueness of tourism resources and service quality. Up till now China has about 20,000 tourist attractions among which were 2,363 graded with A-level classifications. In May 2007,

CNTA announced the first group of national AAAAA tourist attractions in China. A total of 66 tourist attractions were certified, including many well-known sites such as the Forbidden City, the Summer Palace and Stone Forest. The others include 785 AAAA, 521 AAA, 927 AA and 130 A attractions. The rating system is intended to enhance the management of tourist attractions, improve service quality, tourism resources planning and protection. The evaluation criteria addressed a variety of tourism planning and management issues in accordance with industry standards. Specifically, the evaluation criteria for each category included transportation, on-site tours, safety and security, cleanliness and sanitation, postal and telecommunication service, shopping, operation/management, tourism resources and environment protection, tourism resource attractiveness, image and branding, tourist arrivals and tourist satisfaction (CNTA 2004b). Detailed description of each rating category is as follows:

AAAAA Tourist Attractions

1. *Transportation*. The specific criteria include easy accessibility; environmentally compatible designated parking areas; well-designed tour routes within the attraction and environment-friendly on-site vehicles.
2. *On-site tour*. The criteria include well-located visitor centres with comprehensive facilities and well-trained visitor contact staff; well-designed and located signs for direction; various publications about the attraction; sufficient, highly qualified tour guides; and sufficient and well-designed tourist public facilities with distinctive cultural features.
3. *Safety and security*. The criteria include complete safety and security regulations and their implementation; effective fire alarm and ambulance systems; on-site clinic and medical staff; and emergency response planning and reporting.
4. *Cleanliness and sanitation*. The criteria include clean environment; sufficient and clean rest rooms; environmentally compatible and well-maintained trash cans and waste management. All public facilities, restaurants, and foods must meet the specified national standards of sanitation and environment protection and should not use disposable tableware.
5. *Postal and telecommunication service*. The criteria include postal and national/international direct dial phone services; appropriate charges; and a mobile phone signal within the attraction.
6. *Shopping*. The criteria include designated shopping areas, well-designed and environmentally compatible shopping facilities; clean shopping environment; centralized management of all businesses; and the sale of various and unique local tourist products.
7. *Operation and management of tourist attraction*. The criteria include a professional and efficient management system; complete and effective regulations on quality control, safety and security, the maintenance

of tourism statistics; a professional management team; unique destination image and brand, good reputation on service quality; government-approved tourism planning proposals; a professional training department and highly qualified employees; efficient complaint processing system; and a personalized service for seniors, children and the disabled.

8. *Tourism resources and environment protection.* The criteria include specific requirements on air quality, sound pollution record, underground water quality, and sewage; well-maintained and protected natural, cultural and historic attractions; scientific management of carrying capacity; well-designed and environment-compatible buildings and facilities in the attraction; and an appealing overall environment.

9. *Attractiveness of tourist resources.* The criteria include distinctive appeals of the attraction; exceptional historical, cultural and scientific importance; a rich variety of valuable species and habitats, or distinctive or unique scenery; abundant tourism resources; and well-maintained resources retaining their original structure and shape.

10. *Market attractiveness.* The criteria include being a well-known attraction of international significance and recognition; an exceptionally appealing image, strong market influence and uniqueness of the attraction.

11. *Tourist arrivals.* It is required that the total tourist arrivals exceed 600,000 per year, including a minimum of 50,000 international tourist arrivals annually.

12. *Tourist satisfaction.* It refers to high levels of tourist satisfaction based on tourist survey results.

AAAA Tourist Attractions

In 2001, CNTA conferred 187 tourist sites as the first group of National AAAA Tourist Attractions. By the end of 2006, a total number of 785 tourist sites had been designated as AAAA tourist attractions (CNTA 2007).

The criteria of AAAA Tourist Attractions are the same as AAAAA Tourist Attractions, but with less rigorous requirements in each category. For example, AAAA Tourist Attractions are expected to have national significance and recognition, and have 500,000 annual tourist arrivals (including 30,000 international tourist arrivals).

AAA, AA and A Tourist Attractions

These three categories still use the same evaluation items as AAAA and AAAAA attractions. However, compared to the other two ratings, AAA Tourist Attractions are required to have inter-provincial significance and recognition and receive 300,000 annual tourist arrivals. AA Tourist Attractions are expected to have market recognition within the province and receive 100,000 tourist arrivals, whereas A-rating tourist sites are considered local attractions that accommodate 30,000 annual tourist arrivals (CNTA 2004a).

CHINA'S OUTSTANDING TOURIST CITIES

The programme, 'Building China's Outstanding Tourist City' was organized by CNTA, which established the evaluation criteria and started the assessment process in 1998. By 2007, a total of 306 cities have been entitled 'China's Outstanding Tourist City' (CNTA 2008). The tourist arrivals and revenue generated from these cities accounted for 85 per cent of the national tourism industry. Therefore, these cities are major economic contributors and a leading force for China's tourism industry (CNTA 2004a). A primary focus of China's tourism industry is to improve the quality of tourist cities nationwide, and enhance the sustainable development of the industry. The establishment of 'Outstanding Tourist Cities' is part of a strategy to improve various tourism functions of the Chinese cities and stimulate local economic development.

Evaluation criteria and standards were initiated in February 1998 by CNTA, and have since been revised three times in 2000, 2003 and 2007 to meet new situations and problems in China's tourism industry. The most recent criteria published in 2007 included 20 categories with 183 assessment items (CNTA 2007a). Specifically, the evaluation categories include:

1. *Tourism economic development of the city (60 points).* The criteria include the percentage that tourism contributed to a city's GDP (3 per cent or 5–15 per cent) and that international tourism revenue has a value of US$5 million–US$1 billion.
2. *The importance and scale of the city's tourism industry (35 points).* The criteria require that (a) tourism is one of the leading economic sectors of the city; (b) tourism is an important part of the economic and social development master plan of the city; (c) tourism revenue ranks well in the city as an economic factor when compared to other counties, provinces and the nation's economy and when compared with other cities.
3. *Government investment and specialized tourism policy (35 points).* The criteria include favourable tourism policies in financing, taxation, transportation, cultural promotion, local zoning and infrastructure; designated government investment and specialized favourable policy on tourism development; and possessing a guaranteed tourism marketing and promotion fund.
4. *Municipal government's guidance and support on tourism development (35 points).* The criteria include yearly specialized planning for tourism development; a specified department, institution and meetings for tourism issues within the municipal government; policies and

regulations specific to the tourism industry; and established tourism laws and regulations.

5. *Tourism administration (70 points)*. The criteria include established tourism administration functions; centralized regulation and quality monitoring systems for tourism businesses; and the scientific collection and reporting of tourism statistics.

6. *Work ethics and service quality in the tourism industry (60 points)*. The criteria include good work ethics and morale; outstanding employees and companies; and good tourist feedback.

7. *City's ecological environment (45 points)*. The criteria include urban green space (10–30 per cent of the total land); waste management, air quality, sound pollution record, good drinking water quality and the cleanliness of the overall city environment.

8. *Tourism services (100 points)*. The criteria include conveniently located tourist information centre; tourist information and all kinds of tourist-friendly services provided on a sustained basis; an international level convention and exhibition site; recreation and night life activity.

9. *Tourism education and training (40 points)*. Tourism and hospitality education is provided at the levels of technical/vocational high school, college and university; and provision for training employees and managers in tourism businesses of all forms.

10. *Transportation (60 points)*. The criteria include accessibility (airport, railway, harbour, highway, cable car etc.), and in-city transportation (tour route within the city, tour shuttle bus, taxies, cruise/tour boats etc).

11. *Tourist attractions development and management (40 points)*. The criteria require that (a) the local government has active investigation and exploration of tourism resources of the city; (b) the city maintains well-developed A-rating (from A to AAAAA) tourist attractions that receive 500,000 arrivals per year, and tourist resorts with annual arrivals of 100,000; (c) the city has quality control and monitoring of tourist attractions, as well as tourist attraction planning and protection policies.

12. *Tourism promotion and product planning (60 points)*. The criteria require that the city has (a) tourism promotion and sales increase in the past three years; (b) a promotion strategy, plan and its implementation, active participation in international and domestic tourism fairs, and takes part in overseas promotion; (c) local tourism events; (d) distinctive features of tourism products (historic, cultural tourism, theme parks, conventions/exhibitions, and sport events etc.); (e) product exploration and implementation (for example, ecotourism, agriculture tourism, sports/health and fitness tourism etc.).

13. *Accommodation (50 points)*. The criteria include tourist hotels (adequate number and rating of the hotels; 100 per cent annual review of the star hotels; percentage of green hotels); and other types of hotels.

14. *Travel agency (40 points)*. The criteria require that (a) all travel agencies are licensed, with well-presented price list and materials; (b) there are CNTA-approved top hundred travel agencies in the city; and (c) the travel agencies have legal and professional operations following industry standards.

15. *Food and Beverage (40 points)*. The criteria include government regulations over the tourism food and beverage operations, uniqueness and quality of food and beverage (local cuisine, fast food restaurants, western-style fine-dining restaurant, and sanitation rules are strictly enforced).

16. *Shopping (40 points)*. The criteria include the percentage of shopping related revenue in the total tourism revenue (15–25 per cent), and shopping areas and products (centralized shopping areas, souvenirs and other products with unique local features).

17. *Recreation (40 points)*. The criteria include night performance/show, restaurants with performing arts and dances, day-time performance/show, various nightlife recreations, and no illegal activities such prostitution, gambling and illegal drug taking.

18. *Rest rooms for tourists (40 points)*. The criteria include sufficient number and adequate location; quality of rest rooms (cleanliness, environment compatibility, accessibility for the disabled and international-standard signs).

19. *Legal and ethical operation of tourism business (70 points)*. The criteria include standard operation of "one-day tour" business; no fake products, cheating or fraud; qualified tour guides; business ethics; and complaint processing and recording.

20. *Safety/security and insurance (40 points)*. The criteria include safety and security system, emergency response system and insurance system for travel agencies.

CHINA'S BEST TOURIST CITIES

The classification of 'China's Best Tourist City' is an extension to the programme of China's Outstanding Tourist City. The purpose is to improve the quality of China's tourist cities, build international brands and increase the competitiveness of the major Chinese tourist cities worldwide. CNTA and UNWTO collaborated in establishing the evaluation criteria and conduct the actual assessments. In 2006, three cities—Dalian, Hangzhou and Chengdu—were approved as the first group of 'China's Best Tourist City'. The recognition is designed to further improve the development of the infrastructure, environment and service quality, as well as the destination image of the cities in the global tourist market (CNTA 2006b).

Only those tourist cities which have been categorized as 'China's Outstanding Tourist City' can apply for 'China's Best Tourist City'—either in a comprehensive or specialized category—based on their own features and qualification. The title of 'Best Tourist City', after being awarded, can be retained for three years and expires at the beginning of the next cycle of evaluation. Due to the nature and variety of tourism resources, 'China's Best Tourist City' also has nine specialized categories in sightseeing, food and beverage, meeting/conventions, shopping etc. (CNTA 2006b).

The total score of the first nine 'fundamental qualifications' is 800 points, and a city needs to receive the 'pass score' in each individual category and a total of 525 points. It is also the prerequisite for an application in a specialized category of 'China's Best Tourist City'. The detailed evaluation criteria are as following:

1. *Tourist experience and satisfaction (120 points).* Surveys will be distributed to tourists to assess their satisfaction on items such as overall impression of the city, facilities, infrastructure, transportation, natural and built environment and protection and prices.
2. *Tourism economic benefit to the local community, residents' satisfaction and involvement in tourism (80 points).* The criteria include tourist contribution to GDP, employment and business investment, as well as local residents' satisfaction with tourism-generated employment opportunities, salaries, tourist impacts on the local community, infrastructure development and other benefits.
3. *The abundance, quality and uniqueness of tourism resources and attractions (120 points).* The criteria include on-site examination of natural sightseeing tourism attractions; historic and cultural tourism attractions; leisure vacation resources and attractions; food/beverage and shopping facilities; meeting, convention and recreation facilities.
4. *Planning and protection of natural and cultural attractions (40 points).* The criteria include the overall layout of buildings in the city, the environment compatibility of the buildings, streets and public areas with each other; public transportation; parks; walking tracks; direction signage etc.
5. *Environment quality, protection of natural and cultural tourism resources (80 points).* The investigated items include: air and water quality, sewage and waste management, light and sound pollution control, public rest room cleanliness and maintenance, other environment protection solutions, natural ecology protection, cultural heritage protection and maintenance.
6. *Infrastructure and Transportation (40 points).* The criteria include air, rail, highway transportation, bus and coach service, parking, accessibility for the disabled, electricity and gas supply and international telephone service and internet connection.
7. *Availability, quality and uniqueness of tourism facilities and services (160 points).* Items include accommodation and restaurants, entertainment activities; tour guides, bus, taxies, and rental car services, travel

agency services, group tour services, shopping facilities, meeting, convention and exhibition services, language skills and service quality of tourist contacting staff, safety and security facilities, direction signs, other services for tourist convenience, guest complaint handling, emergency response system, tourist insurance policy etc.

8. *Tourism planning, management, development and marketing of the city (80 points)*. This section includes the applicant city's tourism development planning, including policies, strategies and goals of the tourism industry of the city. The evaluation committee reviews the city's tourism development record over the past three years, as well as major tourism projects and planned activities for the next three years, including human resources management, annual tourism planning proposal, annual budget of tourism development and marketing, tourism promotion and market research and so on.

9. *Collaboration and co-operation among governments, tourism business and local community (80 points)*. The assessment committee will arrange meetings with various government offices, tourism organizations, industry professionals and local community representatives to discuss collaboration among various sectors such as city planning, finance, environment, public welfare, education, human resources and so on.

10. *Unique features of specialized category based on the other nine criteria (200 points)*. Cities which meet all the requirements of the other nine criteria are eligible for the specialized category. The total score is 200 points with a minimum requirement of 150. Based on each city's unique features and tourism resources, the assessment committee will examine the following criteria: distinctive attractiveness, accessibility, features of tourism resources and activities; service scope and quality; hospitality/friendliness, international reputation, branding and image; and price. Each of the nine specialized categories has its own survey questions for tourists to answer, due to the distinctive nature of tourism resources and tourist experience.

CHINA NATIONAL PARKS
(CHINA NATIONAL SCENIC AREAS)

Scenic Areas in China are classified at national and provincial levels and are the equivalent of National and State Parks in international terms. A National Park in China (National Scenic Area) refers to a region which has generally large areas of natural landscape, possesses nationally significant natural, cultural or recreational resources, and offers scientific values. The qualifications of China National Parks are established by the Ministry of Construction and certified by the State Council of China. The first group of National Parks was announced in 1982, with a total number of 44 in the list, including Beijing Badaling Great Wall, Beidaihe Resort, Taihu Lake Resort etc. Currently there are 187 National Parks in China, whereas the total number of scenic areas is 677, covering 1 per cent of the total land size of the country (Ministry of Construction 2007).

According to the most recent regulations published in 2007, the assessment focuses on six major categories: administrative structure and function, master planning strategies and implementation, policies on regulation violation, monitoring and information system, tourism signs, management regulations and policies. The total score is 100 covering the standards of 17 essential items. Special attention was given to landscape and natural environment protection, and regulations regarding facility construction and tourism business operations within the National Parks. The categorization of China National Park (Scenic Area) to some extent overlaps with National Preserve, Culture and Natural Heritage and Ecological Park.

CHINA NATIONAL PRESERVES

From 1956 when China's first National Preserve, Guangzhou Dinghu Natural Preserve, was established, to 2004, China's National Preserves have increased to 2,194 in number, covering 14.8 per cent of the total land size of China. Among these, 226 are of national level, 733 are at the provincial level, 396 are at city level and 839 are county level. These areas, appointed by the China National Environment Protection Bureau, were established to preserve nationally significant natural, scenic, historic, archaeological, geological, scientific, wilderness, cultural, recreational and wildlife resources. The evaluation is conducted every five years, and the criteria include: administrative structure and staff, conditions of maintenance facilities, land management, regulation and policies, natural resource protection and utilization, tourism activities within the site etc. (China Environment Protection Bureau 2006)

Since 1979, a total of 26 National Preserve areas have been designated United Nations Educational, Scientific and Cultural Organization (UNESCO) Biosphere Reserves in recognition of their international significance. Additionally 14 have been listed as World Natural Heritage Sites by UNESCO and 27 have been listed as International Important Wetlands. The designations have enhanced the connection and collaboration between China and the world on protection of natural preserve areas.

CHINA CULTURAL AND NATURAL HERITAGE SITES

Since joining the International Convention concerning the Protection of World Cultural and Natural Heritage in 1985, China had in 2007 35 designated world heritage sites; of these 25 are cultural heritage sites, six are natural heritage sites, and four are cultural and natural (mixed) sites. These sites comprise the major part of China's valuable and rich tourism resources. The Forbidden City, the Great Wall and Huangshan in Anhui are some examples of cultural and natural heritage sites (CNTA 2007b).

CHINA NATIONAL GEOLOGICAL PARKS

The aims of the National Geological Park designation are the promotion and protection of geological heritage coupled with public education and the maintenance of culturally and environmentally sustainable development.

The first group of 11 National Geological Parks was announced by the Chinese Ministry of Land and Natural Resources in March 2001, including Zhangjiajie Gritstone Peak Forest National Geological Park in Hunan Province and Karst Stone Forest National Geological Park in Yunnan Province. The geological landscapes were evaluated by a special leadership panel and an appraisal committee on national geological relics comprising administrative officials and experts from various central functional departments. Currently there are 138 National Geological Parks in China (Ministry of Land and Resources 2007). Since 2004, a total of 19 China National Geological Parks have been designated by UNESCO as World Geological Parks, which worldwide total 53 in number.

CHINA NATIONAL VACATION RESORTS

Prior to 1992, there was no contemporary form of resort-based travel in China. The major form of pleasure travel only comprised sightseeing at natural and man-made scenic spots. China National Vacation Resorts were established as extensive resort complexes with a comprehensive leisure product and service mix intended to satisfy the needs of international vacation markets. The State Council of the People's Republic of China government secured the land, provided conceptual planning and development regulations, and offered favourable polices to private commercial ventures for investing in facilities and services (CNTA 1993).

The State Council approved the establishment of 12 National Vacation Resorts on a trial basis in October 1992. Most of the resorts are located at seaside, lakes and rivers, including Yalong Bay Resort in Sanya, Shilaoren (Stone Old Man) Resort in Qingdao, and Suzhou Taihu Resort in Suzhou. Among the 12 National Vacation Resorts, Wuyi Mountain Resort in Fujian is the only mountain resort.

The specific criteria of China National Vacation Resorts were proposed by CNTA and announced by the State Council (State Council 2001). They include:

1. *Environment quality*: favourable climate, high quality environment and air quality for leisure vacation
2. *Tourism resources*: high quality and valuable tourism resources, high potential for tourism development, attachment to tourist attractions, combined functions of sightseeing and vacation, various cultural, recreational and fitness activities in the resorts
3. *Accessibility and transportation*
4. *Tourism market*: market structure (average length of stay, types and source of tourists), market development
5. *Land planning and facility management*
6. *Infrastructure*: sufficient water, electricity supply, sewage, telecommunication
7. *Accommodation*: high-quality resort hotels, vacation village, camping sites etc.
8. *Shopping and recreation facility*: sufficient vacation activities to meet various needs such as recreation, leisure, fitness, sports, physical therapy and other special services
9. *Overall management*: safety and security, clinic service, cleanliness (public areas, rest rooms, waste management, food sanitation), public service facilities (special facilities and services to the elder, children, and the disabled)
10. *Community management system*: professional management team and staff, complete regulations and policies for quality control, safety, and tourism statistics reporting; close connection and communication with local community residents regarding the planning, operation, good environment for community development, information network and promotion network

COMMENTS ON THE CLASSIFICATIONS

Structured local and regional tourism attraction development has been undertaken for only a short period of time in China; yet the Chinese government has made significant efforts to regulate and direct the development and planning of tourist cities, attractions, resorts and parks. Since 1979 China's tourism has been considered an important 'government-led' industry, and the success of the industry is generally seen to be the result of government involvement and support. The government role has been emphasized in the use of evaluation criteria and rating standards for all forms of tourist attractions, resorts, parks and tourist city planning. Government administration and policy are also generally listed as two of the major rating criteria for assessing the quality of tourist cities, attractions and resorts.

Furthermore, it is important to note that, in many cases, the evaluation criteria and rating systems have utilized the tourism industry standards of developed countries as a source of reference. Some categorizations, such as

China's Best Tourist City, National Preserve, National Ecological Park and Cultural and Natural Heritage site, are closely connected to the standards used by UNWTO and UNESCO. It is believed that the use of these international categorizations will enhance understanding of, and collaboration between China and the world, in preserving and promoting China's tourism resources.

Overall, the classifications and benchmarking used in tourism planning represent the variety and complexity of China's tourism resources. The criteria not only focus on infrastructure and facilities, but also include tourist experiences, environmental protection, sustainable development, data collection from tourists, local residents and tourism practitioners and experts. Some criteria specifically address detailed problems in China's tourism operations and service quality (for example, rest rooms and specialized services for older people, children and the disabled).

However, the classifications used for China's tourism resources may appear confusing in that many criteria for different forms of attractions, cities, resorts and National Parks are very similar, which results in an overlapping of categorizations for many tourist cities and attractions. Consequently some places can be simultaneously categorized as an AAAAA Tourist Attraction, National Park, Outstanding Tourist City, National Culture and Natural Heritage and National Geological Parks. Furthermore, it is thought that in some instances the categorizations being awarded to cities and attractions may be influenced by subjective, political considerations as much as the inherent values of the destinations and the qualification of tourist cities and attractions within the same planning category may vary to a great extent. For example, there are differences in performance on the various criteria among the 306 Outstanding Tourist Cities in China, and standards among the 66 AAAAA Tourist Attractions are thought to not always be consistently applied. Certainly, with reference to hotels, the authors have experienced varying standards between properties possessing the same star rating. Yet, having said this, over time, the public nature of the criteria is an important means of improving the quality of China's tourism resources and services.

EXAMPLES AND CASE STUDIES

These issues are illustrated by the following chapters that examine the process of change induced by tourism and its planning in Luzhi, the example of planning at Zhangjiajie National Park, one of the first areas to be subjected to the planning regime established in the last two decades of the last century in China, and the means by which a planned resort–residential area has been developed in Shenzhen.

In one of these chapters Fan, Wall and Mitchell discuss the impacts of tourism on changing retail patterns and the establishment of tourism retail zones in the water town of Luzhi, Suzhou. Borrowing from a phased tourism development concept based on creative destruction whereby old areas are given new purpose through the nexus of heritage, attractive settings and entrepreneurial initiative, they find both similarities and differences with comparative Canadian sites. One of the key differences is based upon population densities and the existing forms of housing found in Luzhi, so that while there is population movement out of the area as forecast by the model, in this instance it has little to do with tourism but government patterns of rehousing as residents are moved to better forms of housing. However, it is implied that, in part, such population movement is due to the regeneration of zones seen as having heritage value that can be commodified for purposes of tourism. The examples reinforce the notion that to understand tourism planning in China it becomes necessary to consider wider frameworks even while it is possible to describe changes solely attributed to tourism.

In their chapter on Zhangjiajie National Forest Park, Zhong Linsheng, Deng Jinyang and Xiang Baohui utilize Butler's tourist destination life cycle to both examine the concept and the changes in the park itself. However, the chapter has been located in this section of the book because it provides an example of how planning, even of a National Park, in China, is often dominated by economic generation as described in various chapters in this book, including that of Ma, Ryan and Bao, or by cultural considerations as indicated by Li and Sofield. One consequence for Zhangjiajie has been that of environmental degradation as the region's population has not only grown in number, but changed in composition as a consequence of economic change and visitation potentially inspired by cultural heritage rather than natural heritage motives. As described in other sections of this book, this is unfortunately a not uncommon experience in China as the country seeks to draw many millions of people out of poverty and reinforce a sense of 'being Chinese'. The issues raised in this chapter are not wholly restricted to tourism but are epidemic in many instances of economic growth, where new industrialization and urbanization has been sought to improve living standards in terms of disposable income, even while the natural environment has been forced to absorb increasing costs. The value of this case study is that, by drawing on secondary data, it explicitly indicates the costs to a National Park of this approach—and in many ways it is a salutary lesson given that Zhangjiajie National Forest Park was established as the first such park in China in 1982. It thus serves to illustrate the predicament facing current planning needs as emerging evidence grows apace of the costs China faces from its use of tourism as a tool of economic regeneration. An associated chapter is that by Ma, Ryan and Bao on the use of resources and their management in Chinese National Parks. They find that external economies of scale due to the proximity of other parks has a role to play, and argue that destination planning requires a portfolio of attractions to both help sustain the park while also possibly generating economic benefits to rural locations. It will also be noted that the Chinese tourism authorities tend to the generation of classifications for purposes of planning; that is categorizations are based on a 'scientific' method for benchmarking, and any visitor to China will be familiar with this practice. Reference to this is made by Hong and Xin, and their work is specifically included to provide an example of the Chinese approach to these issues. This chapter illustrates the strong tradition of 'scientificism' that is found in Chinese cultural life; a tradition partly derived from Maoist socialist empiricism and a belief in scientific method that was used to justify perceived senses of progress.

In their chapter, Zhang and Zhang provide a description and analysis of real estate development by Overseas Chinese Township—and while this model will be familiar to many western readers, it still represents one of the few such successful developments in China.

APPENDIX 2.1

UNWTO (1992) *An Integrated Approach to Resort Development*, Madrid: UNWTO.

UNWTO (1998) *Asian Experiences in Tourism Development*, Madrid: UNWTO.

UNWTO (2005) *Cultural Tourism and Poverty Alleviation—The Asia-Pacific Perspective*, Madrid: UNWTO.

UNWTO (2006) *Poverty Alleviation through Tourism—A Compilation of Good Practices*, Madrid: UNWTO.

UNWTO (1999) *Silk Road Tourism—Current Issues*, Madrid: UNWTO.

3 Tourism Development and the Tourism Area Life-Cycle

A Case Study of Zhangjiajie National Forest Park, China

Zhong Linsheng, Deng Jinyang and Xiang Baohui

INTRODUCTION

The conceptual framework of the Tourism Area Life-Cycle (TALC) has been frequently examined since first proposed by Butler in 1980. The recently edited two volumes on the model (Butler 2006b, 2006c) further highlight its significance as one of the most used frameworks within tourism studies. Lagiewski (2006), in reviewing previous studies, listed a total of 49 major works related to the TALC. These studies have examined the model at different units of analysis, ranging from a single tourist resource (e.g. Niagara Falls; see Getz 1992) to a destination of varied features (e.g. Lancaster County; see Hovinen 1981, 2002; Canada's NW Territories; see Keller 1987; the Greater Yellowstone region; see Johnson and Snepenger 1993; and the Smoky Mountain region; see Tooman 1997). In addition, the model has also been applied to differing types of tourism attractions and resources (i.e. islands, costal resorts/beaches, mountainous destinations etc.) and from different perspectives (i.e. the validity and applicability of the TALC; see Agarwal 1997; social, environmental and/or economic changes as they relate to each stage of the model; see Berry 2001; Hovinen 2002; Tooman 1997; and tourism planning; see Getz 1992; Bao and Zhang 2006; etc.). The TALC, as examined in these studies, was generally proven to be a useful framework in explaining the dynamics of tourism development for a resort, although a universal consensus about its validity and applicability has not yet been achieved.

Obviously, more studies need to be conducted to test the model (Agarwal 1997; Berry 2001), and as Butler (2006: 286) pointed out in his concluding chapter to the two volumes, an 'understanding [of the way that tourist destinations develop] is still far from complete'. The authors of this chapter would argue that special attention should be paid to areas to which the model has been rarely applied (i.e. National Parks or other peripheral areas), particularly those in developing countries with rapid economic growth. A number of reasons prompt this observation. First, the majority of existing studies are of coastal/island destinations with few being focused on National Parks and other protected areas (albeit there are two exceptions, e.g. Boyd 2006; and Weizenegger 2006). Yet parks are important places for the protection of ecological systems and natural resources as well as the provision of rec-reational and tourism opportunities for the public. This is particularly true for those areas assigned World Heritage status (Boyd 2006). Because of the nature of these areas, governmental interventions could affect the direction and speed of tourism development for these places more than in the case of an island resort because of their greater potential vulnerability to the nega-tive impacts of large numbers of visitors (Weizenegger 2006). Second, the robustness of the model has been tested primarily in North America, the UK and Mediterranean areas where tourism development has a long history and has obtained its existing situation over an extended period of development (Baum 1998). The question arises, would the concept apply within a differ-ent type of economy?

Relatively few studies have been undertaken in developing countries where newly emerging tourism destinations might reach the maturity stage within a short time period as noted by Baum (1998, 2006). Thus he writes:

> It is arguable that the model has rather less value, except perhaps of a cautionary nature, to newly emerging tourism destinations, particularly in the developing world. Here the development period is likely to have been much more rapid and, with the added force of globalisation and multi-national investment, may well have 'jumped' one or more stages within the model cycle. (Baum 1998: 169)

Finally, Johnson and Snepenger (2006: 234) also emphasized that more stud-ies should be conducted in 'other settings with . . . different economic transi-tions . . ., which could all provide valuable insight into the changing nature of tourism impacts and destination maturation processes'. China, the world's largest developing country with the fastest economic growth, could provide such a setting in which the development processes of a newly emerging tour-ism destination in a remote and underdeveloped region can be analysed in relation to the macroeconomic change taking place in the whole country.

In response to these concerns, this chapter applies the TALC to China's first national forest park, namely Zhangjiajie National Forest Park (ZNFP). The ZNFP is chosen for three reasons. First, the park is a single tourism destination established in 1982 and located in a remote mountainous area with a single dominant tourism resource, unique sandstones. Also a tourism destination could refer to a resort, a town, city, region or country and no universal consensus has been achieved among researchers with regard to which unit of analysis is the most appropriate. It can be envisaged that stud-ies based on differing units of analysis could generate differing outcomes as suggested in previous studies. Hovinen (2002: 220), in using the TALC to examine tourism development in Lancaster County, a diverse destination, argued that 'Butler's hypothesized stages may fit certain single-attraction destinations well.' Second, while it is recognized that a major challenge in testing the TALC for many destinations is the difficulty in obtaining accurate long-term trend data of visitors to these areas (Butler 1980; Hovinen 2002; Lagiewski 2006), this is not a problem for a park where visitors can only enter through controlled gates. Thus, as with many other parks in China (e.g. see Bao and Zhang 2006), the ZNFP has kept a complete and reliable record of tourist numbers since its inception. Additionally, the park has also documented tourism revenues for each year. Finally, the park has been extensively examined during the past two decades in terms of tourism's impacts on the environmental, economic, and social–cultural aspects associated with the park. This provides a solid basis for the current study.

Agarwal (1997) pointed out that research on the TALC should focus on either one of two aspects: (a) testing the applicability of the model or (b) redeveloping the model to incorporate different issues (as cited in Lagiewski 2006). In reality, most studies combine the two. This approach is also followed in this study. Specifically, three aspects of the model are examined:

1. The validity and applicability of the model as applied to the tourism development evolution of the park.
2. External and internal forces that shaped tourism development in the park and surrounding areas.
3. Environmental, social, and economic changes as they relate to each stage of the model.

In this chapter, the TALC is examined against observed development trends (Getz 1992; Agarwal 1997). To examine these issues, Getz's (1992) approach of using existing documents and previous research findings, combined with direct interviews, field and map observations and questionnaire surveys is adopted.

TOURISM AREA LIFE-CYCLE MODEL

It is not the purpose of this section to provide a comprehensive review of literature on the model (please refer to Lagiewski 2006; Butler 2006b, 2006c, for a detailed summary of findings), rather this section is to provide a foundation for what our study seeks to accomplish. Thus, previous findings on the aforementioned three aspects of the study are the focus of the following review.

Validity and Applicability of the TALC

As is well known, the TALC proposed by Butler (1980) involves a six-stage evolution of tourism, namely exploration, involvement, development, consolidation, stagnation and post-stagnation. This last stage is further characterized by a period of decline, rejuvenation or stabilization. As already noted, the applicability of the model to a given area has been assessed and judged in previous studies by looking at how well the evolution of a tourist destination's development matched the six phases conceptually described by Butler.

According to Butler (1980: 10), 'not all areas experience the stages of the cycle as clearly as others'. This suggests that the model cannot be applied to all destinations in the same uniform manner. This argument was endorsed by previous studies wherein stages experienced by many destinations were not consistently found and findings were very case specific (Agarwal 1997; Cooper and Jackson 1989). On the other hand Meyer-Arendt's (1985) study of the Grand Isle resort of Louisiana, Cooper and Jackson's (1989) study of the Isle of Man and some other studies (e.g. Berry 2006; Smith 1992; Wilkinson 1987) demonstrated that the life-cycle for these destinations matched the model very well. In addition, an interesting study by Boyd (2006) on the establishment and development of Canadian National Parks as a whole also conformed to the six phases of the TALC. However, in this instance the author still noted the difficulty of determining 'where individual cases may be best represented within the model' (Boyd 2006: 138). That is, the six stages were not found to be consistently experienced by every individual park.

As is the case for individual parks in Canada, there are many other destinations that did not entirely conform to the model. For instance, it has been argued that Atlantic City missed the first two stages (Stansfield 1978) and the Cayman Islands in the Caribbean skipped the first (Weaver 2000). In addition, previous studies also found that different stages could coexist for a tourism destination. For example, Lancaster County, as a diverse destination, was characterized by the coexistence of growth, stagnation, decline and rejuvenation (Hovinen 1981, 2002). Similarly, Niagara Falls, as a single tourism resource destination, 'evolved into a permanent state of maturity in which aspects of consolidation, stagnation, decline, and rejuvenation are interwoven and constant' (Getz 1992: 752).

External and Internal Factors Affecting the TALC

Although the TALC is a useful concept for describing the evolution of tourism development, what stages and/or duration of a stage will be experienced by a tourism destination is largely subject to a number of factors, and therefore, no perfect formula can be applied to different areas. Butler (1980: 11) observed that 'the shape of the curve must be expected to vary for different areas, reflecting variations in such factors as rate of development, . . . government policies, and number of similar competing areas'. Additionally, many other factors can affect the shape of the model, including political unrest, terrorism and natural disasters among others. These factors can be categorized into two groups: internal and external factors (Agarwal 1997). Internal factors include those inherent to a destination (i.e. uniqueness of resources and attractions, local residents and their attitudes towards tourism development and gradual deterioration of tourism resources) and associated management, service practices and qualities. Butler (1980: 9) argued that 'only in the case of the truly unique area could one anticipate an almost timeless attractiveness'. That is, a destination with a unique tourism resource can permanently attract tourists so long as external factors detrimental to the tourism industry do not occur (e.g. natural disasters). If this is true for a destination, then it is less likely to decline permanently. In contrast, the rise and fall of a destination (e.g. Niagara Falls; see Getz 1992; Huangshan Mountain and the Great Wall; see Bao 1998) is more likely to be experienced because tourism, as an open system, is highly responsive to many other external factors (Gunn and Var 2002).

Among the external factors that have been examined in previous studies are producers, consumers and regulating authorities (cf. Keller 1987). For instance, individual entrepreneurs usually play an important role in introducing new elements to the life-cycle and in helping to rejuvenate parts of the industry as evidenced in Lancaster County (Hovinen 2002) and the Gold Coast in Australia (Russell and Faulkner 1999) (for a detailed discussion of this issue, please refer to Russell 2006). In addition to entrepreneurs, tour operators can also play a significant role in the TALC. A study of Cyprus by Ioannides (1992) indicated that a few large tour operators contributed significantly to the growth of tourism growth through charter flights and inclusive tour packages. Similarly, another study of Paradise Island, Bahamas, undertaken by Debbage (1990), showed how multinational corporations controlled and influenced visitor flow to the area through imperfect competition and oligopoly.

In the case of consumers, visitors' changing preferences and needs are partial driving forces for the rise and fall of a destination (Butler 1980). For instance, the emergence of the demand for pursuing nature-based tourism or ecotourism in the past two decades has made protected areas and National Parks (which are often peripherally located) into popular tourism destinations in both developed and developing countries. Finally, for regulatory authorities, the rate of tourism development in the three island nations of Papua New Guinea, the Solomon Islands and Vanuatu was significantly influenced by the pre- and post-independent government of these nations (Douglas 1997). Another example of government's role in tourism development is Cyprus (Ioannides 1992) where tourism was used as a means for economic diversification stimulated by the government through economic incentives and loan programmes. In addition, the Greek Cypriot government also took measures to reduce the tourist growth rate and regulate the geographic distribution of lodging development.

Environmental, Social and Economic Changes as Related to the TALC

Environmental, social and economic situations of a tourism area will inevitably change over time as an area moves from the exploration to the post-stagnation stage (Butler 1980). More often than not it is found that the economic gains of a destination are concomitant with an increased decline in an area's environmental quality and an increasing loss in the authenticity of local culture and customs.

In the context of parks, according to Boyd (2006: 125) 'user levels are low and no noticeable impact occurs on the environment' in the exploration and involvement stages. However, a growth in visitor numbers to a park means yet more facilities and services are required, which may in turn pose threats to the ecological integrity of park resources or even result in the transformation of a natural environment into an urbanized area. For instance, Banff National Park in Canada has suffered a serious environmental deterioration as a result of human impacts. The town of Banff with a population of 7,600 in the 1990s is 'located in some of the highest quality wildlife habitat in the park' (Banff-Bow Valley Study 1996: 137). By 2002 the population had increased to approximately 9,000 (Clevenger *et al.* 2002). Likewise, the gateway communities surrounding Yellowstone have experienced rapid and uncontrolled development since the early 1980s (Ansson 1998). Within the 18-million-acre Greater Yellowstone Ecosystem, the population has increased by over 12 per cent since 1990 to more than 322,000 people in 1998. Communities in the Yellowstone region were largely urbanized as a result of tourism development. Environmental problems associated with tourism urbanization such as pollution, smog, crime and overcrowded condition are now not unknown in the Yellowstone region (Ansson 1998).

In terms of tourism development on social change, a plethora of studies have been conducted to examine residents' attitudes towards tourism development (see Harrill 2004 for a literature review on this topic). Butler (1980) noted that attitudes held by local residents towards visitors and tourism development may undergo a process from euphoria through apathy and irritation to antagonism as suggested by Doxey (1976) in his 'Irridex'. This inverse relationship between development of the life-cycle stages and resident impacts was also supported by Martin and Uysal (1990). In addition, Tooman (1997) observed that under circumstances in which tourism becomes the dominant economic sector (regardless of the particular stage), social welfare indicators failed to show significant improvement.

Finally, in the case of economic development across the six stages of the TALC, income tends to increase rapidly from the involvement to the development stage. In the meantime, parallel to the increasing income is an increased tourism income leakage from locals to outside investors (Tooman 1997). During the consolidation stage, the local economy will be dominated by tourism (Butler 1980) and a few large-scale, corporate enterprises become the dominant economic participants (Tooman 1997).

The question thus arises—are these effects noticeable in the case of a National Park in China?

THE STUDY AREA

The ZNFP is located in the north-western Hunan Province, China, approximately 30 km from Zhangjiajie City (in former Dayong County and Dayong City) and 385 km from Changsha, the capital city of the province. The park lies between 110°24′ to 110°28′E to 29°17′ to 29°21′N, covering an area of 4,810 hectares. It was the first national forest park established in 1982 by the State Council of China. It was then combined with two adjacent nature reserves, Suoxiyu Nature Reserve and Tianzi Mountain Natural Reserve, to be attributed status as a World Heritage Site by UNESCO in 1992 under the title 'Wulingyuan Scenic and Historic Interest Area' (WSHIA). More recently, the park was assigned two more titles, National Geological Park in 2000 by the Ministry of Land and Natural Resources and World Geological Park by UNESCO in 2004. Consequently, at least three agencies, State Forestry Administration (forest parks), Ministry of Construction (scenic areas) and Ministry of Land and Natural Resources (geology parks) are directly involved in the area's management (Deng, Bauer and Huang 2003).

The ZNFP is most known for a unique natural landscape characterized by thousands of quartzite sandstone pillars, ranging from 50 to 300 m in height, extending from the valley floor. These pillars are distributed in the park's six scenic zones: Yellowstone Village, Gold Whip Stream (GWS), Kidney Village, Shadao Ravine, Pipa Stream and Yuanjiajie (see Figure 3.1). These pillars and adjacent scenery can be viewed through hundreds of scenic spots and viewing platforms in the park. Currently, sightseeing by the world's highest sightseeing lift, cable cars, bus shuttles or hiking is the most prevalent tourism activity in the park. With its combined charm of uniqueness, elegance, wildness, serenity and sense of mystery, the park attracts millions of tourists from both China and abroad.

The area is also home to a large amount of rare and endangered flora and fauna, and is a valuable place for environmental education and scientific research. In addition, the park and surrounding areas are also rich in cultural and historic resources. The majority of local residents consist of several unique ethnic groups (e.g. the Tujia, Bai, and Miao) who have been living in the area for thousands of years (Gu and Zhong 2005).

RESEARCH METHODS

Triangulation is used for this study, whereby data are collected through personal in-depth interviews, questionnaire surveys and secondary data

Figure 3.1 Geographical location of Zhangjiajie National Forest Park.

sources. Survey results reported in this study are part of a comprehensive study about the park. In-depth interviews were conducted from July 2004 to November 2005. During this time period, we visited the main attractions and surrounding facilities in the park and interviewed 22 individuals with diverse backgrounds. Interviewees were asked to answer questions like 'How has the park been developed since its establishment in 1982?' and 'What are the consequences of tourism transformation in the park?' among others. The interviewed individuals are two Zhangjiajie City officers, five park managers and staff, five professors from three universities, three local tour guides, three local residents and four visitors. Some of their comments are incorporated in appropriate places in the latter part of this chapter.

In addition, two questionnaire surveys were undertaken between March 2004 and November 2005. The first is about visitors' perceptions of tourism development of the park and its impacts on their tourism experience as well as on local communities. This structured and self-completed survey was conducted at the main scenic spots of the park, where visitors were randomly approached. Those who agreed to participate were provided with a questionnaire that asked about their trip characteristics, tourism motivations, perceptions of tourism development and perceptions of the authenticity of local cultures. A total of 480 visitors were approached and 400 were willing to participate. Of these, 73 participants did not fully complete their questionnaires. These incomplete questionnaires were then excluded from analysis. Thus, data analysis was conducted based on 327 usable questionnaires.

This survey found females (55.9 per cent) numbered slightly more than males (44.1 per cent). The majority of participants were young (62.1 per cent aged below 39 years), with the mean of 32.6 years. Most participants were also well educated (67.9 per cent had a university degree). In addition, 33.3 per cent of participants had an annual family income over 25,000 RMB. As Ryan and Gu (2007) comment in a study of domestic visitors usage patterns of the Chinese hotel industry, these apparent skews in socio-demographics in Chinese domestic tourism tend to reflect the actual patterns of visitation that are due to the recent emergence of a well-educated, comparatively afflu-ent middle class in China, and one that tends to be urbanized.

The majority (71.9 per cent) of participants reported a stay in the park for three days or more, with the average length of stay being 3.6 days. Most participants (72.7 per cent) were first time visitors while another 27.3 per cent had visited the park more than twice. This percentage is comparable to recorded intention to revisit, for 32.3 per cent of respondents intended to visit again.

The second survey examined how local residents have been affected as a result of tourism development related to the park. The survey was adminis-tered at two villages under the jurisdiction of the park: Zhangjiajie village and Yuanjiajie village. According to the park census data of 2003, Zhang-jiajie village had 1,527 residents in 455 households, and Yuanjiajie had 404 residents in 136 households. A stratified random sampling method was used to identify participants. A copy of the questionnaire was then given to those households identified using this approach. The survey was conducted face-to-face in these households. Each household was asked to complete the questionnaire by any member over 18 years of age in the household. A total of 29 questionnaires collected from those 200 households were not completed or were left blank, resulting in 171 usable questionnaires for further analysis.

The majority of respondents from this survey have lived in the area for quite a long time with the average length of residency being 15.1 years. The average family members were 4.6 persons, of whom two were working in tourism-related businesses or services. In addition, most household mem-bers had a lower level of education with three members on average holding a high school degree or less. The reported total annual family income in 2004 varied from a minimum of 2,000 RMB to a maximum of 100,000, with the average being 25,958 RMB. In the same year, the average tourism-related income was 15,788 RMB, ranging from to 80,000 RMB. Thus, the average tourism-related income accounted for 60.8 per cent of total family income in 2004.

The main secondary data sources are publications (i.e. peer-reviewed journals, books, news articles, environmental reports etc.) about the park which are largely available in Chinese. Maps used in this study include a topographic map (scale: 1/10,000) published in 1983, a land use map (scale: 1/10,000) published in 1996 and two TM images obtained in 1987 and 1998. Spatial patterns of transformation are analysed based on these two images using an unsupervised classification method. In addition, visi-tor arrivals and other information were obtained from the parks' annual statistic report prepared every year from 1982 by the park administration office. Secondary data cover over 20 years.

RESULTS

Tourism Development Stages and Pushing Factors

Zhangjiajie was famous for its unique scenery long before the area was officially established as a national forest park in 1982. For instance, as early as the Song Dynasty (960–1279), Zhangjiajie was visited by poets who were astonished by the unique beauty of the area as stated in their poems (Gu and Zhong 2005). However, the area was not known as a popular tourism destination until the early 1980s. The following is a description of the tourism-development process for the park based on the TALC stages.

Exploration Stage (1978–1981)

The park was formerly a state-run forest farm established in 1958 under the jurisdiction of Dayong County. Planting and logging were the two main activities of the forest farm before 1978. This practice of forestry gradually changed after 1978 when Chen Ping, a staff member with the Forestry Department of Hunan Province, published a paper, 'Travel in Zhangjiajie' in a national magazine, *China Forestry*, to introduce the natural scenery to the public for the very first time (Zheng 1999; Gu and Zhong 2005). However, the park did not draw wide public attention until 1980 when the *Hunan Daily* published an essay titled 'Zhangjiajie: An Undiscovered Scenic Jewel' by Wu Guangzhong, a famous Chinese painter, who described Zhangjiajie as incomparable when measured against many other famous Chinese mountains (Gu and Zhong 2005). His comments were substantiated by other painters and photographers who were also attracted to the area, making the area even better known and publicized through their publications and photographic exhibitions.

During this period, there was no road/trail construction or facilities developed for tourists although painters, photographers, journalists, scientists and other adventurers and explorers came often to visit the area. According to Liu (1999), a total of 88,000 individuals visited the forest farm during this time. Approximately 1,000 of them visited the farm on a single day: 1 October 1980 (the Chinese National Day), (Gu and Zhong 2005). The annual average arrivals were 29,333 people at this stage.

Local government quickly realized that developing tourism could generate huge economic gains both locally and regionally. During this short time period, several leaders from the provincial government (e.g. the general secretary and governors) and the Ministry of Forestry (renamed the China State Forestry Administration in 1997) visited the farm. They emphasized the importance of a scenic zone plan, the maintenance of natural beauty and the construction and expansion of roads connecting the farm to Dayong, the then county seat of Dayong County. In addition, they examined the feasibility of tourism development in the park. In October 1980, Dayong County decided to increase accommodation facilities, maintain and repair roads and solve the problem of electricity shortages to meet the increasing demands and needs of tourists (Gu and Zhong 2005). In the same year, an administrative sector responsible for tourism development of the area was created in the county. In December 1981, a nationwide forest tourism symposium held by the then Ministry of Forestry in Beijing prepared a 'Memorandum on Forest Tourism Development' which listed the farm as one of seven forest tourism pilot test places in the country. In the same year, officers from the China National Planning Committee and the then Ministry of Forestry proposed the farm be set aside as a national forest park (Gu and Zhong 2005).

Involvement Stage (1982–1988)

Due to the unique landscape and scenery being increasingly publicized through those pioneers, yet more visitors were attracted to the forest farm. The increasing number of informal visitors and the direct economic benefits they brought to the area led the farm to be officially established as a national forest park in September 1982. From 1982 to 1988, a total of 2,215,100 visitors were attracted to the park with annual average arrivals being 316,443, about 11 times the number for the exploration stage.

Facilities began to be built to accommodate increasing visitors. In 1982, the first hotel, Golden Whip Hotel, with 260 beds was opened. In 1984, a major road was built to connect the park with its nearest town: Dayong (renamed as Dayong City in 1985 and Zhangjiajie City in 1994), making the park more accessible for visitors arriving from across the country. In the same year, a railway was constructed to connect the city to other parts of the country through the national railway network. The railway then became the primary access point to the city during this period (Zheng 1999). The construction of the railway was seen as a turning point for after 1984 packaged tours began operating in the destination. Also in this stage of the TALC, more lodging facilities were built within the park boundary (e.g. at Luoguta, Shuiraosimen and Yuanjiajie). As shown in Table 3.1, the number of family hostels/hotels/beds increased from 0, 1, 260 respectively in 1982 to 18, 25, and 2,590 in 1985. Likewise, the number of stalls and stores increased from 20 in 1982 to 45 in 1985.

As with the exploration stage, painters, journalists and other celebrities continued to play an important role in publicizing the park through their visits and publications. However, the difference was that some were now being invited by the park administration and local government. For example, the park invited over 30 journalists and editors from Hong Kong, Macao and Taiwan in 1986. Also, the park administration and local government took a further step in going outside of the park to market the area. Six promotional

Table 3.1 Lodging and Shopping Facilities in the ZNFP from 1982 to 2004

Year	Beds	Hotels	Family Hostels	Stalls, Stores
1982	260	1	0	20
1985	2590	25	18	45
1990	4020	32	60	190
1995	7080	42	76	280
1999	8585	49	196	326
2004	5005	35	125	302

Source: Zhangjiajie National Forest Park Administration (2006).

teams were organized to travel across the country holding photographic exhibitions in 1987. The Tourism Bureau of Hunan Province displayed the park's pictures and photographs in the United States and Japan in the same year as part of its own marketing promotions.

In this period scientists and researchers from the China Academy of Science and other universities began to study the geology, vegetation, climate and landscape of the area. Government officials from the Ministry of Geology, National Tourism Administration and National Environmental Protection Administration also often visited the area. In addition, leaders from the central government including the then prime minister and VIPs from foreign countries (e.g. the Belgian prime minister) visited the park. These visitations increasingly made the park even better known and more popular as a tourist destination.

The government still played an essential role in the development of the park during this stage. A 'Tourism Construction and Development Leadership Group' was formed in 1982 under the Xiangxi Autonomous Prefecture of Tujia and Miao Nationalities, the only autonomous prefecture for minority nationalities in Hunan. In 1983 Zhangjiajie National Forest Park Administration, a government agency at the county level was established to manage the park as well as three villages: Zhangjiajie village, Yuanjiajie village and Xiejiayu village. The park prepared a book, *Interpretation for Tourist Guides*, in 1984. In the same year, a vocational tourism school, Dayong Vocational Tourism School, was established. One year later the first travel agency in the area—Zhangjiajie National Forest Park Travel Agency was created. Subsequently, in 1988, Dayong Travel Agency, affiliated with the China International Travel Agency, was established. The first hotel, Dayong Hotel in Dayong city, commenced construction in 1985.

Government also played an important role in planning and regulating the park in other ways. The first plan, *Zhangjiajie National Forest Park Master*

Plan—1983–1985, was prepared in 1982 by several sectors of the province. Several regulations were enacted during this stage, including the first by Dayong County in 1983. In 1985 Xiangxi Autonomous Prefecture of Tujia and Miao Nationalities approved a foreign co-operation plan of economics and technology, emphasizing a transition of economic activities from being a traditional agriculture-oriented zone to one related to tourism. In 1985 the state council agreed to build a civil airport, Lotus Airport, near the city. In 1988 Dayong City was promoted as a prefecture-level city and Wulingyuan District, which directly manages the park and other nature reserves, was established in the same year.

It is worth mentioning that the construction of the park in this stage was primarily funded by the central and provincial governments with a total amount of 12 million RMB being invested in the construction of the park (Xia 2004).

Development Stage (1989–1999)

This period marked yet another rapid increase in visitor numbers. The park accommodated 381,500 visitors in 1989. Ten years later, this number jumped to 1,187,400 in 1999 (Figure 3.2). The total visitor number reached 7,941,900 during this period with annual average arrivals being 721,991, approximately 2.3 times that of the involvement stage.

Parallel with the rapid increase of visitors was the rapid development of the park and surrounding areas, which was characterized by (a) internationalization, (b) regionalization, (c) modernization and (d) physical transformation (as will be discussed at more length in the next section).

First, the park director was invited by Maine University to visit several National Parks in the United States in 1989. The first international

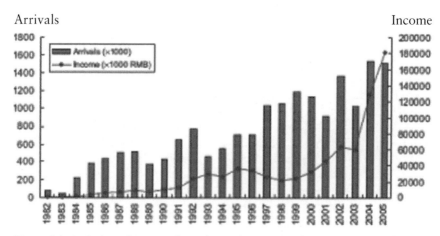

Figure 3.2 Arrivals and income from the park. Source: Zhangjiajie National Forest Park Administration (2006).

co-operation agreement between the city and travel agents of the former Soviet Union was signed in the same year. The first Zhangjiajie International Forest Protection Festival advocated by the then vice-governor of Hunan Province was held in 1991 and has become an annual event since then. Several international academic conferences and several sports festivals were held at the park (e.g. 1991 International Ecology and Forestry Conference, 1992 international Forest Protection Cup climbing contest etc.).

The park and its two adjacent nature reserves gained its international reputation by being listed as a World Natural Heritage Site by UNESCO in 1992. During this period, foreign capital and technology were used to build hotels and other tourism-related facilities. For instance, Dayong Hotel was expanded and revamped using foreign investment and was renamed Xianglong International Hotel. A Taiwanese company invested in a cableway with a horizontal distance of 863 metres that was built in 1997. In addition, the so-called world's highest outdoor sightseeing lift commenced construction in 1999 and was finished in 2002. The project was jointly invested with a total investment of 120 million RMB by three companies including Mackarl in the United States. The lift itself was produced by a German company.

Second, the establishment of the WSHIA with the ZNFP as the core attractive zone created an additive regional effect, in that the whole scenic area covering a total area of 397 square kilometres with a population of 44,954 became one of largest tourism destinations in the country. The existence of the scenic area also promoted the development of many other scenic areas and spots in the western province.

Finally, during this time period, the park and surrounding areas became increasingly modernized. Zhangjiajie airport was officially opened for commercial use in 1994. Dayong City was renamed as Zhangjiajie City in the same year. The city was specifically renamed using the park's name, thereby simultaneously recognizing the fame of the park, and further utilizing it to attract capital to the wider region. The three Chinese characters also imply 'opening family door to outside world'. From that date visitors from China's major cities including Hong Kong and Macao could now fly directly to the city. As of 1999, apart from two airlines connecting to Hong Kong and Macao, another 23 domestic airlines were operating flights to and from the airport. The airport accommodated appropriately 500,000 visitors in that year.

As in previous stages, government continued to play an essential role in speeding the development of the park and surrounding areas. For instance, in 1993 the WSHIA Administration was established. In addition, *The WSHIA Master Plan* prepared by Tongji University and Hunan Construction Commission was approved in 1993. In the same year, the *Dayong City Tourism Industry Management Regulation* was promulgated. The then state president Zheming Jiang visited the park in 1995 and proposed 'to develop Zhangjiajie into a domestically and internationally known tourism destination'. In the same year, the first tour train in the province from Zhangjiajie to Guangzhou, named *Zhangjiajie*, was operated. One year later, in 1996, priority was placed on tourism by the city municipality with 'tourism as the locomotive to drive regional economic development'.

With this said, the role of entrepreneurs, through their multinational investments, cannot be ignored. Without those investments, the park could not have been developed and transformed as it is.

Consolidation Stage (2000–Present)

The park entered the consolidation stage in 2000. During this stage, a total of 7,496,800 tourists visited the park (from 2000 to 2005) with an annual average of 1,124,947 tourists, about 1.7 times as many as that for the development stage (see Figure 3.2). Facility development in the park began to slow (as discussed in the next section) while priorities were placed on external factors including regional ground transportation expansion and improvement for domestic travellers and marketing and advertising for international visitors, which could have long-term impacts on the sustainable development of the park. The area's economy became largely tied to tourism during this time as shown in the following.

A highway connecting Changde, a major city 137 km to the east, to Zhangjiajie was started in 2002 and was completed in 2005. As a result, the travel time from Changde and Changsha to the park was reduced by half. More importantly, the park became even more accessible through ground transportation than before because of the connection of this route to the rest of the national highway network.

Since 2001, continuous efforts have been made by governments at various levels to market and advertise the park to international audiences with South Korea being the principal target. In this period the general secretary of the city led several groups of delegates from travel agencies and relevant sectors to visit Dalian where many Koreans are concentrated. They also visited Korea to directly market the park and WSHIA to the Korean market. In addition, the China National Tourism Administration (CNTA) and the Tourism Administration of Hunan Province also endeavoured to advertise the area to Koreans. As a result, Korean visitors increased in numbers from 102,400 in 2002 to 220,000 in 2004 and jumped to 360,000 in 2005 (Deng 2006; Wan, Du and Tian 2004). It is anticipated that Korean visitors will continue to increase in forthcoming years given that a non-stop flight from Zhangjiajie to Seoul began to operate in 2005. Moreover, in order to impress Korean visitors and make them feel at home, the park hired over 400 tourist guides who can speak fluent Korean. Currently, even the local peasants and peddlers can speak simple Korean in order to sell their products. Korean became one of three languages in the park's interpretation signs (the others are Chinese and English).

Since the Korean market primarily comprises older visitors and that country's total population is not large, the park has a concern that the Korean market will eventually be saturated. With this in mind, the park and the WSHIA began to look for substitute markets with North America as the main target. A recent first ever publication of an article about the area by Winchester (2007) in the *New York Times* marks the potential beginning of a new wave of international marketing or even a new wave of tourist arrivals for the park.

Environmental, Social/Cultural and Economic Changes

Environmental Changes

The park's environment was well protected before the exploration stage. However, the natural appearance of the park has gradually changed since the involvement stage onward. This happened drastically during the development stage wherein the park environment was largely transformed. The spatial pattern of residential houses and built facilities at the exploration phase was dispersed and dotted, but evolved into a linear pattern during the involvement period with more lodging facilities built around Luoguta, Shuiraosimen and Yuanjiajie. The development stage featured further ribbon development. The success of the first hotels acted as a catalyst for more hotels to be built, and these were built behind the first generation of hotels because land close to the roads became less available. Thus the hotels spread out little by little along the road leading to the main entrance of the park and a ribbon development pattern was formed. Some indoor businesses were moved outside as stalls and booths for the visitor's convenience, and of course to permit the retailers to more easily access the tourists. During this period, two more visitor villages emerged centred around Shuiraosimen and Yuanjiajie respectively, while Luoguta as a residential location and visitor village was rapidly expanded. As can be seen from Figure 3.3, the park was fragmented in 1998 when compared to 1987. This is particularly true in Shuiraosimen, Yuanjiajie and downstream GWS.

With the process of physical transformation, the number of beds in the park increased from 4,020 in 1990 to 8,585 in 1999. The rapid transformation of the park environment and the whole WSHIA shocked officers from the UNESCO world heritage commission when they visited the park in 1998. They warned that uncontrolled tourism development in the park was destroying its scenic beauty. Partially in response to this warning, the park has begun to implement a removal plan since the end of 2000. Simply put, it is thought the park has developed visitor facilities to their fullest and it has become time to control this development more tightly than in the past. This control represents a consolidation phase. From 2000 onwards, a number of buildings located in Shuiraosimen and Yuanjiajie that were considered a threat to the ecological integrity of the park environment were demolished and removed. Consequently, the number of beds fell significantly from 8,585 in 1999 to 5,005 in 2004 (see Table 3.1).

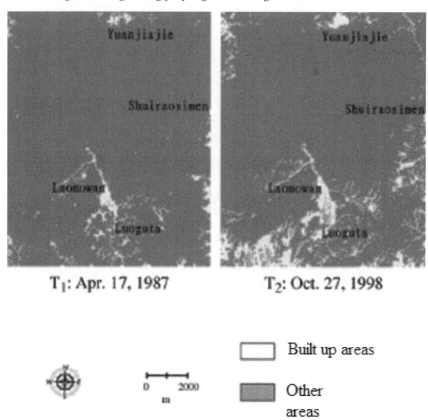

T₁: Apr. 17, 1987   T₂: Oct. 27, 1998

☐ Built up areas

▨ Other areas

Figure 3.3 Increasing urbanization within the park: 1987–1998.

As a result of physical change and commercialization of the park environment, the park's air quality and ground-water quality have deteriorated over time and problems peaked at the development stage. In addition, wildlife and their habitats have been largely disturbed. With reference to the air quality of the park, the main sources of pollution are coal and oil consumption, and gas emissions from vehicles that are concentrated in three locations: Luoguta, Shuiraosimen and Yuanjiajie (Shi 2005). As shown in Table 3.2, coal consumption was 70t in 1981 (the exploration stage). This increased to 1,200t in 1985 (the involvement stage) and jumped to 6,100t in 1998 (development stage). Accordingly, the amount of pollutants for SO_2, NOx and smog increased exponentially from the exploration to the development stage. Shi (2005) also reported, based on a pollution index analysis, that air quality in Luoguta was moderately polluted in 1984 and 1988 (the involvement stage), but was severely polluted each year from 1993 to 1999 (the development stage).

Ground-water in the park has also increasingly deteriorated over time as a result of the increasing consumption of water and discharge of sewage

Table 3.2 Measures of Increased Park Environmental Degradation

Pollution Source	1981	1982	1985	1990	1995	1998
Coal Consumption (t)	70	500	1200	2300	6300	6100
SO$_2$	4.20	30.00	72.00	138.00	378.00	366.00
NO$_x$	0.25	1.81	4.34	8.33	22.81	22.08
Smog	6.44	46.00	110.40	211.60	579.60	561.20

Source: Shi, Q. (2005) *Environmental Impact Assessments of Forest Parks.*

(see Table 3.3). For example, water consumption increased from 6,000t in 1981 (the exploration stage), increased to 20,000t in 1985 (the involvement stage) and then reached 250,000t in 1999 (the development stage), resulting in an increased discharge of sewage from 5,000t in 1981 to 225,000t in 1999. As is the case for coal consumption, Luoguta consumed the largest amount of water and is the main pollution source for the GWS. Before and at the beginning of the establishment of the park, the stream was regarded as unique in Asia (Xia 2004), and was 7,500 m long with transparent, crystal water and well-protected vegetation. However, from 1990 onwards, the water quality in the stream deteriorated with each passing year (Shi 2005). For instance, according to the local environmental monitoring department, the average value of COD (Chemical Oxygen Demand) sampled in the Laomouwan segment of the GWS increased from 0.79 mg/L in 1984 to 2.5 mg/L in 2000 (Quan 2003).

The main reason for deteriorating water quality was the rapid transformation of Luoguta located in the upper reaches of the stream. Luoguta directly discharged its human waste from the hotel and other facilities into the waters of the stream. According to Yang and Zhou (2005), the daily highest discharge of sewage was 2,800t and the annual sewage drainage was approximately 700,000t in 2002, which was discharged into the stream without first being treated (but by 2005 a waste-water treatment centre was finally in operation). Given the relatively small volume of flow of the stream (i.e. daily flow volume is only between 34,000 and 86,000 cubic metres even in an abundant water season), the discharge of sewage in large quantities

Table 3.3 Water Consumption in the Park

Year	1981	1982	1985	1990	1995	1999
Water Consumption (ml)	6	10	20	160	230	250
Sewage Discharge (ooos)	5	9	18	144	207	225

Source: Shi, Q. (2005) *Environmental Impact Assessments of Forest Parks.*

is beyond the self-purification capability of the stream, resulting in the contamination of its water quality (Yang and Zhou 2005). One local guide described the serious situation of water pollution in GWS thus: 'I have not drunk the water directly from the stream at all after the middle 1990s as I did when I was young. I have always seen the white foams on the surface of water. But the colour of stones under the stream has become grey and black'.

Biodiversity in the park has also been adversely impacted by tourism development. Activities such as quarrying rock, building roads and paths, logging, planting and constructing gardens pursued by visitors are among the negative impacts affecting wildlife behaviour and habitats. These activities have destroyed the integrity of the natural ecosystem, reduced the migratory and movement frequency of animals, reduced both the spatial and temporal diffusion of plant seeds and fragmented wildlife habitats. Consequently, the park biodiversity is lower than before. According to the data derived for this study, the number of species and mammal population of the park have decreased since 1980. Significant examples include species such as red dog (Cuon alpinus), musk deer (Moschus berezovskii), goral (Nemorhedus goral), asiatic black bear (Selenarctos thibetanus), clouded leopard (Neofelis nebulosa) and masked palm civet (Paguma larvata), all of which are no longer present within the park. In addition, the population of rhesus monkeys (Macaca mulatta) has fallen from approximately 30 in 1982 to seven in 2003. Shi (2005) also examined the fauna diversity of the park by looking into the extent to which local residents can perceive the existence of a particular species. He found that 87 per cent, 80 per cent and 94 per cent of respondents reported that birds, beasts and reptiles were respectively visible in the late 1970s (the exploration stage). However, this perception rate reduced to 56 per cent, 38 per cent and 39 per cent, respectively, during the late 1980s (the involvement stage), and dropped rapidly to 35 per cent, 16 per cent and 7 per cent, respectively, during the late 1990s (the development stage).

Socio-Cultural Changes

In 1982, there were 1,092 permanent residents in the park, including a small non-agricultural population of 174. Twenty-three years later, the permanent population increased to 3,495 with the majority of the population being non-agricultural residents, totalling about 2,630 in 2005 (see Table 3.4). This growth in total population is a direct consequence of people migrating from the rest of the country to the park seeking better jobs and higher incomes. This drastic increase of a non-agricultural population signifies the transition of the area from being agriculturally oriented to tourism oriented. The survey for this study indicated that 60.8 per cent of local residents' incomes were primarily from tourism. In addition, 87.9 per cent of local residents reported that their quality of life was positively affected by tourism

Table 3.4 Population Change in the Park

Year	Total Population	Non-Agricultural Population (NP)	NP as % of Total Population
1982	1092	174	15.9
1985	1723	643	37.3
1990	2828	1098	38.8
1995	3291	1193	36.3
2000	3310	2590	78.2
2005	3495	2630	75.3

Source: Zhangjiajie National Forest Park Administration (2006).

development. And 93.4 per cent of respondents preferred developing tourism for more income rather than limiting its development for a quiet living environment. Similar findings were also reported by Wu and Liu (2001).

Previous studies have found that tourism developments in the park have generated negative impacts on local residents in terms of the retention of local cultural traditions (Wu and Liu 2001). In our visitor survey, 86.3 per cent of respondents reported that they did not feel a strong ethnic ambience during their stay in the park and only 7.7 per cent of participants regarded the local residents as being hospitable and friendly. This suggests that local residents' friendliness towards tourists has gradually faded over time. Instead, they are more interested in earning more and more income. In the survey, perceptions of visitors' dress was used as one proxy for the tolerance of visitors and 54.5 per cent of respondents reported that they 'liked' or 'liked very much' the way tourists were dressed, while only 2.5 per cent of them expressed a very negative view about the way tourists dressed; the remainder being neutral in their attitudes to this item.

With tourism development in the park, opposition and discontent among local permanent residents increased and peaked in the consolidation stage. One major issue is related to residency relocation, a plan implemented since 2001 in response to the UNESCO warning that the whole WSHIA, particularly ZNFP, has been largely transformed to the point where the values of the park were being threatened. For example, a recent survey conducted by Li (2007) indicated that 56 per cent of residents were willing to move outside of the core areas. However, a significant number of residents remained reluctant to do so, mainly for two reasons. First, they were not satisfied with the relocation compensation that varied between villages. Second, finding a new job in a new place was not easy for many of them. It seems that in the past the local government has not heeded local residents' opinions, which has resulted in several petitions against the local government (Li 2007).

The tourists' experience, particularly nature-based experience, was largely impacted adversely by increasing commercialization and the transformation that took place in the transition from the exploration to the development stage. One tourist interviewed for this study said it felt like he was in a commercial zone of a big city when his party arrived in Luoguta. In addition, many peddlers who are local residents are present in the park. According to a tour guide, although local residents earned extra money by selling souvenirs and other local products to tourists, they also negatively affected tourists' experience by persistent pestering; especially when it was perceived that they were 'forcing' tourists to buy their merchandise. A local tourism company manager also complained that some areas within the park are too commercial, causing the disappearance of local natural and cultural characteristics and so putting at risk the features that attracted tourists in the first place (Zhang and Ouyang 2004).

Landscape change and some destruction have negatively impacted on the aesthetic quality of the park. It has destroyed the integrity and authenticity of the natural landscape that visitors come to see. One study evaluating the park's landscape and its aesthetics indicated that the main negative factor affecting the quartz sandstone landscape of the park was the built environment of accommodation facilities and services (Zhou and Yu 2004). In this study, all 179 interviewees reported that the aesthetic quality has become worse since human facilities became embedded in the natural landscape.

Economic Changes

Although this area was only 30 kilometres from the nearest town, Dayong, in the early 1980s there were no major roads connecting the area to the rest of the country because of its mountainous nature. Local residents who lived in Zhangjiajie village and Yuanjiajie village mainly depended on agriculture and forestry to make a living with the average annual individual income being about 193 RMB in 1981. Local residents' income increased rapidly from the involvement stage onwards. For instance, the annual average income per capita increased to 4,000 RMB in 2002 (Administration of Zhangjiajie National Forest Park [AZNFP] 2004). According to one resident interviewed in Yuanjiajie, before 1982 the village was isolated from the outside world. It took him about four hours to walk from the village to the nearest highway. His life was entirely dependent on crops. The villagers often borrowed foods from other places when shortages of rice occurred, which was not uncommon. After 1983, increasing numbers of visitors came to the village. And the villagers began to make money by providing accommodation and food for tourists. The revenue from tourism in several families now exceeded that from agriculture. The revenue for the park increased, too. The gross product of the forest farm in 1981 was 268,000 RMB and the average annual personal income was 640 RMB for park staff (Wang and Zhang 1996). Since 1982 when the farm became a national forest park, park revenue increased rapidly during the involvement stage from 1981 when it was 558,800 RMB to 1988s 9,101,500 RMB. The total revenue during the period was 2,940,100 RMB with an average being 420,014 RMB. Although revenue fluctuated during the development stage, the general trend was upwards. Revenue increased from 6,844,200 RMB in 1989 to 23,808,000 RMB in 1999 with a total of 250,526,900 RMB with the average being 22,775,172 RMB, approximately 54 times the annual average revenue for the involvement stage. During the consolidation stage, a total of 511,642,300 RMB was generated from 2000 to 2005 with an average being 85,273,717 RMB, about 3.7 times the annual average revenue for the development stage (see Figure 3.2).

Tourism development in the park and the whole WSHIA promoted economic development for Zhangjiajie City; this was particularly the case during the consolidation stage. For instance, in 2002, 59 per cent of the city's tax revenues were from tourism and related sectors while in the early 1990s it was about 20 per cent (Liu 2005). In 2005 the city's tourism-related industry accounted for 54.4 per cent of the city's GDP (Liu 2006). The industry structure has been rapidly transformed from the primary sector and secondary sectors to a tourism-related tertiary sector-oriented with the proportions being 17.9, 27.7 and 54.4 respectively in 2005 (Liu 2006).

While acknowledging the positive effect of the economic development on local residents and the whole region, it appears that income leakage existed, at least from the perspective of local residents, with the increasing development of the park. For instance, because of the construction and operation of the sightseeing lift, fewer tourists took sedan chair rides. As noted by one local resident of Zhangjiajie village, 'My husband is a sedan chair carrier for visitors. His business has become worse since the sightseeing lift was in operation in 2002 because most visitors preferred to take the lift to the top of the mountain instead of hiking along the trail which is more physically and mentally challenging'. In our survey, when asked about the fairness of the distribution of tourism income among villagers, only 18.2 per cent considered that the income distribution was 'very fair' while 81.9 per cent of respondents reported it was either 'basically fair' or 'unfair'.

Entrepreneurs are the stakeholders that have received the most benefit. For example, the Mount Tianzi cable chair concession had sales of about 500 million RMB in 2002. This figure is more than the district's financial income of 300 million RMB in 1998 (Zhang 2002), and only a little less than the district's total revenue from entry fees of 660 million RMB in 2005. According to Zhang (2002), the economic benefits that local government and communities have received from tourism development is certainly less than profits made by the private investor sector.

To summarize, the park has been transformed considerably since its inception. Certainly the changes in physical appearance were the most salient impact, according to local residents in our survey. When asked about the extent to which local environment and socio-economic characteristics have been affected owing to tourism development of the park, 37.1 per cent of respondents considered that the natural environment was affected most, followed by the socio-economic structure (31.4 per cent), local life (18.6 per cent) and local customs (10.0 per cent).

DISCUSSION

The TALC has been rarely utilized in previous studies with reference to the development of National Parks and other protected areas and their associated tourism. Certainly the concept has not been applied to China's protected areas, including forest parks. This study examined its application to China's first national forest park—Zhangjiajie National Forest Park. In addition, factors affecting the park's tourism development and associated environmental, socio-cultural and socio-economic changes were also analysed. Findings indicated that the park has experienced four stages (i.e. exploration, involvement, development and consolidation stages) that confirm Butler's (1980) paper. For example, in the exploration stage the park was visited by pioneers who made the park increasingly popular among the public. At the involvement stage, government and public agencies began to provide or improve transport and other facilities for visitors. A mass tourism market began to emerge at this time. After that, the park entered a development stage characterized by a rapid increase of tourists as well as a decline in local control of the development. In addition, changes in the physical appearance of the park became very noticeable. Finally, at the consolidation stage, the area's economy became largely tied to tourism. Marketing and advertising were wide reaching and efforts were made to attract international tourists and to tap into other tourist markets. International tourism (e.g. South Korean) increased significantly. These four distinct stages that the park has experienced have supported neither Baum's (1998) assertion that the model has rather less value when applied to developing countries, nor Weizenegger's (2006: 137) argument that 'the model can be applied to protected areas to a limited degree only'.

The identification of these four stages was further confirmed when comparing tourist numbers and revenues from one stage to the next. At the involvement stage, the annual average arrivals were 11 times that of the exploration stage, while at the development stage, the annual average arrivals were 2.3 times that of the involvement stage. During the consolidation stage, this figure was 1.7 times that of the development stage. Although tourist numbers have increased from 2000 to 2005, the rate of increase was relatively low during the consolidation stage when compared to the previous stages.

Furthermore, it was found the model applies to the park very well by reference to the relative changes of revenues for each stage. Specifically, at the development stage, the annual average revenue was 54 times that of the involvement stage. However, this trend slowed significantly when the park entered the consolidation stage, wherein the number was about 3.7 times that of the development stage.

With this said, obvious discrepancies still exist. First, visitors outnumbered local resident's right from the beginning of the process and remained so until the park reached the consolidation stage. It was only in the later part of this consolidation stage that visitors again reached or even outnumbered the regional population of 1.56 million. This suggests that the ratio between tourists and local residents as a characteristic for the consolidation stage did not conform to usual understandings of the TALC in this case. However one must note the role of the 'open-door policy' and that, from 1978 onwards, China has seen rapid economic development. The rapid improvement of people's living standards and the formation of positive attitudes towards leisure and tourism have motivated travel, while the Golden Week (i.e. three annual seven-day national holidays) implemented in 1999 by the central government has stimulated travel. In this sense, tourism development of the park mirrors the national development. In addition, in this case, local and regional government policies have condoned if not specifically approved the use of the park as a population growth node in the pursuance of its policy of the use of tourism as a catalyst for economic growth.

A second difference from the standard TALC is that while Butler's hypothesis states local residents are the primary providers of accommodation at the exploration and involvement stages, in this case yet even more residents provided accommodation in the latter stages. For example, in our survey on local residents, 7.3 per cent of respondents reported they provided lodging for tourists before 1982, but this had increased to 29.0 per cent in 2004. This increase is partly explained by the ZNFP and local government encouraging local resident participation in hospitality provision in the form of family hostels as a means of income generation.

A third difference lies in the park having had its fullest infrastructure during the development stage. That is, road construction, artificial facilities and other ancillary facilities had either been completed or commenced during this stage (e.g. the sightseeing lift). No more spaces are now available for further facility development and road construction. As a result, physical carrying capacity has been reached. In addition, the environmental quality of the park has declined to a point below national standards, suggesting environmental capacity has been exceeded. This suggests that it was during the development, not stagnation stage as described by Butler (1980), that carrying capacities were reached. In contrast, environmental quality was improved during the consolidation stage because of the use of environmentally friendly buses, recycling of sewage and reduction of coal consumption (Bu 2006). Carrying capacities are not fixed. Rather, they are dynamic. Physical capacity in the park increased rapidly during the development stage as described earlier. For example, the construction of paved road networks, the sightseeing lift and the cableway system sped up the rate of tourist movements in the park, thereby increasing physical carrying capacity in terms of coping with visitor numbers, but equally the catalyst was laid for reductions in environmental quality. This distinction in itself indicates the complexity of measuring carrying capacities.

Fourth, tourist numbers have not fallen as a result of environmental decline and the loss of opportunity for an 'authentic' nature-based tourism experience. This could be explained by four reasons. First, the park's landscape is unique and second to none in China or even worldwide. Second, although the construction of the sightseeing lift is controversial, its main transport function may also be an attraction for tourists. Third, sightseeing is the prevalent activity in the park. Tourists visiting the park may not come for the purpose of being in close contact with nature. Rather, they come to marvel at the wonder of nature, and then be whisked away to other places. These tourists can be considered as 'soft' ecotourists who have a high requirement for convenience and comfort, e.g. in transport. This is particularly the case for Korean visitors, many of whom are elderly and in packaged tours with low levels of physical challenge. Such tourists are less likely to be critical of artificial facilities in the park. In this study the majority of respondents (54.4 per cent) were highly supportive of the construction of the cable trams while only 19.9 per cent were strongly against. Likewise, local residents' attitudes were also very supportive. For instance, 49.0 per cent of respondents were either 'supportive' or 'very supportive' of the existence of the lift and 58.0 per cent of them either 'agreed' or 'strongly agreed' with the construction of the trams. Finally, the majority of tourists (72.7 per cent in our survey; 84.2 per cent in Luo 2006) are first-time visitors. Thus, while crowding is reported by most visitors to be a major issue in peak seasons (Luo 2006) and visual disturbance adversely affects visitor's satisfaction, given that many are one-time visitors such negative experiences will only very slowly negatively impact on future visitor numbers. Thus, Plog's assertion that 'destination areas carry with them the potential seeds of their own destruction, as they allow themselves to become more commercialized and lose their qualities which originally attracted tourists' (cited in Butler 1980: 6) seems not to apply to a destination that is mostly visited by non-repeat visitors, as evidenced by this park.

A fifth difference is that while sightseeing of the unique scenery is and will continue to be the dominant tourism activity in the park, it can be anticipated that even when the park enters its stagnation stage, 'heavy reliance on repeat visitation' (Butler 1980: 8) will continue to be unimportant. This challenges Lundtorp and Wanhill's (2001) argument that the TALC model can only apply to those destinations whose visitors are 'repeaters'.

The park was made public by pioneers such as journalists, painters and photographers at the exploration stage who have continued to visit the park through the consolidation stage and beyond, and have in fact been joined by others of similar interests. In this instance the park displays a feature of visitor segments not serial in nature, but additive.

The evolution of the park's tourism development has been a function of many factors, including consumers, government, entrepreneurs and tour operators/agencies. First, the increase of visitors in a large number prompted the increase in facilities and services in the park to meet the needs of tourists. As noted, the park is 30 kilometres from Zhangjiajie City, and it takes at least two days to visit the most attractive scenic spots within the park. In addition, living in or adjacent to the park can offer more chances to view the park scenery. Thus, many visitors choose to stay in or adjacent to the park during their visitations. This has stimulated investors to build more facilities in or adjacent to the park, and consequently increases the transformation of the park environment.

Government has played a crucial role in planning, regulating and co-ordinating the park's development. The planning process involved several sectors from solely within the province at the involvement stage, but later development involved regional consideration with experts from other provinces becoming engaged in further planning. The planning was then carried out at the national level at the consolidation stage with policymakers and experts from the central government being the major actors. With this said, irrational planning at the early stage of the park's development has contributed to the environmental deterioration. In 1984, a master plan of the park adhered to the planning principle of maximization of visitor satisfaction, but ignored the fact that Luoguta lies in the upper area of the GWS watershed. Luoguta in the plan was regarded as the reception area of the park. Thus more and more waste-water flowed into the GWS with the increase of overnight visitors in this area. In 1992, the master plan of WSHIA including the park was approved by the Ministry of Construction of China. The idea of constructing a sightseeing lift was proposed in that plan. Since this park was designated a UNESCO world heritage site in December 1992, many environmental experts opposed this idea because of its magnitude and the fear it would destroy the integrity of the landscape. According to the Convention Concerning the Protection of the World Cultural and Natural Heritage adopted by the General Conference of UNESCO in 1972, protecting the integrity of the landscape is one of the goals for world heritage designations (Xie 2000). However, the sightseeing lift was subsequently completed due to the expected economic benefit.

This leads to a further lesson, namely, that with the evolution of the park's development even the government began to lose control of tourism development. Instead, entrepreneurs, through multinational investment, have largely impacted politicians' decision making. The aforementioned construction of the controversial sightseeing lift is a good example. The lift was operated in 2002, was then forced to stop operating a year later for the reason that it was not a project approved by local government, then the lift was reopened for business in 2004 after being appraised and approved by experts to be a safe and environmentally friendly construction.

Under the central government's policy of alleviating poverty through tourism, development tourism has come to be regarded as a new economic growth point by government officers at all levels (Zeng 2006). The tourism industry has been developed with more attention paid to its economic aspects and less to its negative environmental impacts. In order to promote the development of the regional tourism industry, local governments have adopted a policy of privilege for investors, including low prices for leasing land, tax exemption and rapid approval of applications. Furthermore, there is no legislation that deals with forest parks despite the fact that forest parks are China's most heavily impacted protected areas. For example, the existing Forest Park Management Regulation allows commercial development to occur within the park other than in rare natural/cultural spots or core scenic zones (Clause 11). Legislation is also inadequate for controlling tourism development within China's nature reserves (Han and Zhuge 2001). Similar shortcomings exist for scenic areas. For instance, Clause 2 in the Provisional Act of Scenic Area Management provides the basis for overlapping and multiple administration of protected areas because any nature reserve or forest park could be assigned the designation 'scenic area' so long as they meet given conditions. This gives rise to conflicting controls, while the legacy of most existing legislation and policies lies in the context of a centrally planned economy that is currently being slowly dismantled, thereby again generating mixed planning outcomes. Consequently, today China is increasingly assigning national scenic areas to the operation of concessionaires and private owners. Therefore, new laws or policies are necessary to deal with this issue to ensure environmental protection is not only legislated for, but also implemented.

Finally, tour operators/agencies have been playing an important role in bringing more tourists to the park since the involvement stage onwards, particularly at the consolidation stage when Korean tourists swarmed to the park as packaged tours. Without the co-ordinated operation between tour operators/agencies in both countries, the Korean inflow in large numbers would not be possible.

CONCLUSION

The park has experienced the first four stages as described by Butler in his seminal article. Since its establishment in 1982, the park has been transformed over the past two decades from a less accessible forest farm to a world-renowned tourism destination now accessible by trains, vehicles and aircraft. The macro-social and economic changes, since 1978 when China adopted an open-door policy, have seen governments at all levels, celebrities and entrepreneurs all playing essential roles as catalysts in this rapid transformation. Government officials, at all levels, have been major players from the very beginning in planning, regulating and directing where the park should go and how, through personal visitations and comments, policy-making, granting privileged land use for concessionaires, events organization and domestic/international marketing.

Although theoretically the park will enter a stage of stagnation and then a post-stagnation stage, it seems that dramatic and total decline is unlikely to occur because such a decline is unacceptable both economically and politically (see Agarwal 1994) given the region's heavy reliance on tourism and the public nature of the destination. Government intervention and regulation is inevitable. However, this does not apply to any given specific segment of the tourism market. For example, the Korean market, which started and rapidly increased during the consolidation stage, may experience its own micro-life-cycle, in that the market will become saturated and then decline or even exit from the destination. In foreseeing this possible future, the park has taken a proactive measure to start focusing on the North American market. This potential market may emerge before the Korean market declines and may also experience its own cycle from growth to stagnation and even to exit. Thus, in the long run, the area may be described by Handy's (1994) sigmoid curve, wherein old products may exist in tandem with newly introduced ones (also see Baum 2006). The sequential entry of new market segments may also follow the shape described by Haywood (1986), wherein a standard S-curve shape may not be followed in the same way by each segment.

This micro-life-cycle may also be experienced by any given segment of the domestic market. For example, repeat visitors may become less likely to visit if they have visited more than once because of the nature of activities being largely related to sightseeing. In addition, those who come to seek the authenticity of being close to nature may turn to other places after their first visitation because of the park's loss of naturalness and increased commercialization. It is hard to say that such micro-life-cycles, as experienced by a given market segment for the park, will significantly affect the long-term sustainability of the park, given its uniqueness and the fact that domestic visitors are and will remain as the main market component.

Butler (1980: 9) argued that 'only in the case of the truly unique area could one anticipate an almost timeless attractiveness'. With this said, the park should take the TALC as a warning by which long-term sustainability can be achieved or enhanced through constant adjustments, corrections and proactive actions. The park is a core component of the WSHIA, which belongs to the Category V (Protected Landscape) in the IUCN system of protected area categorization. Category V protected areas emphasize the interaction of people and nature. According to the IUCN, Category V protected areas should be 'models of sustainability' (Phillips 2002). This is in parallel with the TALC, in that maintaining or enhancing sustainability of a destination requires understanding the process of tourism development in any area (Butler 2006a).

The main limitation of this study lies in the fact that the park has not completed its cycle of development. Thus, it may be too early to say that the model entirely applies to the park. However, since the park has experienced the first four stages, it may enter its stagnation sooner or later. Therefore, from the perspective of sustainable development, the park and other government sectors should take pre-emptive measures to prevent its stagnation or decline so that long-term sustainable development can be achieved. It seems that the park and the municipality have anticipated this problem by extending marketing efforts to other international audiences such as North America. Nonetheless, it can be anticipated that the consolidation stage will be maintained for a long time considering the improvement of regional ground transportation which makes 'do-it-yourself' tourism possible for those who own cars.

Future research may need to be conducted to monitor the change in given segments of both the domestic and international markets. Also, the dynamic changes of local population should be monitored and examined in relation to the TALC model. For example, youth retention is an issue in the area. Owing to the low-paying jobs offered to local residents, young adults tend to migrate to other places (mainly large cities) to find better paying jobs. In the meantime, well-educated outsiders come to the area seeking good jobs and more incomes. These outsiders may not be as concerned about the environment as existing locals because they have not developed an emotional or place attachment to the park. This increasing emigration of young adults and immigration of outsiders may destabilize the basis upon which long-term sustainability can be built. It is also clear from the evidence cited that water, air, vegetation and animal life have suffered during the park's development, but the failure to consistently use any form of cost-benefit analysis has tended to place these concerns secondary to the economic, and it is this omission that perhaps contains the seeds of eventual decline of the park.

ACKNOWLEDGEMENT

This chapter originally appeared as an article in *Tourism Management* and has been reprinted through the kind permission of Reed Elsevier Company.

4 Chinese National Parks— Resource Usage Efficiencies, Spatial Proximity and Roles
An Application of Data Envelopment Analysis

Xiao-Long Ma, Chris Ryan and Ji-Gang Bao

INTRODUCTION

Chinese National Parks are not only concerned with conservation and recreation, but also the economic development of rural areas through a contribution to tourism as stated in Chapters 1 and 2 of this book. They are often characterized by higher levels of infrastructure development (including accommodation, retail and transport development) than their western counterparts; partly on the premise that such development creates high levels of employment, as evidenced by Zhong, Deng and Xiang in the previous chapter. In addition Chinese concepts of harmony also permit greater human intrusion in National Parks than would be found in, for example, a North American National Park, where an ethic of wilderness untouched by human presence is perceived as an ideal. Given these differences, within the Chinese context the efficiency of resource usage becomes a key component of park management. Consequently the purpose of this chapter is to first discuss the nature of National Parks paradigms in China, second to assess resource usage effectiveness in parks through the use of data envelopment analysis (DEA) and third to note the role of park proximity as a source of externally generated economies of scale. Finally the results are discussed with reference to their implications and the limitations of the study. This chapter therefore extends previous studies in three ways. First, DEA is applied to Chinese National Parks, and no prior study using this approach in this context has been found. Second, the initial findings gave rise to considering geographical clustering of parks as a determinant of scale economies, and this too is a new consideration, although it may be specific to the Chinese context. Third, the chapter also extends the work of Zhong, Deng and Xiang reported in

Chapter 3 by indicating that within the tourism area life-cycle the progress made by any one destination is linked to the patterns of demand associated with other sites and attractions, and that proximity or clusters of attractions, sites or destinations also need to be considered.

The chapter is thus located within a quantitative paradigm, which represents both a strength and a weakness. Given the complexity of purposes associated with a National Park (economic, conservation and recreational) it is problematic to determine measures of efficiency. This chapter draws upon a readily available set of data relating to physical assets and manpower that are used as input resources, and utilizes visitor expenditure in terms of park fees as output to develop a measure of efficiency based on input–output analysis. Visitor expenditure might be deemed to be a proxy measure of a park's success as a recreational resource, a generator of visits for local economic gain and a measure of dissemination of knowledge about natural and cultural heritage. It can be argued that the measures used in the analysis are crude variables—an argument the authors recognize—but models by definition are abstractions from reality that seek to identify key relationships that aid understanding or management practice. In this case the data point to the importance of tourism clusters of attraction, and it is argued that in a context of Chinese rural economic management as well as the contingencies of park management, this is an important finding.

LITERATURE REVIEW

Chinese National Parks

It has become a cliché to note the emergence of tourism as a rapidly growing sector of economic importance in a process of globalization. Many researchers have noted that the tourism industry has a significant impact on regional economic development (Wagner 1997; Deying *et al.* 1997; Dwyer and Forsyth 1998; Dwyer *et al.* 2003; Narayan 2004; Briedenham and Wickens 2004), especially in developing countries and regions (Durbarry 2004). China is no exception to this (People's Daily 1998). Of the 31 Chinese provinces, no less than 25 mainland provinces have identified tourism as a major industry in their economic system and tourist activities and plans have mushroomed nationwide (Sofield and Li 2007). Given this context the manner of use of tourism resources, including National Parks (see Arlt and Xu in Chapter 10 of this volume), will become increasingly important to the economic development of China (Ma and Yang 2003; Ao and Wei 2006). However, as noted by Li, Wu and Cai (2008) with reference to China's World Heritage Sites, Chinese sites face three great challenges to sustainability, namely, population pressure, development policies of local governments and lack of financial support. Given this lack of financial support various issues arise including: Are current resources well used and would an increase in resource actually achieve more efficient usage of park resources?

National Parks, by their very nature, are repositories of outstanding natural scenery and are cultural and/or historic resources both popular and significant as sites of tourism development. In China 80 per cent of domestic tourist trips are to National Parks (Qiu 2006), meaning the parks hosted more than 1 billion domestic travellers in 2006. However, the concept of a National Park in China is relatively new. The first National Parks, such as Ba Da Ling Great Wall and Ming Shi San Ling, were designated as such only in 1982. Since that date the number of designated National Parks and nature reserves in China has increased to the point that by 2007 there were nearly 1000 such parks of different classifications with a combined area of more than 96,000 square kilometres, accounting for 1 per cent of the Chinese land area (National Department of Construction 2007).

In western societies National Parks are regions for sightseeing and carrying out scientific and cultural activities, with the primary intention of preserving natural resources and a secondary use as recreational assets (Hall and Page 2002), but in the Chinese context they also have an important role in the economic development of rural areas (Gu, Ryan and Zhang 2007; Arlt and Xu in Chapter 10 of this volume). In turn the parks use human and capital resources in order to achieve these objectives and thus there is a requirement to use resources in as optimal a manner as possible (State Department 2006). From a western perspective, the objectives of a National Park revolve around a tension between two values, preservation and recreation, and thus they must resolve this tension in ways pertinent to each case as determined by not only the nature of the park but the social, political, economic and legal frameworks within which they operate (Barros 2005a). Preservation values include the sustenance of biological species and cultural diversities (Greenley, Walsh and Young 1981) and usage value is related to consumer surplus benefit derived from actual recreational use. Often this latter value translates into a purpose of satisfying tourism demands and supplying services for tourists, which, as noted, in China is also important in terms of developing infrastructure for general economic development. However, given that approximately 56 per cent of the Chinese population still resides in rural areas, the role of National Parks assumes an economic role not wholly comparable with that found in the West (e.g. see Arlt and Xu and Li and Sofield in this book).

It can also be noted that the view of leaving parks wholly as wilderness areas has rarely been voiced in the Chinese literature. This has much to do with the Chinese cultural perspective of the relationship between nature and humans. Located in Confucian and Taoist thought, the desired relationship is one of harmony—one translation of 'nature' in Mandarin is *da-ziran*—a coming into being that encompasses all things, living and inorganic. This world-view is, as noted by Sofield and Li (2007), both anthropocentric and anthropomorphic and thus human intrusion enhances the natural especially when consistent with a valued heritage and cultural inheritance. The attribution of names to natural features that are then enhanced by landscape features like gardens, pagodas, temples, and by extension hotels, serves as a fusion of the natural and man-made to create a harmony made all the more important through the literary tradition of *shan shui* that was established by the eleventh century (see Figure 4.1). *Shan shui* is a literary aesthetic shaping perceptions of nature and human harmony through concepts of tranquillity as in, for example, the building of places for meditating on the natural or the shaping of tree forms through bonsai. It still filters the Chinese planning mind when considering the scientific value of a National Park—the park derives its value from the cultural inheritance as much from, if not more, any scientific rationale (Li 2005).

Another factor affecting Chinese National Parks that differs from normal western practice lies in planning regimes. Early studies in China primarily regarded National Parks as a form of tourism resource, and concentrated on their development and planning within the context of general economic

Figure 4.1 Enhancing natural settings by buildings (above) and inscriptions (opposite)—Stone Forest Park, Yunnan Province.

and infrastructure development (C.G. Zhao 1982; Yu and Gu 1988; Chen, Gu and Yu 1990). While top-down planning remains the norm, in China's transformation into a more open economy such planning also paradoxically incorporates 'planned competition' which is made yet more complex by incomplete application of regulations. For example, Ren Zhuge (2000) found in a sample of 83 nature reserves that 68 permitted at least one practice prohibited under the Regulations of the People's Republic of China on Nature Reserves 1994. These banned practices permitted at local levels included quarrying, hunting, mining and logging. In short, a combination of planning, competition and inconsistent application of planning regimes has not inhibited a sense of establishing market competition strategies that include National Parks as part of tourism policies; which policies may be subjected to needs for economic development. Additionally, the pressure induced by the size of the human population is also a key factor (Li, Wu and Cai 2008).

This pressure leads to a further way by which Chinese National Parks can be differentiated from their western counterparts, and this is the high degree of 'urbanization' permitted within National Park boundaries to cater for the demand for accommodation and accessibility as well as the notion of enhancing nature by man-made artefacts. Indeed, as in the case of Ba Da Ling, the National Park is premised on human constructions (in this case the Great Wall of China) and the main site from which the visitor accesses the Wall is one of hotels, restaurants, souvenir shops and a cinema (see

Figure 4.2 Retail developments at Ba Da Ling, Great Wall of China (above and opposite).

Figure 4.2). At another popular National Park, that of Huangshan, access to the complex of hotels is via a cable car and concrete footpaths are constructed to take visitors to see trees and views. Many of these constructions fall within the remit of National Park management and add to the resource base of the park, and such construction is considered important in creating local employment (Wang, Lu and Liu, 2003).

Faced with these issues, the Chinese government has established strict laws and rules to maintain, preserve and, where possible, regenerate natural areas and biodiversity within the parks. From one perspective, therefore, to preserve the National Parks is a legal and not solely an economic action, much less a conservation one. Theoretically, every National Park must seek to achieve conservation and regeneration removed from a system of economic rewards and sanctions. On the other hand, the value of Chinese National Parks as tourism destinations that attract and serve tourists serves an economic function requiring efficient use of the assets made available to park managements. It is this latter perspective that legitimizes assessment of resource efficiency even while bearing in mind conservation needs, albeit such needs are framed within a cultural perspective determined by Taoist and Confucian philosophies.

Wang and Yang (1999) argued there was an overuse of resources that threatened the sustainable development of parks, even while So, Dong,

Yang, Wang and Jiang (2006) identified the continuing role of National Parks for economic development, cultural protection and the maintenance of heritage. Other studies have assessed the modes of usage, protection and management (Jin and Ding 1994; Wang, Zhou and Li 2001; Zhao 2001; Zhang, Li and Dong 2003; Xu, Liu and Bai 2004). Certainly the evaluation of performance is a critical component of the management process in any type of organization. Business performance is recognized as a multidimensional construct, as it covers diverse purposes and types of organizations/levels (Lewin and Minton 1986).

DATA ENVELOPMENT ANALYSIS

The evaluation of performance is a critical component of the management process in any type of organization, yet it is a multidimensional construct because it covers diverse purposes and types of organizations/levels (Lewin and Minton 1986). The use of multiple criteria lends itself to the application of DEA for reasons discussed in the following. In the wider tourism literature most efficiency studies appear to be of the hospitality sector (e.g. Baker and Riley 1994; Morey and Dittman 1995; Anderson *et al.* 2000; Tsaur *et al.* 1999; Hwang and Chang 2003; Wang *et al.* 2006; Chiang 2006; Barros 2005a, 2005b, 2006), travel agencies (Köksal and Aksu 2007; Barros and Matias 2006) and transportation operations (Anderson *et al.* 1999; Charles and Paul 2001; Sarkis and Talluri 2004). Other studies include ecological efficiency (Stefan *et al.* 2005), sporting venue efficiency (Preda and Watts 2003), tourism destinations (Fuchs 2004), tourism advertising programmes (Wőber and Fesenmaier 2004) and human resource management (Hwang *et al.* 2005). Nature-based tourism resources, such as National Parks, are receiving growing attention from academics, but with few exceptions, such as Lee and Han's (2002) study on use and preservation values of Korean National Parks' tourism resources (estimated by using a contingent valuation method). Little research has been conducted to estimate resource efficiency of National Parks and their sources of such efficiency. This is not to deny that a rich literature exists on National Parks. Within countries such as the United States, Canada, Australia and New Zealand studies have been directed towards issues of crowding, visitor satisfaction, flora and fauna regeneration, noxious weed control, types of visitors, role of zoning and other forms of visitor control (e.g. see Newsome, Moore and Dowling 2002) but studies within China are still embryonic (see Li and Sofield 2007).

For comparative purposes such as benchmarking, or in the use of multiple criteria for efficiency measurement, the usual modus operandi adopted by scholars is to select some input and output variables related to an organization's outputs and techniques such as DEA, Stochastic Frontier Approach (SFA) or ratio analysis are used to evaluate the efficiency of different tourism sectors. However, there is no definitive study as to which methodology is superior (Anderson *et al.* 1999). This chapter adopts DEA because it permits ad hoc modelling and enables use of secondary data published by the National Department of Construction as described in the following. The technique also lends itself to proximity analysis as previously used by Wőber (2007). DEA is a non-parametric methodology used to estimate efficiency based on linear programming techniques (Charnes *et al.* 1978, 1981). An important feature of DEA is that performance is judged in a comparative fashion, so it is effective in comparing the efficiency and productivity of the Decision Making Units (DMU) that possess the same classes of inputs and outputs (Wei 1998). Generally speaking, DEA is an extension of traditional ratio analysis that identifies an organization as optimally efficient when no other unit is capable of producing higher outputs with the same inputs (output orientated) or, alternatively, of producing the same outputs with less inputs (input orientated).

Consequently, in the arithmetic of DEA, the DEA efficiency score of any DMU is derived from comparison with the other DMUs included in the analysis, considering the maximum score of unity as a benchmark. The score is independent of the units in which outputs and inputs are measured, and this allows for a greater flexibility in the choice of inputs and outputs to be included in the study. In this study Chinese National Parks are considered DMUs with reference to their usage of human and capital resources. As described by Halme *et al.* (1999) the construction of efficiency frontiers diagrammatically produces a cone within which different organizations can be located, while most preferred solutions (MPS) are determined by reference to either weights or preference information (Golany and Roll 1994). In this instance, an input-orientated model that aims to produce the same outputs with minimal inputs will be used because the recreational use outputs of National Parks are subject to market demand while the inputs are under the control of park management.

Suppose there are J homogenous DMUs (DMU$_j$; j = 1, ... , J), that is all Chinese National Parks have the same function as tourism attractions to greater or lesser degrees, all seek tourism revenue and all utilize a similar capital resource in buildings, walkways, accommodation, retail units and human resources. Given this, it becomes possible and feasible to evaluate resource usage efficiency. Each DMU can be characterized by a vector of *m* inputs $X_j = (x_{1j}, x_{2j}, \ldots, x_{mj})$ and a vector of *s* outputs $Y_j = (y_{1j}, y_{2j}, \ldots, y_{sj})$. This chapter uses an output-oriented model, that is, the organization seeks the highest output for any given input level and can be subjected to constant returns to scale (CRS), variable returns to scale (VRS) or non-increasing returns to scale (NIRS).

From the perspective of CRS, one can obtain the DEA model of CRS from the j$_0$-th DMU $_{j0}$, which satisfies the condition of inputs and outputs being efficient. For each DMU, the following linear programming model (Banker *et al.* 1985) must be solved:

$$\min \{\theta\},$$

$$s.t. \sum_{j=1}^{n} X_j \lambda_j \leq \theta X_{j_0},$$

$$\sum_{j=1}^{n} Y_j \lambda_j \geq Y_{j_0}, \tag{1}$$

$$\lambda_j \geq 0, \ \theta \text{ free。}$$

When two congestion variables S$^-$ and S$^+$ are embedded into the upper formula, then it can be modified as:

$$\min \{\theta\},$$

$$s.t. \sum_{j=1}^{n} X_j \lambda_j + S^- = \theta X_{j_0},$$

$$\sum_{j=1}^{n} Y_j \lambda_j - S^+ = Y_{j_0}, \tag{2}$$

$$\lambda_j \geq 0, \ S^- \geq 0, \ S^+ \geq 0, \ \theta \text{ free。}$$

where X_{j0} and Y_{j0} are respectively input and output vectors of the j$_0$-th DMU, θ is the ratio of input shrinkage, λ is the linear combined coefficient of DMU, and the character with (*) is the optimal solution. When the optimal solution exists, $\theta^* = 1$, $S^{-*} = S^{+*} = 0$, that is the j_0th DMU is efficient; if $\theta^* = 1$, S^{-*}, S^{+*} possess non-zero value, that means the j_0th DMU is inefficient. Under this kind of condition, if the non-zero value is near to 1, the efficiency of National Parks is deemed to be close to 'efficient' (e.g. see Gregoriou *et al.* 2005).

According to normal procedures of DEA-based efficiency research, when different restrictive conditions are embedded into the equation, three DEA models with optimal solutions based on CRS, VRS and NIRS can be obtained (Coelli 1996). If a DMU is not efficient, one can judge whether it is operating at decreasing or increasing returns to scale through comparing the usage efficiency of CRS with that of NIRS. If they are equal, the National Park can achieve more efficiency through enlarging and increasing inputs; but if the usage efficiency of the CRS is less than that of NIRS, the National Park is experiencing decreasing returns to scale, and operators and management must reduce inputs to obtain more efficiency. (It can be noted that implicit in this analysis lies a potential limitation whereby, for example, management would resist reductions in budget for non-economic factors such as prestige, patronage and similar reasons. This is discussed in the final section of this chapter). Finally, this technique usually permits disaggregation of data into three classifications, namely: (a) technical efficiency, transforming efficiency from actual inputs to outputs; (b) congestion efficiency, the optimal distribution and combination of inputs with lowest cost and (c) scale efficiency, where combining different inputs and the respective usage efficiencies are equal to the product of all decompositions. (For discussions of this see Färe *et al.* 1994; Charnes *et al.* 1994; Coelli 1996; Coelli *et al.* 1998; Cooper *et al.* 2000; and Wöber 2007).

Careful assessment of the nature and types of inputs and outputs is required, but such consideration of the inputs and outputs types is obviously not without problems, both conceptual and practical. Other studies indicate various criteria can be used (e.g. Barros 2006b). One possible empirical criterion is availability of measures. A rigorous model requires that the measures of inputs and outputs must be relevant, adequate, and appropriate archival data must be available. Second, a review of past literature survey is itself a way of ensuring the validity of the research and therefore constitutes another possible criterion. A further test for measure selection can be the professional opinion of managers.

Achabal, Heineke and McIntyre (1984) suggested, in the sphere of retailing, that the true value of outputs in a distribution consist of a series of services resulting from the activities developed, and this can be applied to the case of National Parks. One indirect monetary output, the revenues gained by a park, can be identified as the output measure of efficiency. This has, for example, been used as an output index for travel agencies and hotels (e.g. Anderson *et al.* 1999, 2000; Tsaur *et al.* 1999; Barros 2005a, 2005b). Given the common practice of charging for entry to Chinese National Parks, revenue becomes a proxy for visitation and tourism 'attractiveness' within an economic system/approach. Similarly parks utilize as inputs the land area, number of employees, the value of capital assets and operating expenditure. Such variables are consistent with traditional economic theory, where land, labour and capital are regarded as essential production factors (Say 1963). The inclusion of land area can be justified in terms of crowding and perceptual carrying capacities given both the huge population of China and the numbers of visitors that parks attract—often far greater in the most part than their western counterparts. It is also a measure of the degree to which a park protects areas. Following the National Parks regulations of 2006, and conceptually Achabal *et al.* (1984) in perceiving outputs emanating from resources producing a series of services, it is justifiable to use employee numbers as an input factor. Shao (2006) notes the role of tourism industry employees, and among these are not only those directly employed in National Park services but also those working in accommodation, catering and retail services within the parks. Given the role of parks and their urbanization as actualizing the 'gaze' of visitors who come to stare rather than experience wilderness experiences as understood in, say, the United States, (e.g. Zhao 2008) the retention of employee numbers within the parks in an efficiency assessment can be justifiable. Finally, lagged effects are ignored and simple input–output ratios for any given year are used.

Unless specifically stated otherwise, all input and output data used in this chapter originate from the China Urban Construction Count Annual Report of 2005 issued by the National Department of Construction, China. To ensure uniformity, data from 171 Chinese National Parks of the same grade as examined and approved by the State Department at the end of 2005 form the research population. (Other parks are approved by local governments, so the data of their inputs and outputs cannot be compared to each other.) The area of these National Parks covers 69,000 square kilometres and host 370 million visitors a year.

For each input and output variables, DEA requires all data must include some non-negative and non-zero scores. Examination of the data set revealed that data were incomplete for 35 parks, leaving a final sample numbering

Table 4.1 Descriptive Statistics of Inputs and Outputs

Variables	Styles	Minimum	Maximum	Mean	Standard Deviation
Area (Square Km)	Inputs	3.00	10000.00	378.99	979.83
Expenditure (10k RMB)	Inputs	30.00	67700.00	4909.59	10335.43
Investment(10k RMB)	Inputs	1.00	270000.00	7227.34	28379.85
Employees (Person)	Inputs	18.00	8300.00	857.32	1441.77
Revenues (10k RMB)	Outputs	44.00	293000.00	9844.88	28533.61

136 parks distributed in 25 provinces, cities and autonomous regions. Basic information on the input and output variables of these Chinese National Parks are shown in Table 4.1.

RESULTS

Usage Efficiency—Mean Scores

Calculations showed four DMUs with a score of 1, which locates these parks at the usage efficiency frontier. They are Pu Tuo Mountain in Zhe Jiang province, Wu Ling Yuan in Hu Nan province, Huang Guo Shu in Gui Zhou province and Ming Sha Mountain-Yue Ya Quan in Gan Su province. They account for fewer than 3 per cent of the 136 parks, meaning there are 132 parks with a score less than 1. This means that the relative usage efficiency of these National Parks is located at varying distances from the usage efficiency frontier. The mean park usage efficiency score is 0.17, which means that on average one National Park in China could produce the same level of outputs with 17 per cent of the current inputs, or it could reduce current inputs by 83 per cent. The standard error of all Chinese National Parks usage efficiency equals 0.21, implying some variability in the scores, but reinforcing the tendency to low efficiency usage of resources. Of the parks, only 36 parks score more than 0.17—the remainder scoring considerably lower scores. The distribution of scores is shown in Figure 4.3, and it can be concluded that on these measures the resource usage efficiency in Chinese National Parks is low.

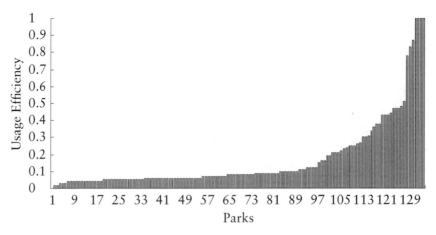

Figure 4.3 Distribution of scores of Chinese National Parks' resource usage efficiency.

Disaggregated Scores

Based on DEA theory, Chinese National Parks' usage efficiency can be classified as comprising scale, technical and congestion efficiencies. Disaggregation of data reveals that each of these respectively accounts for 4.41 per cent, 26.47 per cent and 46.32 per cent of total efficiency computations. Assessing individual park scores indicates that the scale efficiency score is 0.44, the congestion efficiency is the highest equalling 0.90 and technical efficiency is 0.53. Table 4.2 provides a summary of the data. It also became evident that the ability of the parks to combine different inputs was poor, and the features contributing most to variability were scale and technical efficiencies.

Through comparing the usage efficiency of the CRS with those of the NIRS, we can see that, except for Tian Zhu Mountain and Lu Nan Shi Lin, the other 134 Chinese National Parks are operating at increasing returns to scale and that by enlarging and increasing inputs higher returns and efficiencies can be achieved than is currently the case.

The question thus arises, in which of the classifications of scale, technical and congestion efficiencies might the best returns be achieved? In order to answer this question three scatter diagrams based on overall usage efficiency and disaggregated scores were established where the x-co-ordinate denotes overall usage efficiency and the y-co-ordinate denotes the sub-classification. The inherent characteristic of efficiency decomposition indicates that the nearer co-ordinates are to the diagonal, the greater is the impact of the sub-classification on overall usage efficiency (see Figure 4.4).

Most co-ordinates in each graph are far from the diagonals, implying none of the sub-classifications impact much on overall usage efficiency and that little difference exists between scale, technical and congestion efficiencies as to their impact on overall mean efficiency. Closer examination of the data reveal that scale efficiency was relatively more important than technical and congestion efficiency, while technical efficiency was slightly

Table 4.2 Characteristics of Chinese National Parks' Usage Efficiency

Efficiency Compared	Number of Efficient Parks	Percent of Total	Minimum	Maximum	Mean	S.D.
Overall Efficiency (Eff.)	4	2.94	0.00	1.00	0.17	0.21
Scale Efficiency (S.E.)	6	4.41	0.01	1.00	0.44	0.34
Technical Efficiency (T.E.)	36	26.47	0.07	1.00	0.53	0.34
Congestion Efficiency (C.E.)	63	46.32	0.00	1.00	0.90	0.21

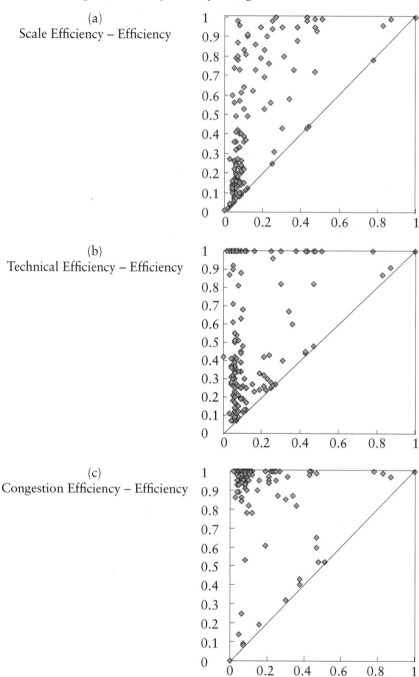

Figure 4.4 Correlation between Chinese National Parks' usage efficiency and disaggregation (horizontal axis is overall usage efficiency index, vertical axis is sub-scale as indicated).

more important than congestion efficiency. This conclusion was supported by taking overall usage efficiency and one of its components as variables while holding the other two as constants to calculate the partial correlation analysis under conditions of two-tailed significance. The results show that the partial correlation coefficient between overall usage efficiency and scale, technical and congestion efficiency are respectively 0.82, 0.73 and 0.53 ($p<0.01$). This confirmed the suggestion that Chinese National Parks' overall usage efficiency is more influenced by scale than by technical and congestion efficiencies.

These results are significantly lower than those recorded in other related studies, and the apparent inefficiency of resource usage is a matter of concern. Given that scale efficiencies appear to have some importance in potentially contributing to overall efficiency, the researchers began to question whether economies of scale might be achieved from external sources based upon proximity to other parks. Consequently, the remaining part of this chapter is devoted to examining this issue.

SPATIAL PROXIMITIES OF CHINESE NATIONAL PARKS AND USAGE EFFICIENCIES

Nearest neighbour analysis is a well-established technique (Smith 1983; Zhang and Yang 1991) and the three main patterns of spatial distribution are equilibrium, random and agglomeration, although Ryan (1991) presents a more detailed categorization of proximities. Following Smith (1983) and Zhang and Yang (1991), when the shortest distance of each National Park with its nearest neighbour is measured as r_i, then the mean of all Chinese National Parks can be denoted as $\bar{r}_1$. When all Chinese National Parks are randomly distributed (Poisson distribution), their theoretical nearest distance can be denoted as stated by Smith (1983, 1989).

$$\bar{r}_E = \frac{1}{2\sqrt{n/A}} = \frac{1}{2\sqrt{D}} \tag{3}$$

Where, $\bar{r}_E$ is the theoretical nearest distance, A is the regional area, n is the number of Chinese National Parks and D is the density of Chinese National Parks.

The nearest neighbour index, R, can be denoted as the ratio with the actual nearest distance and the theoretical nearest distance:

$$R = \frac{\bar{r}_1}{\bar{r}_E} = 2\sqrt{D}\bar{r}_1 \tag{4}$$

So, when $R = 1$, $\bar{r}_1 = \bar{r}_E$, then Chinese National Parks belong to random distribution; when $R > 1$, $\bar{r}_1 > \bar{r}_E$, Chinese National Parks belong to an equilibrium distribution; when R < 1, $\overline{r_1} < \overline{r_E}$, Chinese National Parks belong to conglomeration distribution.

Based on the formula (4), the theoretical random distribution distance of Chinese National Parks can be computed based on China's land area of 9.6 million square kilometres.

$$\overline{r_E} = \frac{1}{2\sqrt{n/A}} = \frac{1}{2\sqrt{171/9,600,000}} \approx 118 \text{ (Km)}$$

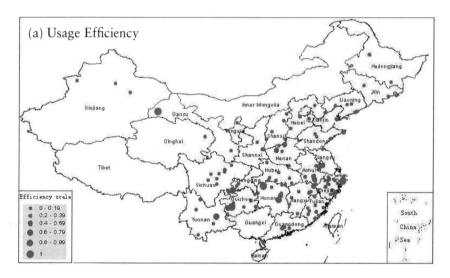

Figure 4.5 Distribution of Chinese National Parks' usage efficiency and its components (above and opposite).

Based on geographical co-ordinates, the actual nearest distance between one Chinese National Park with another can be calculated, denoted as r_i, (i = 1 ~ 171), then the mean of all can be estimated.

$$\bar{r_1} = \frac{1}{171}\sum_{i=1}^{171} r_i \approx 89 \text{ (Km)}$$

Given $\bar{r_1} < \bar{r_E}$, Chinese National Parks belong to conglomeration distribution.

In Figure 4.5 the actual spatial distribution of Chinese National Parks is indicated. In several provinces, such as Zhe Jiang, Si Chuan, Gui Zhou and

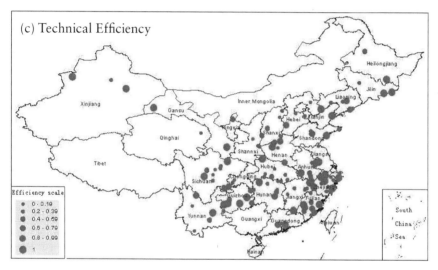

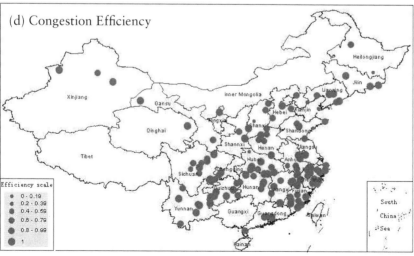

Fu Jian, there can be found more National Parks than in others (17, 16 and 13 respectively). In other administrative regions such as Ning Xia, Qing Hai and Tian Jin there might be found only one National Park. Additionally, along the Chang Jiang River, there is a dense cluster of National Parks with more than 100 found in 11 provinces, whereas the remaining parks are found in 20 other mainland Provinces.

Taking the data derived from the first part of the chapter, and using a score of 0.2 as a cut-off point, the consequent distribution of all Chinese National Parks based on these scores and the disaggregated scores are shown in Figure 4.5, when the size of the dots represent the level of efficiency—the bigger the dot the more efficient is the park based on the earlier calculations.

From this it can be seen that several provinces, such as Gan Su, Hu Nan and Gui Zhou, have a high mean efficiency score, respectively 0.53, 0.33 and 0.32, but other provinces, such as Ning Xia, Qing Hai, Liao Ning, Xin Jiang and Tian Jin, all have parks with very low mean usage efficiencies, the lowest in Tian Jin is only 0.05. What becomes obvious is that parks with a mean score higher than 0.17 (the national average) are all distributed along the Chang Jiang River with the exception of Gan Su and He Han. These parks are in Hu Nan, Gui Zhou, Jiang Su, An Hui, Yun Nan and Zhe Jiang provinces. It was also found that National Parks with high scale, technical and congestion efficiencies also cluster in the provinces along the Chang Jiang River. Can it be concluded that the proximity of parks to each other might be an enabling factor contributing to these high efficiency scores? Neoclassical economic theory proposes external economies of scale as a contributor to operating efficiency, and units cluster to enhance their competitive ability and efficiency though saving costs and obtaining more revenues than those more widely dispersed (Marshall 1920; Weber 1929). From the computation of the nearest neighbour position index and their spatial distribution, it is shown that while National Parks are geographically immobile they still show similar cluster or dispersive characteristics.

There are two kinds of spatial interactive relations, (a) complementary and (b) substitutive effects between clustered destinations. Further destinations may be separately described as being heterogeneous or homogeneous tourism destinations (Bao and Chu 2001). When heterogeneous National Parks are clustered (that is parks based on different morphology or biodiversities) complementary effects will arise and the attractiveness of a cluster of National Parks as one destination will be improved through an 'external economy'. So, under any given input level, they can achieve higher outputs than can one single National Park, thereby improving their mean overall usage efficiency score. But when homogeneous National Parks are clustered (that is parks with similar features of geography, fauna or flora) substitutive effects may occur. Under these conditions, and given a need to generate revenues to finance activities, parks will compete for market share and try to maximize outputs from given inputs levels thereby improving productivity and perhaps reducing transaction costs and using resources properly.

Finally, those National Parks without a near neighbour, while perhaps enjoying a local monopoly, may have a poorer ability to use resources efficiently because of the lack of externally induced economies, thereby leading to consequent low overall usage efficiency scores.

Obviously, the more National Parks are clustered, the more there exist opportunities for complementary effects and substitutive effects to arise with their subsequent implications for resource use efficiencies. It is thought such clustering in the Chang Jiang River accounts for the higher efficiencies being recorded. However, the existence of exceptions such as Ming Sha Shan—Yue Ya Quan indicates that this theory is not wholly sufficient to explain the outcomes noted by this research, thereby implying the role of other factors as possessing importance.

CONCLUSIONS AND DISCUSSION

National Parks are one of the more important resources for regional tourism development in China. Based on their economic, recreational and conservation functions, the pursuance of efficiency and resource usage is important in terms of justifying expenditure on parks. The very complexity of their task raises problems about how one might assess return in investment in resources, and it is recognized that the nature of the variables used have the virtue of being easily accessible, but are at best proxy measures of efficiency. However, the results possess importance in answering questions such as: What might be the government response to improving the resource usage efficiency of Chinese National Parks by increasing expenditure on the parks' capital structures? The main role of government is the establishment of policies and goals to better the nation. Within a still centralized planning regime, one possible implication of the study is that resource usage might be improved through the creation of not simply clusters of National Parks, but rather clusters of tourism resources around National Parks; clusters which might generate required flows of visitors to help justify capital expenditures and asset formation in the parks. In short, external economies of scale may be increased not solely by the proximity of other parks but also by proximity to other tourist attractions. Parks thus become part of a portfolio of tourism product. Government can therefore prompt more tourism projects and enterprises to be established near these National Parks and encourage participation in the tourism market through the administration of tax concessions, the provision of finance, grants and destination planning policies etc., which together potentially could improve the usage of park assets in a more efficient manner. The National Parks thus can serve as nodal points of tourism development that support income and employment-generating infrastructure that is important in generating wealth for rural areas. The counter-argument from a western perspective is that increased recreational usage might endanger the conservation role of the parks, and thus one needs to recognize that, as in other parts of the world, zoning, quotas and other management techniques will assume greater importance than in the past.

Another question is: What can the management of National Parks do in order to improve efficiency? Mostly, managers can control their inputs such as the manner of use of operational areas, employees, investment and expenditure and improved operational conditions. However, an important consideration, if one is seeking to increase the intensity of an operation, is an understanding of the limits of acceptable change from environmental and psychological carrying capacity perspectives. Site planning and management are affected by technical and resource mix considerations (which currently are, according to the earlier data, poorly managed) and need very careful consideration to resolve tensions between better usage of resources and recreational usage on the one hand, and conservation necessities on the other. Improving the state of operations must not imperil natural environments and biodiversity if species of fauna and flora are to be saved from extinction. It can also be noted that the results imply an increasing rate of return to inputs and thus better financial support would address one issue identified by Li, Wu and Cai (2008)—namely, better funding may ease pressures that threaten longer term sustainability.

In the process of destination planning and assessment of National Parks, the results in this chapter possess value in terms of identifying other management issues. It appears that the parks in the provinces along the Chang Jiang River have more opportunities to gain higher than average resource usage efficiency. The manner of inter-park co-operation, possible sharing of resources and joint marketing (or de-marketing) is highlighted by the results. Without closer examination of actual management practices, and a wider holistic consideration of regional characteristics, economic and tourism planning policies it is difficult to indicate more precise measures, but the study shows the potential for more gain than is currently being achieved. Even allowing for the uncertainties, it is argued that the research can be used for future benchmarking.

From a conceptual perspective the study indicates that DEA is a technique applicable to National Parks and the results can generate new ways of assessing patterns of business. It is recognized that if measures of conservation effort were introduced into the output measures, different results could well accrue—but such measures would probably need to be based on cost-benefit analysis techniques to provide monetary values. This highlights a major limitation to this study—namely, the choice of visitor expenditure as an output variable against which to measure resource efficiency. On the other hand, the study has indicated the significance of scale and the role of factors external to the parks. It is evident the study possesses implications for regional policy planning, and it is equally evident that the study exists within a paradigm of increased visitor expenditures being the output criterion—in short, this criterion is both the strength and the weakness of the study in terms of assessing National Park performance. Nonetheless, the study shows how a technique like DEA can inform discussion and produce results that are meaningful when considering the role of National Parks as tourism assets and their place in destination and tourism planning.

ACKNOWLEDGEMENTS

This chapter originally appeared as an article in *Tourism Management* and has been reprinted through the kind permission of Reed Elsevier Company. The photographs were taken by Chris Ryan.

5 Overseas Chinese Town
A Case Study of the Interactive Development of Real Estate and Tourism

Zhang Ning and Wen Zhang

INTRODUCTION

There has always been a link between tourism and property development. Within European literature works such as Frank's (1920) book on the economics of the Roman Republic leads one to believe that Roman property speculators built villas on the Italian coast-line to sell to the emergent merchant classes in the second century BC. However, it was in the nineteenth century, as part of the Industrial Revolution that created both a new management and administrative class and ease of access through rail transport, that one began to see the development of purpose-built coastal resorts combining property development in locations such as Bournemouth and Nice (Meller 2001). In modern China, such a development model has come to be regarded as the ideal choice for city planning. This chapter examines the characteristics of the real estate and tourism interactive development model by taking the Overseas Chinese Town in Shenzhen as a case study.

With the economic development that has taken place in China generating significant increases in many people's income, increasing numbers of Chinese have become more concerned about the comfort and aesthetics of their living environment rather than simply the physical expansion of housing stock. It should be noted though, that for many people the housing stock remains a pressing problem due to urban redevelopment as described in the Introduction to this book. But for both a growing number of purchasers, and the construction industry, these changing economic conditions have generated an interactive development of real estate and tourism. The Overseas Chinese Town (OCT) took this stance as a long-term development strategy. The OCT model has proved to be successful for that company in developing its portfolio of properties and themed attractions in China. The OCT Corporation has used tourism demand as the basis for residential property development around its theme parks, and in carefully designed environments the company has incorporated residential areas for both permanent and holiday/recreational use. Combining these functions has led to pleasant living environments, comfortable housing, easy access to leisure activities and convenient transportation. The company also claims that in addition to the initial economic impetus generated by the construction of the site, the provision of theme parks has added economic impetus to local employment and income through tourism while providing a show-place for Chinese culture and theatre. This model of combining tourism with property development has been replicated in other parts of the world, sometimes on a very large scale as is evidenced by the construction of Dubailand in the United Arab Emirates, which upon its completion will host not only theme parks, retail provision and a new central business district, but also a population of 2.5 million people.

The OCT model has thus been copied by other regions in China. In this chapter the authors, based on survey data derived from interviews conducted in Shenzhen, seek to analyse the characteristics of joint tourism and real estate development, the problems that might be involved, and in that way contribute to a better understanding that might aid others who seek, in China, to duplicate this model.

The chapter therefore has the following structure. First it offers a brief review of literature pertaining to real estate and tourism development. Second, it provides the context of the study by offering a description of OCT before indicating the methodology used in the study. The remaining parts of the chapter then describe and discuss the results of the study and their implications.

CHINESE OVERSEAS TOWN

Shenzhen Overseas Chinese Town Holding Company is a publicly listed company on the Hong Kong Stock Exchange and is primarily engaged in the tourism and properties industries, but also has some interests in the sale of commodities. Its accounts report that for the year that ended 31 December 2006, the Company obtained approximately 53 per cent, 32 per cent and 10 per cent of its total revenue from respective sales in its theme parks, park tickets and travel business. In 2006, approximately 60 per cent and 40 per cent of the Company's total revenue was from Beijing and Shenzhen, China, respectively. The Company is based in Shenzhen, Guangdong Province where its theme parks are located. In the same year the company had a net profit margin of 27.55 per cent. The value of its sales was 1,679,800,000 RMB. At the end of 2007 the Company announced a new joint venture with Yunnan Metropolitan Real Estate Development Co. Ltd. The name of the corporation comes from its connections with overseas Chinese business interests. Within the tourism industry it is best known for its operation of Chinese Overseas Town in Shenzhen that incorporates a residential area with the theme parks China Folk Culture Village, Splendid China and Window of the World. Within the total complex a number of hotels can be found, including

Figure 5.1 Show at China Folk Culture Village.

a Crown Plaza Property, (a five-star hotel themed in the Venice cultural style of China), and several other major properties. Currently the company is exploring the development of a further site on the east side of Shenzhen—its current complex being on the west side of the city.

A major draw is the theme parks, and these are of a high standard and among their features are the spectacular nightly shows as shown in Figures 5.1 and 5.2. These are very popular among both domestic and overseas tourists and feature in many tour operators' programmes.

The Joint Development of Real Estate and Tourism

In the modern period the initial joint developments of real estate and tourism can be discerned in the latter part of the nineteenth century when the development of the railways, a retiring managerial and administrative class and discovery of the value of seaside locations began to create middle class enclaves along the coast-lines of Britain (e.g. Bournemouth and Torquay) and France (especially in the Brittany and Mediterranean coast-lines). Villas were built as properties for holidaying and subsequent retirement, and as notably happened in Nice and Bournemouth, were specifically constructed as real estate developments and suburbs. In part this was associated with a growing confidence in urban planning as the newly created local authorities sought to improve living conditions with the building of water supplies,

Figure 5.2 Show at Window of the World.

sewage treatment plants, and the provision of gas and then electrical lighting of newly paved streets being patrolled by police forces. The late nineteenth century in Europe saw the introduction of planning controls and regulations seeking to govern the orderly development of expanding cities complete with parks, schools, retail areas and residential zones separate from the fumes of industrial areas (Meller 2001).

In the 1960s the villa resorts around the Alps region in France based upon skiing and summer rambling primarily explored the time-share marketing approach to attract consumers to help fund building and resort development, a pattern quickly replicated on the Spanish *costas*. Nowadays, real estate for the tourism market has matured with the emergence of time-share holiday hotels, golf course resort complexes and country resorts as well as leisure villas. Other contemporary examples include harbour redevelopment as the old shallow water ports of the nineteenth century give way to marinas and historic enclaves as demonstrated at Sydney, Baltimore, Boston, Cleveland, Liverpool, London, Cardiff and Swansea—each with their retail zones, maritime museums and residential areas of apartments with views overlooking sea and boats. Many of these developments are based on zones and phased development, with earlier built developments being sold to private purchasers, thereby funding the next stage of development. Not all have been successful, with the Japanese example of Huis Ten Bosch at Sasebo near Nagasaki being one such example where the initial property developer was almost bankrupted by the costs as apartments and houses remained unsold for long periods of time.

The pattern continues. When the centre areas of a city become built-up and costs of building are high, the real estate developer moves building projects to the suburban areas. Where possible they usually choose tourism destinations as profitable regions in which to develop. These sites include tourism resorts, recreational facilities, 'loosely' located hotels as well as villas combined with retail services. Around the world such developments can be identified, including China. In some instances though, supply has outstripped demand. The Australian market, for example, is currently oversupplied. Units worth just under AUS$23 million were sold in 2004, which was well short of AUS$38 million projected for the year (Zou 2004).

OVERSEAS CHINESE TOWN

Within China, the projects associated with OCT are often credited as being the most successful. OCT commenced operations in the mid 1980s in Shenzhen, which at that time had commenced its growth but did not resemble the modern large metropolis of some 8 to 9 million people that it is today. Indeed, at the commencement of the 1980s Shenzhen was still a comparatively quiet back water, but its position within the Guangdong Provinces Economic Free Enterprise Zone was soon to change that. OCT embarked on a scheme involving more than 1 billion RMB capital expenditure in the establishment of the China Folk Culture Village, Splendid China and Window of the World theme parks as well as the Happy Village, instead of more traditional projects such as holiday hotels and time-share hotels. With its growing prestige, increasing numbers of people pay a visit here, including overseas tourists who often access it from Hong Kong. Meanwhile the commercial, education, health care and cultural institutions were established one after another in phased developments. From 1985 to now, the OCT has grown into a modern sea-shore area with a nice environment and a wide range of social functions (Ming and Ping, 2004).

Tourism, commerce and real estate were associated together from the outset to develop step by step. OCT thus served as a new model in China for real estate and urban planning. The main characteristics of such a model were that it paid careful attention to a comprehensive and inter-regional development programme within the province and also focused on connections with associated industries, for instance, a series of investments by other companies in the hotel sector, commerce, finance and recreation. It saw these not so much as competitors, but as contributors to a critical mass that it too could take full advantage of by offering products in a niche not being built by others—which were the aforementioned theme parks. Such industries series and their location formed the core the development of whole city. Diagrammatically this is shown in Figure 5.3.

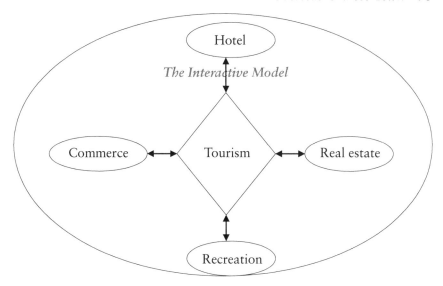

Figure 5.3 The interactive model.

Such an approach can integrate different resources together. It comprises tourism, accommodation, business and residential properties that together create a synergy. All can attract large investments in a short time and the associated resources can each play to their own advantages. Tourism can draw the attention of tourists from home and abroad, and create a base demand that helps the financial viability of the other developments. The complex also spins off benefits for the surrounding city, helping that city to develop a stronger brand image that enables it to become better known, and therefore tourism brings more opportunities for the city's development. The high quality of the available hotel accommodation combined with its recreational assets can also serve as the base for conventions, conferences and a meetings market—all of which are necessary functions for a modern city's development (Feng Wun Wei, 2006).

This approach has worked well for both OCT and Shenzhen. Another factor has been the attention to sustainability, for as the company's own publications indicate, it has sought a 'development driven by tourism, and an environment focusing on culture—profit created by environment'. Thus today 53 per cent of its land area is covered by green plants, and that includes, where possible, the use of green plantings on roof-tops. These have the effect of cooling the ambient temperatures by reducing radiant temperatures from tarmac and concrete, thereby reducing need for air conditioning by a few percentage points. At the same time it permits shade, fragrances from plants in the night air and the sound of birds. With such a pleasant environment, combined with a subtropical climate, the OCT has become a 'paradise for living' in Shenzhen and even in China. Over time the level of investment has increased from first 1, then to 3 and currently the present complex represents a value of 5 million RMB (Xiu, 2005).

Possibly where OCT differs from other theme parks in the regions that involve hotel accommodation, such as Janfusun in Taiwan, is the very scale of the undertaking. The city authorities permitted the property developers to have ownership of the land prior to development and to permit phased introduction of the complex so that each stage aided the financing of subsequent stages—thereby reducing the need for initial capital and reducing the operating risk. Table 5.1 illustrates the area and initial investment of OCT across three of its theme parks.

The other facet in which the project differs from others elsewhere is the linkage with residential properties. Indeed, on opening the theme park Splendid China in 1991, the price for OCT real estate increased by 131.41 per cent within 12 months. The OCT is known as an 'affluent area' in Shenzhen (China.org.cn 2007a), and its presence has many spin-offs, even when the residential area does not always succeed in attracting new investment. For example, in October 2007, a new international school was developed in Shekou district of Shenzhen by the Japanese Chamber of Commerce, who preferred that area after considering OCT—yet the very fact that there is a Japanese presence in Shenzhen is due to the commercial and industrial hub that is Shenzhen, to which OCT makes an important contribution. Indeed, it has been estimated that half of the city's 20,000 Koreans (who stay on average five years in the city) are residents in OCT, indicating another function of the complex which is the development of an environment that can attract international families who come to work in the city (China.org.cn 2007b)

The complex continues to develop, and additional resources appropriate to residential and tourism sectors are added, one such example being the He Xiangning Art Museum that was established in OCT on 18 April 1997, having been approved by the central government as part of its urban and cultural promotion policies. Perhaps it is almost unnecessary to add that under the Chinese classifications of destinations, OCT is graded as a 4-A top-quality tourist destination (Yan Liu, 2004).

Table 5.1 Areas and Capital Investment—Overseas Chinese Town

Project	Invest	Area
Splendid China	0.1 billion RMB	450km2
The Window of the World	0.3 billion RMB	660km2
The China Folk Culture Village	0.11 billion RMB	300km2

RESEARCH DATA

Face-to-face personal interviews and a completion of a questionnaire were used to collect primary data which were supported by secondary sources. The study instrument was developed through a two-stage process. First, an examination of the literature on topics related to real estate and tourism was undertaken. Second, respondents who lived in OCT were interviewed to generate different perspectives on the mix of tourism and estate management. These interviews were based on a semi-structured interview technique as proposed by Minichiello *et al.* (1995) using open-ended questions. Interviews averaged 45 minutes and were recorded and transcribed for analysis. Prior to analysis, responses to each question were entered into appropriate cells of a spreadsheet. In addition, the authors used archival data to gain additional insight, thereby increasing the validity and reliability of the evidence.

The respondents were approximately equally divided between females (47.7 per cent) and males (52.3 per cent), with the majority (46.5 per cent) between the ages of 25 and 44. In total they numbered 43. In terms of the education level, 44.2 per cent of the respondents reported that they had bachelor degrees and above, followed by college degrees (37.2 per cent), high school degrees (14.0 per cent), and less than high school leaving qualifications (4.6 per cent). In terms of duration of residency in the OCT, 72.1 per cent had lived in OCT for less than 10 years, but generally in excess of two years. Of the remaining 27.9 per cent all had lived there more than 11 years and indeed one respondent reported living in the area for 35 years and thus had seen the whole complex develop from its origins.

A five-point Likert-type scale was used on the written questionnaire and a summary of results are shown in Table 5.2, although it must be noted the sample was small and the primary purpose of the questionnaire was to get respondents thinking about the issues. The score of '5' represented the most positive response.

Three main areas of influence were identified—the economic, social/cultural and the impacts on the environment. With reference to the economic sphere, most respondents held a positive attitude. They agreed that the tourism and real estate complex attracted more investment ($m = 4.3$, $sd = 0.56$) as well as increasing employment ($m = 3.74$, $sd = 0.98$). As noted from the open-ended questions, these undoubtedly contributed to a better living environment. With reference to the social/cultural components, the mix of tourism and residential areas was generally seen as a positive experience, and also of interest. There was a greater diversity of views on the impact of traditional culture, as evidenced in the standard deviation, but overall the mean score erred more to the positive. Hence the scores for communicating with outsiders were $m = 4.12$, $sd = 0.63$, and impact on culture, $m = 3.28$, $sd = 1.01$. With reference to the environmental issues, while mean scores tended to be slightly above 3.0 as shown in table 5.2, the standard deviation with

Table 5.2 Views on Complexes Involving Residential and Tourism Zones

(Mean Scores)		
Statement	*Mean*	*SD*
Attracted more investment	4.30	0.56
Improved the infrastructure	4.28	0.66
Increased employment	3.74	0.98
Increased the price of goods	3.19	1.00
Communicated with more outsiders	4.12	0.63
Changed the traditional culture	3.28	1.01
Caused heavier traffic	3.35	0.90
Caused more pollution and noise	3.44	1.10

a value of approximately 1.0 indicated quite differential experiences. That these scores emerge is not wholly surprising given the location and nature of the site. Shenzhen is a fast growing metropolis of the last two decades. But having been built from 'green field' sites it has not had the problems of urban renewal and population displacements from existing residential areas associated with many of China's other major cities. The city is characterized by high-rise, modern sky-scrapers of office blocks and apartments linked by motorways. It has not, to a large degree, had to graft the new onto the old. Indeed, while being in the Cantonese-speaking south of China, Mandarin is in effect spoken more widely in Shenzhen because of the influx of migrants from other parts of China. The city is indeed new—and thus for its inhabitants, a scenario of change, of new constructions and a growth in traffic has been the norm of their lives in Shenzhen.

With reference to the qualitative data, the following extracts provide some examples of text.

A Retired Engineer, 70 Years Old, Living Here for Two Years

I have retired for more than ten years. I have been to many places in my work years. Comparatively speaking, the Overseas Chinese Town has been developing at a striking speed. The environment and surrounding buildings were improved step by step. Its theme parks were a big attraction to the tourists from outside. It attracted many people to invest here and buy houses here. Of course, some crimes appeared with the development of tourism. And also the heavy traffic influences our life to some extent.

[But] totally speaking, we get more benefits from model of estate development here.

A Lady, 35 Years Old. Worked Here for Seven Years

I am from Hebei Province in the north China. Seven years ago, I came to Shenzhen for work. During the time here, I found its rapid developing speed was at a leading position in China. Just take the theme parks and real estate for example, from the establishment of the first theme park, the company maintained construction year after year. The total appearance of the Overseas Chinese Town would grow beyond my recognition if I did not live here all the time. As to the negative influences, I found that the price of goods has generally gone up.

A Local Resident, 62 Years Old, Living Here All His Life

I was born here and grew up here, so I saw the great changes taken place in Shenzhen, especially in OCT. What impressed me most was its design and development principles which contributed a lot today's prosperity. Under the guidance of sustainable development principles, the original environment was carefully protected and improved. After 10 years of research, the plan was put into practice. Therefore, scientific and feasible planning played a vital role in OCT success.

With its reputation growing, people from outside gathered here. Those who could not find a job easily committed crimes such as stealing, robbing and fighting. That was because many manufacturing firms that produced pollutions were moved out from here and the employment was greatly decreased.

To sum up, most residents here are basically satisfied with the ideal environment and harmonious development.

From these findings, we can arrive at various conclusions. Generally speaking, the development of OCT had a positive influence for the current residents living there, although some evidence was found of occasional unpleasant experiences happening to tourists. Today this is controlled much better by the government.

DISCUSSION

The findings in this chapter have led to a general conclusion that the combined tourism and real estate development model of OCT has achieved success in Shenzhen. It established a trend for developers to study and learn valuable experience. But, in the light of subsequent experience, whether it is suitable for other cities to copy requires careful thought. After the success of OCT, many developers invested in joint tourism and real estate projects in other places in China, such as Shanghai, Fujian, Dalian etc. The number of all such projects is thought to total about 100. Yet only a few developers succeeded. One main reason was that they failed to build tourism projects attractive to the domestic market. Without these successful tourism resources, the real estate projects lacked their core asset and there was no substantial difference when compared with traditional real estate projects. A key determinant was the 10 years OCT spent in market research and planning. The investment in constructing the theme parks at the beginning was significant. With a decade's efforts, the tourism function was fully integrated into the plans, which added much value to the real estate. This provided an advantage important to the long-term success of the development.

Given these characteristics, the following suggestions are made:

- Strategies aimed at long-term sustainability count, and should be given special consideration. Operating within such a sustainable strategy has enabled OCT to achieve success.
- Have comprehensive and detailed research prior to development so as to reduce the risks associated with investment. Before opening a new project, developers must undertake thorough research to secure profits.
- Inherit and sustain traditional culture and highlight the culture of the local area. Shenzhen is a modern prosperous city but its attractions reflected a unique cultural characteristic, thereby providing a point of difference from competing destinations.

ACKNOWLEDGEMENTS

The photographs in this chapter were taken by Chris Ryan.

6 Heritage Retail Centres, Creative Destruction and the Water Town of Luzhi, Kunshan, China

Chennan (Nancy) Fan, Geoff Wall and Clare Mitchell

INTRODUCTION

In North America, investments in landscape and commodity reproduction have become a common activity since the early 1970s, while, as Zhang and Zhang write in the previous chapter, the concept of estate management being tied to tourism and property development is certainly not unknown in China. Such developments range in size from those associated with 'green field' sites such as Overseas Chinese Town to a smaller, newer type of post-modern community that has emerged known variously as a tourist, recreational or heritage shopping centre or village (Mitchell 1998). Such centres are small in scale and are not restricted to conventional villages. Such communities are centres of consumption that specialize in the provision of hand-crafted products reflecting local or regional heritage and they also often sell products brought in from farther afield.

This chapter continues the theme of property development commenced by Zhang and Zhang in the prior chapter, but does so within the context of a well-established older town, Luzhi, bound by a canal system, and thus where available land space is more constrained. The case illustrates a number of the issues raised in Chapters 1 and 2, showing how tourism has become a part of economic planning and place creation complete with changing land use patterns and implications for both local residents and visitors. The study concentrates on two main *leitmotivs*—the development of small-scale retail provision based on commodified heritage goods or ambience, and conceptually the notion of creative destruction, by which old forms of place give way to new, each associated with different patterns of usage and meaning.

The purpose of this chapter is to determine if the model of creative destruction can be applied successfully to the water town of Luzhi, China. To meet this objective, three variables that interact at each stage of the model are examined: business composition, visitor numbers and expenditures, and resident attitudes towards tourism. Also the behaviour of provincial, municipal and Luzhi local governments is documented. Interviews, surveys and on-site observation were employed to gather the necessary information. The structure of this chapter is first to describe briefly how land usage patterns change with various impacts for local communities. The next section details the stages of 'creative destruction', which is a model describing how business units, especially retail outlets, can change as a destination matures over time and changes its focus. The next section then describes the context of the research, namely Luzhi, which forms part of Suzhou near Shanghai. Finally, the concept of creative destruction is assessed with reference to the area and the implications of the research are discussed.

LAND USE CHANGE AND COMMUNITY REACTION

Researchers have found that heritage shopping centres share three common characteristics (Dahms 1991; Getz 1993; Mitchell, Nolan and Hohol 1993). First, they are accessible to a large and relatively affluent population. Second, an amenity heritage environment must exist before commodification begins. The elements required to create such an environment include an attractive physical setting, such as a river, an appeal to local culture and heritage, and the presence of a rural or small-town atmosphere, associated with an image of a life-style that is happy, healthy and problem free. Third, an entrepreneurial spirit must be present and, with it, the availability of capital.

The heritage shopping centre is one means of observing land use changes at various stages of tourism development, and perhaps begins to demonstrate processes of 'creative destruction'. Such centres combine the carefully crafted products of a pre-industrial society (e.g. quilts, pottery or stained glass), with the modern conveniences (e.g. washrooms, parking and air conditioning) that contemporary consumers have grown accustomed to. The appearance of heritage shopping centres can be traced both to consumer demand and entrepreneurial response. From the demand side, many people now desire to purchase 'signposts' (Urry 1990) of the past, as both tangible products and aesthetic experiences. From the supply side, entrepreneurs are the first group who contribute to the development of these postmodern communities. To generate profits, entrepreneurs invest in the purchase and restoration of historically significant buildings and in the construction of venues where heritage products are sold. Through marketing, these communities are sold as packaged experiences (Mitchell 2003).

This process is not confined to the western world. As both international and domestic tourism in China has grown exponentially in the last quarter century, increasing pressure is being placed on communities with interesting heritage attributes and they are being transformed by tourism development as is evidenced by Arlt and Xu's contribution in this book about Ganzi. This chapter explores this transformation through the application of a model that has not yet been widely used by tourism researchers (but see Tonts and Greive 2002), and has been applied previously in a very different cultural setting.

THE MODEL OF CREATIVE DESTRUCTION

Several models have appeared in the literature over the past two decades to describe the evolution of tourist destinations and to analyse the impacts of tourism development in a community: Butler's Tourist Area Cycle of Evolution (1980), Bjorklund and Philbrick's matrix of Host Attitudinal/ Behavioural Responses to Tourist Activity (Mathieson and Wall 1982) and Doxey's Irritation Index (1976) are examples. However, these models have either dealt primarily with resort towns (e.g. Butler 1980) or viewed places largely in isolation, divorced from exogenous economic and social forces that drive their development (Mitchell and Coghill 2000). Like Butler's evolutionary model, which expounded upon the relationships between tourism numbers and time, the model discussed in the following is a stage model, albeit one that focuses on the roles of key actors and incorporates a greater number of variables. While such models have many deficiencies, particularly the establishment of thresholds between changes and their assumption of unidirectionality, which have been discussed at length in the papers in Butler (2006b, 2006c), they continue to receive wide use in a diversity of tourism contexts, ranging from aboriginal cultural performances to disaster management where the interest is in the process of change. Accordingly, a stage model is proposed and applied in this chapter.

Building upon the economic research of such as authors as Schumpeter (1943) and Harvey (1987, 1988, 1999), and more recently articulated by Cowen (2002), Mitchell (1998) designed a new model by drawing on an established body of literature in economic and rural geography and linked the evolution of heritage shopping centres to (a) the entrepreneurial drive to accumulate profit and (b) the consumer desire to accumulate nostalgic experiences. This model is called 'the model of creative destruction' and it describes the evolutionary process of communities whose development has occurred around the commodification of heritage. There is a long history on research on creative destruction on which this research draws. For example, Harvey (1989: 167) has written as follows:

> Its (capital's) internalized rules of operation are such as to ensure that is a dynamic and revolutionary mode of social organization, restlessly and ceaselessly transforming the society within which it is embedded. The process masks and fetishes, achieves growth through creative destruction, creates new wants and needs, exploits the capacity for human labour and desire, transforms spaces, and speeds up the pace of life.

The model was first tested in a retail tourism context at St Jacobs (Mitchell 1998), a heritage village in Ontario, Canada, and then in another two Ontario small towns: Elora (Mitchell and Coghill 2000) and Niagara-on-the-Lake (Mitchell, Atkinson and Clark 2001). In the case of St Jacobs,

Mitchell (1998) stated that the creative destruction model is based on a relationship among three variables: entrepreneurial investment, consumption of commodified heritage and destruction of the small-town atmosphere (which she called the rural idyll). The premise underlying the model is that the desire to accumulate capital drives investment in the production and sale of heritage. These investments entice the postmodern consumer in search of nostalgia for a vanishing, simple small-town or rural life. While the resulting consumption of heritage provides entrepreneurs with profit for reinvestment, the creation of a commodified landscape will ultimately result in the destruction of the small-town or rural atmosphere. However, Mitchell *et al.* (2001) found in a later study of Niagara-on-the-Lake that the entrepreneurial desire for profit is not the only dominant force driving the creation of such heritage landscapes. Preservationists and producers of heritage are the other two stakeholders whose activities contribute to the creation of a heritage shopping centre.

Two groups of preservationists typically can be found within these communities; first, those whose main concern is the restoration of architecturally significant structures and they include local council, residents and interest groups among their number. Second are those who fight against development initiatives to prevent destruction of the original buildings and atmosphere. The activities of these two groups are enhanced by the work of producers; the visual, performing or literary artists who contribute to the stock of both tangible and intangible heritage within a community (Mitchell 2003). It is the combined actions of these three agents (entrepreneurs, preservationists and producers of the heritage) that foster the creation of the heritage landscape, although their goals may be different and there may be tensions between them. While investment is important to create the businesses that cater to tourists, the preservationists strive to protect the ambience of the community that also attracts the tourists while the producers create items, such as crafts and theatre that animate the place and enhance the visitors' experiences through the provision of activities in which they can participate.

There are five stages in the process of creative destruction. In the first stage, 'early commodification', the marketing and sale of tradition or heritage is initiated. Investments that are made generate financial benefits and cosmetic improvements and are viewed very favourably by local residents as enhancing the economic well-being of the community. The stage of 'advanced commodification' is marked by acceleration in the level of investment. An increasing number of entrepreneurs introduce new merchandise to meet the demands of the visiting population. The community is marketed extensively, resulting in an escalation of visitor numbers. Those involved in the tourism industry will extol its merits, whereas a number of residents, particularly those who do not benefit financially from commodification, will perceive an erosion of the quality of their environment; in other words, a partial destruction of the landscape idyll occurs.

'Early destruction' sees the reinvestment of surplus value into businesses that provide for the needs of the expanding number of visitors. While some are in line with the heritage atmosphere of the locale, others stray significantly from this theme (e.g. a restaurant serving fast food). While numbers of visitors continue to escalate, a growing number of residents will perceive an erosion of their community ambience as problems of crowding, congestion or crime become apparent. The 'advanced destruction stage' begins when residents become resigned to the inevitability of ensuing change. If left unchecked, significant investment will be made, such as a major hotel development supplementing existing small-scale accommodations. An outmigration of local residents may occur, while those who remain come to accept change and may opt to seek involvement in the lucrative tourism market. Thus, by the end of this period the destruction of the idyllic landscape and atmosphere has occurred.

The 'post-destruction stage' is an indefinite stage in that several scenarios are possible. One is that the heritage shopping centres will be replaced by a recreational/tourist shopping centre, appealing not to the heritage-seeking consumer, but to the mass tourist market. In this case, consumption levels will again rise and the hope of regaining any vestige of the original atmosphere is completely lost. Another possibility is that the community has not developed other venues for tourism consumption and consumers seek out another landscape where the desired atmosphere is still intact, and then the result may be a decline in visitor numbers at the initial location. With less congestion, noise and other ills associated with overcrowding, a partial return of the sought-after atmosphere may occur, which will be reflected in the attitudes of local residents whose perceptions of their landscape become more positive. A state of equilibrium may be achieved.

In later work, Mitchell, Atkinson and Clark (2001) found that the model could be applied to many types of heritage communities: One based on rural tradition as in the case of St Jacobs, or one based on historical/cultural heritage, as in the case of Niagara-on-the-Lake. To determine whether a community has evolved along the path of creative destruction as predicted by the model, three variables that interact at each stage are examined: business composition, visitor numbers and expenditures, and resident attitudes towards tourism.

Before testing the model in a Chinese context, one thing should be noted: In China, as elsewhere, governments play an important role in promoting tourism or coping with the results of tourism. As Dogan (1989: 227) noted, 'whether the touristic development is encouraged or blocked, and whether the effects of tourism are effectively dealt with, depend very much on the policy of the government'. The government's influence is very obvious in China. The State Council of China regulates the economy and implements industrial and financial policies directing economic development and establishing the fundamental conditions aimed to guarantee China's lasting economic growth and social stability (Wang 2003). Similarly, tourism planning and promotion in China are controlled primarily by the central government (Xie 2001) and implemented by different provincial and local governments. Thus, in addition to examining the performance of the three main players in the model of creative destruction, the behaviour of local government, as producer, protector or promoter of heritage should be emphasized.

THE CONTEXT OF THE STUDY—LUZHI

Geographically, water towns are concentrated in the Jiangnan area, including the present Jiangsu province, Zhejiang province and Shanghai area, to the south of the Yangtze River (Figure 6.1). These river-based settlements appeal to tourists for two main reasons: First, the historic and cultural heritage of the ancient towns is very attractive; and second, the atmosphere of the water towns represents harmony between nature and its residents. In particular, the idyllic small-town setting enables urban residents to escape from the noise, pollution and pressures of cities (Ruan and Shao 1996).

Figure 6.1 The location of the study area.

Water towns are widely advertised as tourism destinations but little tourism research on them has been undertaken. There is a small amount of Chinese research on tourists' images (Li, Zhang and Chen 2006), planning and management problems (Xiong, Zhang and Zhou 2002; H. Xu 2003) but nothing that we are aware of on the processes of change that they are experiencing.

Luzhi is a typical water town, surrounded by five lakes and crossed by six canals. It is located in the Wuzhong district 25 km east of Suzhou city in Jiangsu province, China. Figures 6.1 and 6.2 show the position of Luzhi. It borders on the Singapore industrial park to the north-west and is only 18 km from the core of Suzhou city, 48 km from Shanghai's Hongqiao Airport and 60 km from Shanghai city. The Su Hu (Suzhou to Shanghai) Airport Expressway and the Round Suzhou Express both cross Luzhi town. The Hu Ning Expressway is 8 km to the north of Luzhi and connects Shanghai, Suzhou and Nanjing (the capital of Jiangsu province). This makes transportation very convenient and Luzhi highly accessible to a large population.

Luzhi is about 50 sq km in size, with a population of 50,000 and comprises four regions: the ancient town, the economic development region, the villa region and the resort region. The ancient town is about 1 sq km in size and it has a permanent population of around 6,280 in about 2,300 households.

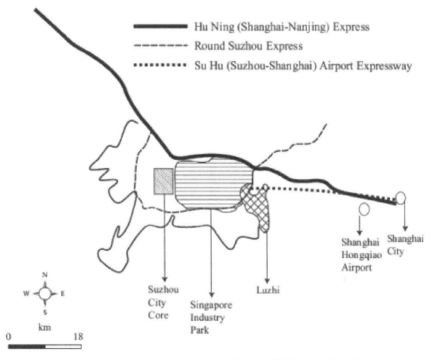

Figure 6.2 Location of Luzhi town (after Suzhou Luzhi Tourism Development Corporation, 2004).

The resort region is now just being constructed and is not addressed in this paper although, once in place, it could have relevance to the later stages of the creative destruction model. The layout of the town is shown in Figure 6.3. It has a history extending 1,400 years and Baosheng Temple was initially constructed in AD 503 and parts of the current temple include a Buddhist stone pillar from the Tang Dynasty (AD 618–907). In addition to many historic buildings and areas, another feature is that from time to time the visitor may see a local traditional costume of tops and trousers woven from different cottons and colours topped off with colourful shawls and shoes. Yet, in spite of these distinctive advantages, Yabing, Zhang and Chen (2006), based on a survey of 1,619 visitors to Zhouzhang and Tongli (other water towns in the area), find weak images of the towns are held: one contributing factor being the commercialization of the streets.

Although originally outlined within the context of a developed country and a postmodern society, the model of creative destruction may also be applied to Luzhi. As described earlier, heritage landscapes emerge when three criteria are present, each of which can be found in the study site.

First, a community must be accessible to a large and relatively affluent population. Luzhi is readily accessible to metropolitan Shanghai and is part

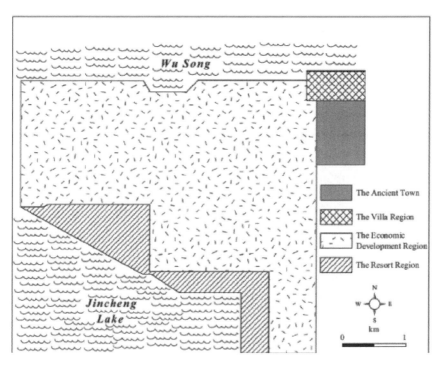

Figure 6.3 Layout of Luzhi water town (after Suzhou Luzhi Tourism Development Corporation, 2004).

of the affluent city of Suzhou. Indeed the wider Suzhou municipal area now has a population of over 6 million people, with a fast growing per capita income. Zhu (2005: 2) states that 'It is fair to characterize the grand change in Suzhou since the mid-1990s as nothing but urban renewal.' Luzhi is thus within one of the most developed areas of China—the Yangzi River delta area.

Second, some elements of a heritage environment must be in existence before commodification begins. With its wealth of canals, rivers and surrounding lakes, Luzhi has a very distinctive natural environment. Luzhi also has an exceptional history that is perhaps the most important contributor to its amenity environment. In spite of a settlement history of about 2,500 years, Luzhi still maintains a traditional layout and living environment and its residents are proud of their traditional Chinese culture. The image of Luzhi water town—'small bridges, flowing canal water and rest houses'—is deeply rooted in Chinese people's hearts. Traditional artefacts, food, architecturally appealing buildings and ancient forms of transportation (boats) are abundant in Luzhi. Luzhi appeals to tourists as it exudes an image of a happy, healthy and problem-free life.

The third criterion is the availability of capital and an entrepreneurial spirit. Luzhi is located in Jiangnan area; one with a long tradition of encouraging businesses and entrepreneurship. In modern Jiangnan, economic reforms have released a tremendous amount of productive and entrepreneurial energy in rural areas. For example, in 1984, the former commune and brigade enterprises were renamed town and village enterprises (TVEs), which motivated many government officials and party members to become entrepreneurs. Marton (2000: 77) noted that 'official sanction of township and village industries, individual and institutional entrepreneurship and a commensurate rise in commercial and trading activities in the countryside, have led to a revitalization of the marketing and commercial functions of small towns'. Moreover, since 1992, the main feature of China's economy has been the growth of the private sector. Deng Xiaoping's call in his southern tour in 1992 for an acceleration of market-oriented reforms signalled to entrepreneurs that party-state attitudes to private enterprise were becoming more welcoming. During the 1990s the party-state acknowledged more completely the virtues of entrepreneurship. In 1997, at the fifteenth Party Congress, for example, private enterprise was recognized as an important component of the economy. This reform rapidly accelerated in Jiangnan due to the fact that 'the increased entrepreneurship in the modern Jiangnan area is not new, but is consistent with historical patterns of commercialization' (Marton 2000: 78). Therefore, the Jiangnan area is featured by its continual and deeply rooted entrepreneurial spirit. When combined with its heritage base, and accessibility to a large market, Luzhi appears to be an ideal site to test the model of creative destruction in a Chinese heritage context. Figures 6.4 and 6.5 indicate the nature of the locale.

Figure 6.4 Canals of Suzhou.

Figure 6.5 Canals of Suzhou.

RESEARCH METHOD

Interviews were adopted in this research for two reasons. First, business and residential records about the tourism development of Luzhi before the mid-1990s could not be found. Second, interviews are a good way to understand the attitudes and actions of different governments towards developing tourism in Luzhi and water towns in general. Face-to-face interviews were conducted with 10 key informants, including four government officials, five businessmen in the tourism industry and one university scholar. Different questions were directed as follows:

> *For government officials*: the process of tourism development in Luzhi and several other water towns, the status of the tourism industry in Luzhi, the status of ancient water towns in Suzhou's tourism industry, the policies and plans for promoting water-town tourism, tourism impacts on Luzhi and the sustainable development of Luzhi.
>
> *For businessmen*: the business operation situation, efforts to promote tourism in Luzhi, tourism impacts on Luzhi, the relationship between Luzhi and other water towns, attitude towards tourism's future development in Luzhi and understanding of sustainable development.
>
> *For the scholar*: the relationship between Luzhi and other water towns, tourism impacts in Luzhi, sustainable development and opinions concerning future development.

These interviews were undertaken in the form of casual conversations that began with an introduction to the author's research purpose. Several informal interviews were also conducted to gather information on and to understand the population structure and living standards of Luzhi.

In addition, questionnaire surveys were used to obtain local residents' perceptions of tourism impacts in Luzhi and to examine the local business community. The resident questionnaire consisted of two sections. The first section was comprised of 15 attitudinal questions designed to measure perceived tourism impacts of different types. These included economic, environmental and socio-cultural aspects and residents' overall evaluations of and attitudes towards tourism development. The questions were developed following an exhaustive review of the literature towards tourism and, with reference to residents' perceptions and attitudes, it can be seen as being part of the long tradition of such research in tourism that has been summarized by such authors as Perdue, Long and Allen (1990); Long, Perdue and Allen (1990); and Teys, Sonmez and Sirakaya (2002). However, until quite recently, little such research has been undertaken in China (e.g. see Gu and Ryan 2008). Participants were asked to indicate their degree of agreement with statements of tourism impacts on a five-point Likert scale, in which '–2' meant strongly disagree, '–1' disagree, '0' neutral, '1' agree and '2' strongly agree. The second section of the resident questionnaire dealt with demographic and socio-economic data. The business questionnaire asked for such information as type of enterprise, number of employees, length of operation and seasonality of the business.

After a successful pretesting, the questionnaire survey was conducted in the ancient town, which is about 1 sq km in size. There were several reasons for choosing this region as the study area. First, the aim for the survey was to examine Luzhi local residents' perceptions of and attitudes towards tourism. In the economic development region, the villa region and the resort region, newcomers constitute a very high proportion of the local population and it would have required much extra effort to differentiate them. Second, the ancient town is the place that local people and visitors generally refer to as 'Luzhi Ancient Town' and the majority of tourism activities take place in this area. Residents and entrepreneurs in the ancient town region have direct contacts with tourism. Their perceptions and attitudes are being given priority in this research for the time being. Although some tourism activities occur in the resort region, the scale is very small and the resort has yet to be constructed completely. Third, the research was restricted by time and funding constraints so that it was necessary to focus efforts on a manageable area.

The ancient town of Luzhi consists of four districts: Nanhui, Jianxin, Zhengyang and Zhonghe. They are of fairly equal size and population density, and separated geographically by canals. A geographic cluster sampling design was utilized to acquire data from residents of the four districts. Only permanent residents of the community, defined as those persons who had lived in the community for at least one year, were surveyed. Based on previous studies of tourism impacts and the ancient town's permanent resident population (about 2,300 households), 220 questionnaires were assigned to residents, which represented about 10 per cent of the households. Because the four districts were fairly equal in number of households, 55 questionnaires were allotted for each district. No lists of households existed to form a sampling frame. Therefore, in each district a street was chosen randomly and the first house and every tenth house was visited. Then a second street was selected randomly and so on until the quota of interviews was completed. A questionnaire was left with the individual answering the door and collected the next day. Only adults aged 18 years and above were invited to answer the questionnaire.

There were about 170 businesses in total in the ancient town region; however, the businesses were distributed very unevenly. After on-site observation of the distribution of businesses in these districts, it became evident that tourism businesses were concentrated in Nanhui and Jianxin districts, while Zhonghe district and Zhengyang district had far fewer businesses. There were only 10 businesses in Zhengyang district and 13 businesses in Zhonghe, whereas there were about 150 businesses roughly evenly divided between Nanhui and Jianxin districts. Therefore, 10 questionnaires were allocated for Zhengyang and Zhonghe respectively, and 30 questionnaires were handed out each in Nanhui and Jianxin.

Due to the fact that most of the local residents were seniors and they tended to speak dialects, as well as possible safety issues, help was requested from the Luzhi Residents Committee to facilitate undertaking a door-to-door survey. Four officials, each responsible for one district, accompanied the first author to hand out questionnaires. These officials were familiar with door-to-door surveys because in 1999 they had helped to collect data in a similar way for the Luzhi Town Plan.

There was no list of all households. Following the random identification of a starting point, potential respondents were selected in a systematic manner. Only adults (aged 18 or above) were invited to complete the questionnaire. Interviews were conducted after 6 PM when most people were expected to be at home. The questionnaire was dropped off and collected at the same time on the following day. In the case of businesses, all the 10 businesses in Zhengyang district were visited. In the other districts, potential respondents were selected in a systematic manner following the random identification of a starting point. As for the household interviews, questionnaires were dropped off and collected the next day.

The surveys were conducted in July 2005, and of the 220 households contacted, 199 questionnaires were completed, for a response rate of 90.5 per cent. Non-respondents included people who refused to participate as well as those who were absent when the questionnaire was handed out or collected. Of the 80 businesses contacted, 69 questionnaires were completed, for a response rate of 86.3 per cent. The responses were numerically coded, entered into a computer and analysed using SPSS.

In addition to interviews and surveys, detailed observations and notes, including the drawing of maps, were made on the land use pattern and the types and numbers of businesses in the ancient town of Luzhi. Also, academic literature in both the West and in China was reviewed to establish a conceptual context for this research. Government documents such as Luzhi Ancient Town Protection and Restoration Plan (1999) and Luzhi Tourism Development Plan (2001) were photocopied with the permission of the local government. Tourism brochures and magazines, local newspapers, books and videos were searched for relevant information. Photographs reflecting the living environment and tourism attractions were also taken.

THE CREATIVE DESTRUCTION OF LUZHI

In the following sections the development of tourism in Luzhi will be described according to the stages of the creative destruction model using information of a variety of types as discussed earlier.

Early Commodification

The stage of early commodification in Luzhi began approximately in the late 1980s and continued to the mid-1990s. After 1978, China emphasized market economic development and, as a result, the entrepreneurial spirit was given a chance to reappear. In Luzhi, the industrial and service sectors developed quickly. During this period, Luzhi Township started to initiate tourism mainly by maintaining and promoting Baosheng Temple as a national historic site for domestic Chinese visitors.

> Luzhi did not develop tourism with great efforts but focused on industrial economic development before the early 1990s. (Interview with the deputy secretary of Luzhi Township)

It is difficult to gauge resident attitudes towards tourism during this stage since no surveys were done. No local newspaper or government documents have been found that show the attitudes of local residents towards tourism development. However, interviews with local government officials and businessmen and casual conversations with local residents suggest that before the mid-1990s tourism was at a small scale and was viewed very favourably by local residents as enhancing the economic well-being of the community.

Advanced Commodification

In the middle of the 1990s, Luzhi Township began to promote tourism widely, particularly to develop Luzhi's ancient town as the main tourism attraction. In the early 1990s, Luzhi's economy shifted to an industry-centred and export-led economic mode, abandoning the previous agriculture-centred and closed mode of operation. The new economic orientation motivated more local government officials and local entrepreneurs to increase investments related to tourism development.

In 1995, Luzhi was listed as one of the four pioneering cultural and historic towns by Jiangsu Province (Luzhi Ancient Town Protection and Restoration Plan 1999). In 1998, Luzhi launched the Luzhi Water Town Costume Tourism Festival to increase Luzhi's reputation as a tourist site and to attract more tourists. Since 1999, the local government has invited famous officials and scholars from within and out of China to visit and comment on the ancient town. For example, both Li Lanqing (former vice premier of the State Council) and Helmut Kohl (former prime minister of Germany) were invited to visit the area. Also, Luzhi Ancient Town was praised as 'The First Ancient Town of Chinese Watery Regions' by Fei Xiaotong, vice president of the Standing Committee of the National People's Congress and a famous scholar in China. This praise, 'The First Ancient Town of Chinese Watery Regions', has become a slogan for advertising Luzhi and is printed in numerous tourism brochures. Also after 1999, the Wu county year-book began to include tourism development information on Luzhi. In 2001, Luzhi was recorded in the preliminary list for world heritage status by UNESCO and as one of Jiangsu Province's relic protection units.

In 1997, Luzhi received 82,500 tourist arrivals and a tourism income of 3.5 million RMB (Luzhi Tourism Development Plan 2001). In 2000, there were 233,200 tourist arrivals, further increasing Luzhi's ability to attract outside investment (Wu County Year Book 2000). In 2001, a tour bus from the train station of Suzhou to Luzhi was made available. As the deputy secretary of Luzhi Township stated, 'The GDP of Luzhi Town in 2003 was RMB 2,500–3,000 million'. In 2005, from interviews with a Luzhi government official, the general manager of Suzhou Luzhi Tourism Development Company and casual conversations with local tour guides, it was learned that the major tourism product in Luzhi is sightseeing and visitors tend to stay in Luzhi for just half to one day. Additionally, Luzhi targets Shanghai as its most important market. Travel agencies in Shanghai organize Shanghainese to take a one-day cuisine tour in Luzhi to taste its local aquatic products. The majority of visitors to Luzhi are from Shanghai or nearby provinces.

Suzhou Luzhi Tourism Development Company has been responsible for promoting and marketing tourism in Luzhi. This corporation is a local government company, which has two functions: (a) planning and maintenance of the ancient town; and (b) marketing the ancient town and selling Luzhi's tourism products. Suzhou Luzhi Tourism Development Company mainly promotes one-day tours in Luzhi to potential customers, such as the one-day tour in Luzhi Ancient Town, the one-day combined tour of Luzhi Ancient Town and Zhangling or Tingling Park which is about 2 km away from the ancient town, the one-day agricultural tour in Luzhi which combines the ancient town and vegetable green houses and model production fields and the one-day industrial tour which combines the ancient town and the new industrial area.

From on-site observation, the authors found that businesses are concentrated in the 'ancient town' of Luzhi. Gift stores, restaurants, tea-houses and grocery stores are crowded along the narrow lanes and canal banks in the ancient town (Figure 6.6). About two-thirds of businesses in the ancient town of Luzhi had been opened in the past five years (Table 6.1). The only hotel that was surveyed was built in 2003. Five of six restaurateurs that responded to the survey indicated that their businesses had opened in the preceding five years, with two in 1999, two in 2000 and one in 2001. All the clothing/shoes/hat stores appeared after 1999; one had been open for six months, and the other two for two years. Five of the six surveyed 'other businesses' had opened in the preceding five years.

Of the 64 surveyed businesses, 51 businesses (80 per cent) had only one or two employees. Only four businesses had six or more employees and all of them were restaurants; the other two restaurants had three or four employees. Therefore, the scale of surveyed businesses tended to be small

Figure 6.6 Tourism businesses in Luzhi.

Table 6.1 Businesses Appearing After 1999

Business Type	N	%	Opened Since 1999
Grocery store	20	31.3	12
Handcraft store	12	18.8	9
Special food store	10	15.6	6
Restaurant	6	9.4	5
Antiques store	6	9.4	5
Clothing/shoes/hats store	3	4.7	3
Bookstore	2	3.1	2
Tea-house	1	1.6	—
Hotel	1	1.6	1
Other	3	4.7	2

and the slightly larger ones among them tended to be restaurants. Tourism also created more business opportunities for local entrepreneurs. A local entrepreneur opened a four-story hotel in May 2003. He reported that he saw the great potential of developing tourism in the ancient town. After consulting other hotel owners, he found that their businesses were making profits and so he decided to set up his own business. To date, his hotel on average maintains a 70 per cent occupancy rate, which is evidence that this business is doing well.

With the continual growth of tourism, businesses are expanding rapidly. A café that is not in harmony with the traditional architecture of the ancient town has appeared (Figure 6.7). Also, a specialized business district outside of the ancient town, facing the entrance to the ancient town, is being constructed for tourists to buy souvenirs. Local government officials, as policy-makers, realized the economic potential of the ancient town and have made

Figure 6.7 A café in Luzhi.

efforts to make local residents realize that the ancient town is precious, not obsolete and outdated. The following quotations illustrate this point.

> Protection is the precondition for developing the ancient water-town tourism. (Interview with deputy director of Suzhou Municipal Administration of Tourism)

> Making a viable plan and protection are the premise of developing tourism in ancient water towns. Not only the legacies should be well protected, but also the old and the modern should be harmonized, creating a living environment approved and welcomed by local residents. (Interview with Director of Planning and Statistical Division of Jiangsu Provincial Tourism Bureau)

> The local government put a lot of effort into protecting the ancient town, inviting relevant planners and experts to develop Luzhi Ancient Town Protection and Restoration Plan, and then invested a huge amount of money to implement this protection plan. (Interview with deputy secretary of Luzhi Township)

> Protection of the ancient town is the first and foremost task. In order to realize effective protection, proper planning should be carried out. (Interview with general manager of Suzhou Luzhi Tourism Development Company)

> The local government of Luzhi and Suzhou Luzhi Tourism Development Company worked closely to protect the natural environment in the ancient town, trying to avoid over-commercialization and improving the living environment of the old town. (Interview with chief officer, Suzhou Luzhi Tourism Development Company)

Luzhi Township leads in repairing and maintaining historic sites and protecting its ancient town. In 1998, Luzhi Township invited planning experts from Shanghai Tongji Urban Planning Institute to devise the Luzhi Ancient Town Protection and Restoration Plan. After the plan was finished and approved by relevant government departments in 1999, Luzhi government began its implementation. The working strategy proposed by this plan is 'protecting the ancient town, constructing new zones out of the ancient town, developing tourism and promoting the economy' (Luzhi Ancient Town Protection and Restoration Plan 1999).

In recent years, the local government has allocated more than 20 million RMB for the renovation of historic buildings in the ancient town, repairing seriously damaged old buildings to prevent them from collapsing, repairing ancient bridges and dredging and cleaning the river course. A chemical plant and other factories were moved out of the ancient town. New roads were

Figure 6.8 Attractive scene in Luzhi.

built bypassing the ancient town rather than going through it. All these efforts have resulted in a beautified environment in the town of Luzhi (Figure 6.8).

The government officials and residents interviewed by the first author all have a favourable attitude towards tourism development. Specific comments on the tourism development in Luzhi are as follows:

> Without tourism development, local communities lack funds to repair and maintain heritage buildings; local residents lack initiative and motivation to protect heritage. Tourism enables the community to protect the heritage on its own. (Interview with deputy director of Suzhou Municipal Administration of Tourism)

> Tourism benefited local residents, provided more employment opportunities, improved the local economy and protected the local heritage in Luzhi. (Interview with deputy secretary in the Luzhi Township)

> Tourism development improved the sanitation situation in the ancient town, cleaned the canal river and improved residents' living environment. (Interview with chief officer, Suzhou Luzhi Tourism Development Company)

> Tourism increased local residents' household income. (Interview with general officer, Luzhi Ancient Town Resident Committee)

I hope the tourism in Luzhi can be further developed and bring more tourists. (Casual conversation with a restaurant owner)

We hope that the local government will promote further tourism and expand tourism activities to our district. (Casual conversation with residents of Zhonghe district)

Tourism has improved the infrastructure facilities, built sewage drainage, provided cooking gas, increased local residents' enthusiasm for protecting heritage in the ancient town and contributed to the fiscal revenue. (Interview with general manager of Suzhou Luzhi Tourism Development Company)

Luzhi residents' attitudes towards tourism development are also revealed by the questionnaire survey, the results of which are summarized in Tables 6.2 and 6.3. As can be seen from Table 6.2, on a five-point Likert scale ('-2 strongly disagree' to '+2 strongly agree'), with respect to the local economy, employment opportunities and standard of living, the mean responses were above 0.54, with the highest mean of 0.85, indicating that local residents were inclined to have favourable evaluations of tourism impacts in these three areas. Specifically, 79 per cent, 75 per cent and 58 per cent of participants believed that tourism improved the local economy, employment situation and living standard respectively. However, prices of goods and land were perceived to have increased as a result of tourism, with 47 per cent of respondents confirming this and 38 per cent being unsure.

Regarding environmental aspects, the majority of respondents believed that tourism development had protected local historical and cultural heritage (90 per cent), beautified the living environment (82 per cent) and improved public utilities' infrastructure (82 per cent). There was not a single respondent who strongly disagreed with the assertion that tourism had protected Luzhi's historical and cultural heritage. However, although most respondents (52 per cent) perceived that the canal water quality had been improved as a result of tourism development, a substantial minority (30 per cent) did not agree. Residents also varied in their perceptions of tourism impacts on local traffic and air and noise pollution. About half (47 per cent) of respondents disagreed with the statement that tourism had worsened local traffic (although this was probably outside of the ancient town which vehicles cannot penetrate, although motorcycles can be very annoying in the ancient town) and the quality of air, but 23 per cent agreed, with the rest of the respondents being unsure. Therefore, respondents as a whole believed that the environmental impacts of tourism were beneficial, despite the fact that almost one-quarter of residents perceived negative impacts of tourism related to traffic jams, and air and noise pollution. However, these problems were concentrated outside of the tourism area for vehicular traffic cannot penetrate the ancient town.

Table 6.2 Residents' Perceptions of Tourism's Impacts

Statement	N	Mean	Std. D.
Economic			
Improves the local economy	194	0.85	.71
Increases employment opportunities	193	0.76	.89
Improves residents' living standard	192	0.54	.89
Increases price of land and commodities	192	0.36	.85
Environmental			
Protects historic and cultural heritage	194	1.12	.69
Beautified the living environment	194	1.04	.82
Improves public utilities infrastructure (water, electricity and road)	193	1.03	.90
Improves the quality of canal water	195	0.24	1.21
Leads to traffic jam and overcrowding	194	−0.26	.93
Increases air and noise pollution	193	−0.32	1.05
Socio-Cultural			
Strengthens local residents' cultural identity	192	1.08	.76
Destroys friendly atmosphere in the neighbourhood/ interpersonal relationships	190	−0.42	.96
Increases stealing	194	−0.55	1.11
Overall			
Positive impacts of developing tourism outweigh negative impacts	190	0.78	.72
The local government should continue to promote tourism	193	1.14	.74

(−2 = strongly disagree, −1 = disagree, 0 = neutral, 1 = agree, 2 = strongly agree)

From a socio-cultural perspective, the majority of respondents (83 per cent) agreed that tourism has strengthened their cultural identity and not a single respondent strongly disagreed with this statement. However, although most (65 per cent) respondents disagreed, about one-quarter (24 per cent) perceived that tourism destroyed the local neighbourhood's friendly atmosphere and increased the incidences of theft. Thus, the respondents' overall

Table 6.3 Residents' Perceptions of Tourism's Economic Impacts (%)

Statement	Strongly Disagree	Disagree	Neutral	Agree	Strongly Agree
Economic					
Improves the local economy	0.5	5.2	15.5	66.5	12.4
Increases employment opportunities	3.1	6.7	15.5	60.1	14.5
Improves residents' living standard	1.5	11.5	28.6	47.9	10.4
Increases land and commodity prices	1.6	13.5	38.0	40.6	6.3
Environmental					
Protects historic and cultural heritage	0	4.1	6.2	63.4	26.3
Beautified the living environment	0.5	5.7	11.3	54.6	27.8
Improves public utilities infrastructure (water, electricity and road)	2.6	3.6	11.9	51.8	30.1
Improves the quality of canal water	10.3	20.0	17.9	39.0	12.8
Leads to traffic jam and overcrowding	3.6	46.4	25.3	22.2	2.6
Increases air and noise pollution	12.4	34.7	29.5	19.2	4.1
Socio-Cultural					
Strengthens residents' cultural identity	0	4.2	12.5	54.2	29.2
Destroys friendly atmosphere in the neighbourhood/interpersonal relationships	4.2	60.5	11.1	21.6	2.6
Increases stealing	14.9	54.6	5.7	20.1	4.6
Overall					
Positive impacts of developing tourism outweigh negative impacts	0	2.1	15.0	49.7	33.2
The local government should continue to promote tourism	.5	5.3	20.5	63.2	10.5

perceptions of the socio-cultural impacts of tourism were positive, although a minority of respondents recognized some negative aspects.

The mean scores in Table 6.2 indicate that respondents believed that the positive impacts of tourism outweigh its negative impacts and they supported development of additional tourism. As Table 6.3 shows, almost all respondents (98 per cent) believed that the positive impacts of tourism outweighed the negative impacts and that the local government should continue to promote tourism (94 per cent).

In summary, the survey revealed that Luzhi residents as a whole not only supported the current magnitude of the tourism industry, but also favoured its expansion. Despite this overall positive feeling, some Luzhi residents identified some negative tourism impacts, such as the increasing price of land and goods, and about one out of four respondents perceived that tourism did not improve the quality of canal water and that tourism led to traffic issues, air and noise pollution and stealing and destroyed the former friendly neighbourhood atmosphere.

DISCUSSION AND CONCLUSIONS

Application of the Model of Creative Destruction

The model of creative destruction was developed to describe the evolution of communities whose development has occurred around the commodification of heritage in North America. This chapter has endeavoured to determine if this model can be applied to Luzhi, China, in the context of a developing society. This study first identified that Luzhi satisfies the three criteria necessary for land use patterns to change, including the emergence of a heritage shopping centre. Then, a detailed analysis of the three variables that drive the model was conducted and a description of governments' behaviours was presented.

Results suggest that a substantial number of businesses in Luzhi now cater to visitors. Within the tourist area in the ancient town, more than 100 small businesses have been developed. Even so, as a relatively new destination, Luzhi retains much of its character and there are other water towns, such as Zhujiajiao, and many other tourism destinations, such as Liyang, that have seen greater changes (Li, Zhang and Chen 2006). Most businesses are currently in keeping with the traditional atmosphere of Luzhi's ancient town; however, initiatives that stray significantly from this theme have appeared, such as a café with discordant architecture. Moreover, new businesses initiatives are expected. A business district catering to visitors is under construction in the area just opposite the entrance of the ancient town. These investments, and their accompanying marketing strategies, have had a huge impact on visitation levels. Although visitor numbers are high, local residents support their presence and favour expansion of the industry. Despite this generally favourable situation, residents are aware of the negative impacts caused by tourism. Considering all of these factors, Luzhi currently is in the latter stages of advanced commodification, and is moving towards the next stage of early destruction.

This study has argued that the model of creative destruction is applicable in a Chinese context. Results of this study show that the model should be modified in two ways to enhance its relevance for application outside the developed world. First, in its original form, the model identified entrepreneurial activity as being the catalyst for development. As described in this study, although the private sector has contributed to the development of the heritage landscape, it is government that has invested large sums in producing the heritage landscape. This needs to be acknowledged when applying the model in a Chinese context. Second, the model of creative destruction presents a number of indicators that can be used to determine when a community is moving from one stage to the next. These indicators may not all be useful outside a North American context. For example, it is suggested that when a community is in the final stages of creative destruction, an outmigration of local residents may occur as crowding and congestion become unbearable. Although this indicator may be useful for Canadian heritage centres, which tend to be very small, it may not be valid in China, where population densities are considerably higher and where government can directly influence migration decisions. In the case of Luzhi, for example, outmigration has already occurred at the request of local government who have encouraged people to relocate to reduce densities and provide additional room for tourism development. Residents who undertake such moves do so not to escape from tourism, but to improve their living conditions. Therefore, it is questionable whether the outmigration of local residents is a good indicator of development stage in the Chinese context.

Implications for Policymakers and Managers

The model of creative destruction is useful for guiding development in land usage patterns, perhaps particularly retail sites, in the absence of effective controls. It presents a number of indicators, such as business and visitor numbers and resident attitudes that can be collected to describe a community's passage along its evolutionary path. Although the model does not provide specific numbers or thresholds that symbolize a particular stage of development, information collected in different periods can be compared, thus allowing the process of development to be tracked.

The model, therefore, should be used by local policymakers and managers to monitor the status of tourism development within their local municipalities. Although it does not permit the prediction of the future in a precise, mechanistic way, the model does suggest how such an asset as a heritage shopping centre might evolve, if no actions are taken to modify the trends. Such actions are necessary if authorities wish to create a tourism product that is sustainable over the long term; one that does not degrade the environment in which it exists to such a degree that it prohibits the successful development and well-being of other activities and processes (Butler 1993). Thus, the model should be used to inform decisions that contribute to the sustainable development of Luzhi and, by extension, other small-town heritage tourism destinations. In their work on Elora, Mitchell and Coghill (2000) found that if local residents continued to express their concerns, and the government remained receptive, developments that deviate from a specific theme might be restricted, thus preventing the community's entrance into the model's next stage. This implies that local policymakers and managers should pay more attention to local residents' opinions about tourism development and make efforts to direct the planning and managing of businesses in the area. They further argued that the magnitude of destruction, in other words, the evolution of tourism in the direction suggested by the model, would depend on the extent of domination of a profit-motivated discourse. If profit-driven forces are accompanied and tempered by those who work on preservation and protection, a community's advancement to the next stage is less unlikely. On the other hand, in the absence or unsuccessful intervention of groups other than businesses, the destruction of the original landscape may be almost total. Therefore, local governments should regulate businesses development in heritage settings such as the ancient town of Luzhi to ensure that the highly valued atmosphere is not destroyed by over-commodification.

ACKNOWLEDGEMENTS

The authors would like to thank the officials and residents of Luzhi for their great help with data collection. The research was funded, in part, by the Canadian International Development Agency through the Ecoplan China project in collaboration with the Hainan Department of Land, Environment and Resources, Nanjing University, Guelph and Wilfrid Laurier Universities, under the leadership of the University of Waterloo. A previous version of this chapter was also published in *Tourism Management* and is here replicated through the courtesy of Reed Elsevier. The photographs used in this chapter were taken by Chennan Nancy Fan, with the exception of figures 6.4 and 6.5 that were taken by Chris Ryan.

7 Comparing Scenic Zones in China
Comparisons and Methodology

Li Hong and Zhao Xin

INTRODUCTION

The purpose of this chapter is to describe an approach for the evaluation and ranking of tourist scenic zones in China. It is based on both quantitative and qualitative data and assessments, and the purpose is to establish classifications and rankings that can be used in tourism planning and benchmarking. The chapter begins by indicating the types of data required for this exercise and then indicates the method of constructing the index. The main purpose of this chapter is therefore methodological in that it concentrates on the means of establishing ranking patterns rather than seeking to describe the results. Within China these methodological issues are important as large numbers of classification and rating/ranking systems are used with reference to perceived international best practice and benchmarking, as indicated in other chapters of this book. The questions that thus have to be answered is the degree to which any ranking system of cities, events and scenic areas is based on good scientific assessment, thereby aiding the future development of China's tourist infrastructure by better informing planning decisions about site investment, location synergies and promotion. Such approaches are common and represent an important component in Chinese thinking that informs policies when planning and comparing different destinations.

ESTABLISHING INDICES

Establishing scenic areas as tourist zones or attractions involves many considerations apart from the intrinsic aesthetic values of any given region. These considerations involve the strategic, the technical, the speed of implementation of policies and acceptance by the tourist industry, and the costs involved in establishing such zones. Strategic considerations relate to wider infrastructure planning and the relationship of the proposed zone to existing tourist scenic areas, and the degree to which social and economic benefits are being sought, while the technical consider-ations relate to site development and planning, such as establishment of accommodation, restaurants, footpaths, visitor centres and the like. All of these involve costs, and these in turn can be described in various ways such as total costs, costs per visitor etc. This chapter suggests an evalua-tive index system to assess the international competitive ability of tourist scenic areas based on three class-I level evaluative indices and 14 class-II level evaluative indices. The class-I level indices include the 'objective', potential and consumer competitive characteristics and are divisible into the class-II indices that are listed in the following. Indicative measures are shown for each index.

Class II Indices of Competitiveness

Objective Classification

Classification of tourist scenic area
Visitor satisfaction
Promotion and advertising costs as proportion of total costs
Liquidity ratios
Rate of return on capital invested (profitability)
Labour turnover
Proportion of sales revenue due to new product
Market share

Potential Classification

Costs of capital maintenance and incremental capital accumulation as
    proportion of total costs
Research and development costs as a proportion of total costs

Consumer Classifications

Tourist visit patterns between competing locations
Spatial proximity to markets
Image and importance ascribed to the region by the market

In this chapter both qualitative and quantitative data are used to evaluate the relative position of each of a number of Chinese tourist scenic areas. The final index value was determined by collecting and assessing data to build a model of scenic site evaluation. Both a qualitative and quantitative index were standardized and non-dimensionalized to figure out the contributory weights of the class I indices to a total score for each scenic area , and these were then ranked on the basis of the scores.

DATA REQUIREMENTS

To undertake this exercise the following statistical data for the whole year were needed: the rate of visitors' complaints, the advertising expenditure, the total amount of business income, a description of product characteristics, current assets, current liabilities, value and levels of profitability, the value of initial and term-end funds, the total number of the staff at the beginning of the period, number of the staff at varying times during the period of analysis, sales income from new products, values of initial and term-end capital, value of research and development expenditure, a trend analysis of tourism consumption, location data and data about the surrounding area and its infrastructure and assessments as to the quality of that infrastructure and the means by which it supported the tourist scenic zone in question.

ESTIMATING INDICES

Tables 7.1 and 7.2 indicate the potential bases for estimating indices for ranking sites/destinations, and comprise three sets of data: (a) secondary data of an empirical nature, (b) qualitative comparative rankings (Table

Table 7.1 Indices for Scenic Area Performance Assessment

Index – Level I	Index – Level II	Measure
The real competitive ability of tourism scenic areas	The visitors' degree of satisfaction	Complaint books, surveys of satisfaction
	Advertising expenditure	Advertising spending /advertising as % of total expenditure/income
	Asset ratios	Assets/Liabilities
	Rate of return	Annual Profit/capital ratios
	Labour turnover	Labour turnover measures
	New product success	New products' business income/the total business income
	Market share	Revenue/Visitor numbers of a given area as % of all Revenue/Visitors in a system
The potential competitive ability of tourism scenic areas	The ratio of capital maintenance and capital accumulation	Capital expenditure as % of all capital
	Research and Development Expenditure	Annual additions in R&D budgets R&D budget as % of total expenditure

Table 7.2 Comparative Benchmarking of Scenic Areas

Index Name	The Score Interval and Standard of the Indices			
	[0, 30]	[30, 60]	[60, 80]	[80, 100]
The ranks of tourism scenic areas	A	AA	AAA	AAAA
The characteristics of the products	Local Status	Provincial Status	National Status	Global Status
Market demand	Local	Regional	National	International
Nature of the region where scenic area is located and attractiveness to desired market	Ordinary	Strong	Stronger	Strongest
Resource status	Local	Regional	National	International

7.2) and (c) benchmarking data based on comparisons derived from expert opinion (Table 7.3). For illustrative purposes Table 7.3 may be based on marks awarded out of a total of 100, requiring experts to allocate marks where 100 represents the highest score. Researchers may wish to avoid bias by removing the highest and lowest scores (or a given percentage of such scores) before calculating the mean value. Its formulation is thus:

$$T_{ij} = (\Sigma X_{ij} - \min X_{ij} - \max X_{ij}) / (n - 2) \qquad (1)$$

In which: jth is the stationary index value; i is the travelling scenic area variable; i = 1 . . . . . . m, m are the senate comments the travelling scenic area integer; T_{ij} is the jth qualitative index scoring points of the ith

Table 7.3 Qualitative Indices—Comparative

The Name of Indices	The Evaluated Scenic Areas			
	Tourism Scenic Area Z_1	Tourism Scenic Area Z_2	. . . .	Tourism Scenic Area Z_n
The level of the index				
Area characteristics				
Measure of market demand				
Evaluation of ability to meet demand				
Resource status				

tourism scenic area; Xij is the value of the jth qualitative index of the ith tourism scenic area; and i = 1 . . . . . . n is the number of the tourism scenic areas involved in the evaluation.

ESTABLISHING THE INDICES

The investor, the creditor, the owner, the business operator and the country's economic planners have all paid close attention to the international competitive ability of tourist scenic areas from different angles and to achieve different goals. An objective, reasonable evaluation of the international competitive ability of tourism scenic areas could provide relatively accurate and scientific information for the tourism scenic area itself and other stakeholders because it relates to the development of those areas and the economic interests of other stakeholders while providing a basis for the country's economic and social planning. It enables scientific decision-making on the strategic management of tourism scenic areas, provides quantitative information of the areas' competitiveness, analyses their successes and failures, and enables the competitive ability of tourism scenic areas to be gradually optimized in this dynamic process. Therefore, the authors examined the competitive ability of scenic areas from the perspective of multiple levels and angles using measures derived from international studies, and established the following lateral evaluation model to assess the comparability of tourism scenic areas. This model takes a tourism scenic area as the evaluated system and seeks a comprehensive comparison with other scenic areas in an international context during specific time periods to determine the status of each scenic area in the system and their competitive ranking. The steps involved are:

1. *The determination of the system.*
   Suppose there were *n* independent tourism scenic areas, and the evaluation system could be readjusted according to the subjective assessments. Then the more extensive the system, the more objective and accurate the model results, and the greater the stability and the smaller the volatility. This chapter seeks to establish a model of dynamic evaluation of tourism scenic areas' international competitiveness based on a modified system to the multi-objective II level.
2. *Establish the evaluative indices system as shown in Table 7.4.*
   First determine the number of tourism scenic areas in the evaluation system, (*n*) so that $Zj(j = 1 \sim n)$. Next, collect the value for every index at the level I and II stratum based on available information for each tourism scenic area as shown in the Tables 7.5 and 7.6. Among them, Xij indicates the value of the Z tourism scenic area in the ith item of I level index, Yij(m) indicates the nth modified value in the ith item of I level index. It must be noted that the qualitative analysis index was

Table 7.4 The Corresponding System for Basic Indices and Modified Indices for Comparability of Tourism Scenic Areas

Basic Indices (1st Level)	Modified Indices (2nd Level)
	Index T_{11}
Index T_1	. . . . . .
	Index T_{1t}
. . . . . .	. . . . . .
	Index T_{p1}
Index T_p	. . . . . .
	Index T_{pt}

enumerated as T_0, which means the 0th item of I level index, after quantification and standardization, and the value and weight of the index will then be elaborated as a subsystem.

3. *Define the index's contribution attribute to the total score.*

The index's contribution to the total score can be divided into three categories: the positive, negative and the ideal or optimum point. The positive contribution may be called 'the benefit type'; it creates a positive contribution to the total score, namely the higher the indices, the better for the overall score. Its general set was recorded as $\Omega1$, namely, $\Omega1 = \{Ti|Ti$ is in the category of positive contribution$\}$ $\{Tij|Tij$ is in the category of positive contribution$\}$, where the set comprises the visitor degree of satisfaction, the proportion of expenses accounted for by advertising, the ratio of profit to capital, the proportion of the sales volume due to new products, the share of the market; the potential competitive ability indices including the ratio of capital maintenance and increment, the proportion of expenditure due to research and development, the consumption preference revealed at the tourist

Table 7.5 The Value Table of Basic Evaluative Indices of Each Tourism Scenic Areas' Comparison

Sample Indices	Z_1	Z_2	Z_3	. . .	Z_n
T_1	X_{11}	X_{12}	X_{13}	. . . . . .	X_{1n}
T_2	X_{21}	X_{22}	X_{23}	. . . . . .	X_{2n}
. . . . . .	. . . . . .	. . . . . .	. . . . . .	. . . . . .	. . . . . .
T_p	X_{p1}	X_{p2}	X_{p3}	. . . . . .	X_{pn}

Table 7.6 The Value Table of Modified Evaluative Indices of Each Tourism Scenic Areas' International Competitive Ability

Sample Indices	Z_1	Z_2	Z_3	......	Z_n
T_{i1}	$Y_{i1}^{(1)}$	$Y_{i2}^{(1)}$	$Y_{13}^{(1)}$	......	$Y_{in}^{(1)}$
T_{i2}	$Y_{i1}^{(2)}$	$Y_{i2}^{(2)}$	$Y_{23}^{(2)}$	......	$Y_{in}^{(2)}$
......	......	......	......	......	......
T_{iti}	$Y_{i1}^{(ti)}$	$Y_{i2}^{(ti)}$	$Y_{13}^{(ti)}$	......	$Y_{in}^{(ti)}$

source, the scenic region's referent power to the goal market and the resource status. The negative contribution was also called the 'cost type' as this kind of index has a negative contribution to the total score. Namely the smaller the index, the better is the total score. Its general set was recorded as Ω2, namely, Ω2 = {Ti|Ti is of the negative contribution category}{Tij|Tij is of the negative contribution category}, and this set will be comprised of items such as labour turnover and the loss of knowledge due to those staff leaving. The ideal point index will have more contribution to the total score if it closely approaches the ideal point, and its general set was recorded as Ω3, namely, Ω3 = {Ti|Ti is of the ideal point category}{Tij|Tij is of the ideal point category}. For example, the ideal point of current ratio is 2.

4. *Engage in reiteration of non-dimensional and standardization of the level I and the level II indices.*

   a. When $T_i \vee T_{im} \in \Omega_1$, calculate the sample average value of the various indices in the system first.

$$\bar{X}_i = \sum_{j=1}^{n} \lambda_{ij} X_{ij} \qquad i \in \left\{ i \,\middle|\, T_i \in \Omega_1 \right\} \qquad (2)$$

Among them, λ_{ij} and $\lambda_{ij}^{(m)}$ is the index weight, established according to characteristics and data derived from financial information and the index formulation. If T_i is the profit rate of capital index, then:

$$X_{ij} \begin{pmatrix} \text{(the profit rate} \\ \text{of capital of area j)} \end{pmatrix} = \frac{P_j \begin{pmatrix} \text{(the current} \\ \text{net profit of area j)} \end{pmatrix}}{Z_j \begin{pmatrix} \text{(the average total} \\ \text{amount of capital of area j)} \end{pmatrix}} \qquad (3)$$

And its weight $\lambda_{ij} = U_j / \sum_{j=1}^{n} U_j$ (the determination of $\lambda_{ij}^{(m)}$ could be the same) $\qquad (4)$

Then calculate the sample standard deviation of basic indices and modified indices in the system:

$$S_i = \sqrt{\frac{n}{n-1}\sum_{j=1}^{n}\lambda_{ij}(X_{ij} - \bar{X}_i)^2} \qquad i \in \left\{i \,\middle|\, T_i \in \Omega_1\right\} \tag{5}$$

$$S_i^{(m)} = \sqrt{\frac{n}{n-1}\sum_{j=1}^{n}\lambda_{ij}^{(m)}(y_{ij}^{(m)} - \bar{y}_i^{(m)})^2} \qquad m \in \left\{m \,\middle|\, T_{im} \in \Omega_1, i = 1 \sim p\right\} \tag{6}$$

After that, calculate the normalized index value of the level I and II indices:

$$\xi_{ij} = \frac{X_{ij} - \bar{X}_i}{S_i} \tag{7}$$

$$\eta_{ij}^{(m)} = \frac{y_{ij}^{(m)} - \bar{y}_{ij}^{(m)}}{S_i^{(m)}} \tag{8}$$

b. When $T_i \vee T_{im} \in \Omega_2$, deal with X_{ij} or $Y_{ij}^{(m)}$ by continuing absolute processing and make $\tilde{X}_{ij} = -X_{ij}$ or $\tilde{Y}_{ij}^{(m)} = -Y_{ij}^{(m)}$, take the value of $\tilde{X}_{ij}$ instead of the value of X_{ij}, and place it on the ith line in Table 7.4, or take the value of $\tilde{Y}_{ij}^{(m)}$ instead of $Y_{ij}^{(m)}$ list in table 7.5 and then the standardization processing can be carried out adopting the standardization steps of the indices in Ω_1.

c. When $T_i \vee T_{im} \in \Omega_3$, deal with X_{ij} or $Y_{ij}^{(m)}$ by carrying on the absolute processing, make $X_{ij} = -|X_{ij} - X_i^{(0)}|$ or make $Y_{ij}^{(m)} = -|Y_{ij}^{(m)} - Y_{i0}^{(m)}|$, among them $X_i^{(0)}$ is the ideal value of the ith index, $Y_{i0}^{(m)}$ is the mth modified value of the ideal value of the ith level I index. And then take the X_{ij} instead of the ith line in table 7.4, take $Y_{ij}^{(m)}$ instead of $Y_{ij}^{(m)}$ in table 5, and then finish the index standardization steps of the indices in Ω_1. Through synthesizing this theory,

Table 7.7 The Standardization and Non-Dimensionalization of Basic Indices

Sample Indices	Z_1	Z_2	....	Z_n	Standardization/non-Dimensionalization	Sample Indices	Z_1	Z_2	....	Z_n
T_1	X_{11}	X_{12}	....	X_{1n}		T_1	ξ_{11}	ξ_{12}		ξ_{1n}
T_2	X_{21}	X_{22}	.....	X_{2n}		T_2	ξ_{21}	ξ_{22}		ξ_{2n}
......	......	......		......		......	...	...		......
T_p	X_{p1}	X_{p2}	....	X_{pn}		T_p	ξ_{p1}	ξ_{p2}		ξ_{pn}

Table 7.8 The Standardization and Non-Dimensionalization of Modified Indices

Sample Indices	Z_1	Z_2	Z_n	Standardization/non-Dimensionalization	Sample Indices	Z_1	Z_2	Z_n
T_{i1}	$Y_{i1}^{(1)}$	$Y_{i2}^{(1)}$	$Y_{in}^{(1)}$		T_{i1}	$\eta_{i1}^{(1)}$	$\eta_{i2}^{(1)}$	$\eta_{in}^{(1)}$
T_{i2}	$Y_{i1}^{(2)}$	$Y_{i2}^{(2)}$	$Y_{in}^{(2)}$		T_{i2}	$\eta_{i1}^{(2)}$	$\eta_{i2}^{(2)}$	$\eta_{in}^{(2)}$
......	......	......	......		......	......	......	...
T_{iti}	$Y_{i1}^{(ti)}$	$Y_{i2}^{(ti)}$	$Y_{in}^{(ti)}$		T_{iti}	$\eta_{i1}^{(ti)}$	$\eta_{i2}^{(ti)}$	$\eta_{in}^{(ti)}$

the transformation of two matrices can be carried out, as Table 7.7 and Table 7.8 show.

d. Calculate the contribution weight of the level II index in the total score. This requires calculation of the weights of each level I index using information entropy.

(i) First, divide the interval $(-\infty, +\infty)$ into several subintervals, for example, it can be divided into eight subintervals: $(-\infty, -3)$, $(-3, -1.5)$, $(-1.5, -0.5)$, $(-0.5, 0)$, $(0, 0.5)$, $(0.5, 1.5)$, $(1.5, 3)$, $(3, +\infty)$, the division of subintervals could be readjusted according to the size of sample capacity n. Record P_k as ξ_{ij}, as it is the frequency of $j = 1 \sim n$ fall in the kth subinterval $(k = 1 \sim 8)$.

(ii) Carry out the information entropy of the ith level I index.

$$H_i = -\sum_{k=1}^{8} P_k^{(i)} \ln P_K^i \qquad (9)$$

Note: if one of the $P_k^{(i)} = 0$, then $P_k^{(i)} \ln P_k^{(i)} = 0$

(iii) Calculate the weight of the level I index in the total score.

$$\omega_i = \frac{H_i}{\sum\limits_{i=0}^{P} H_i} \qquad i = 0 \sim p \qquad (10)$$

5. *Calculate the total scores of the competitive ability of tourism scenic areas in the evaluated system.*

a. Theoretical distribution total score is $LF_j = \sum\limits_{i=1}^{p} \omega_i\, \xi_{ij} \varepsilon(-\infty, +\infty)$, indicating the distribution total score of the tourism scenic area Zj in the relative evaluated system. However, LF_j usually falls in the interval of $[-3, 3]$.

b. Direct comparison total score JF_j, indicates the direct comparison of total score of the tourism scenic area Zj in the relative evaluated system.

$$JF_j = 100 \times \varphi(LF_j) = 100 \times \frac{1}{\sqrt{2\pi}} \int_{-\infty}^{LF_j} \exp(-\frac{x^2}{2}) dx \in (0, 100) \qquad (11)$$

Among them, the sort of LF_j and JFj are absolutely unified in the evaluated system, where $LF_j = 0$ and $JF_j = 50$ represent the system's average level. The authors have divided the status of international competitiveness of tourism scenic areas into the following ranks: The strongest standard of international competitiveness of tourism scenic area is $JF_j \geq 75$, nominated as rank A; the stronger standard of international comparative ranking is $75 > JF_j \geq 55$, classified as B; the medium standard is $55 > JF_j \geq 45$, ranked as C; the weaker standard $45 > JF_j \geq 25$, ranked as D; the weakest standard is $JF_j < 25$, ranked E.

c. The complementary introduction of quantification and standardization of qualitative indices. Suppose there were q qualitative analysis indices, each qualitative index has been divided into five ranks, and also suppose that the score value of the mth qualitative analysis index of jth tourism scenic area Zj, which was called T_{om}, is X_{0j}, as the Table 7.9 shows.

   After the standardized processing, $X_{0j}^{(m)}$ was changed into $\xi_{0j}^{(m)}$, the total score of the tourism scenic area Zj's qualitative index is:

$$\xi_{0j} = \Sigma \omega_m^{(0)} \xi_{0j}^{(m)} \qquad j = 1 \sim n \qquad (12)$$

Among them, the determination of $\omega_m^{(0)}$ was carried out separately in the qualitative analysis subsystem using the entropy value method, after ξ_{0j} is returned to the evaluated system to participate in the total calculation as the level I index T_0.

6. *Establish ranking of scenic areas.*
   The goal is to establish a ranking of tourism scenic areas' international competitiveness to establish a benchmarked and comparative position between areas competing for not only visitors but also resources

Table 7.9 Reference Table for Quantification and Standardized Processing of Tourism Scenic Areas' International Competitive Ability

Sample Indices	Z_1	Z_2	Z_n
T_{01}	$X_{01}^{(1)}$	$X_{02}^{(1)}$	$X_{0n}^{(1)}$
T_{02}	$X_{01}^{(2)}$	$X_{02}^{(2)}$	$X_{0n}^{(2)}$
. . . . . .	. . . . . .	. . . . . .	. . . . . .
T_{0q}	$X_{01}^{(q)}$	$X_{02}^{(q)}$	$X_{0n}^{(q)}$

Table 7.10 The Arrangement of the Main Factor Matrix's Scores of Each Tourism Scenic Area's International Competitiveness

Index Sample	T_1	T_2	......	T_n	Synthetically Score	Arrangement	Rank
Z_1	X_{11}	X_{12}	......	X_{1n}			
Z_2	X_{21}	X_{22}	......	X_{2n}			
......	......	......	......	......			
Z_m	X_{m1}	Xm_2	......	X_{mn}			
	Weight number ω_1	Weight number ω_2	......	Weight number ω_m			

from government and other funding bodies. It is possible to create, as described earlier, five prime categories from A to E, indicating strong to weaker levels of competiveness. It is suggested that such rankings can be beneficial in informing, comparing, analysing, controlling and developing scenic areas for purposes of tourism. These considerations include not solely economic ones, but also a consideration of geo-morphic and botanical sciences, technology requirements, patterns of demand and statistical data relating to visitor perceptions, needs and motives. The end-result is that the rankings created by systems such as those described earlier create more structured thinking and thus better decision-making. It permits rankings of the nature illustrated in Table 7.10.

DISCUSSION

In a word, the calculation of ranking systems of scenic areas is complex. However, the systems derived from relevant theories of scenic area manage-ment and expert assessments suggest that three main characteristics need to be considered. These are: (a) the real market structural capabilities, (b) the potential competitive capabilities and (c) the environmental competitive abilities. These can be described at a domain optimization level by using respective concrete indices that can be largely empirically determined, with the results being placed in an order of ranking. Several explanations should be noted. First, the final order was arrived at on the foundation of a quan-titative model and analytical system; second, the absolute score of a scenic area is not the most important score, rather the score (no matter whether high or low) only reflects a relative position compared to a chosen reference set. Third, the final result acquired after comparison represents a relatively fine level of judgement of each tourist scenic area that is included in the reference set, and hence any estimation of economic contribution that might be provided by any given scenic area is bound by the chosen parameters of the set of selected places being considered in the analysis. One can also add that in order to facilitate the data processing and the generalized analysis, computer online data acquisition systems and programming can be used as an auxiliary method for processing and evaluating data. In short, this evaluative system lays the foundation for a convenient and continuous monitoring of the relative positions of scenic areas, one against the other, as changes are made to those areas in terms of performance, structures and developments. The system thus permits a scientific approach to better utilization of resources when seeking to enhance scenic areas or to improve their economic contribution to regional development.

Part II

Destinations and Cultural Representations

8 Destinations and Cultural Representations

The Importance of Political Context and the Decay of Statism

Chris Ryan and Gu Huimin

INTRODUCTION

The purpose of this chapter is to provide a context for specifically Chapters 9 to 13 that represent contributions that discuss aspects of cultural tourism within China. By their nature they tend to represent developments at specific sites, and thus represent issues at a destination or regional level. This approach has several advantages in that they illustrate the 'on the ground' concerns of peoples and administrations as they seek to cope with growing numbers of visitors and design products to appeal to those tourists. But their very specificity effectively silences other debates, and thus this chapter seeks to establish a framework for those debates. The chapter is thus based around an unconventional approach to cultural representation and tourism in that it tries to offer explanations of such issues within a framework of an eroding statism or central command economy in current Chinese developmental policy. It is thus divided into sections that first note long-established themes in tourism and its impacts on communities and their culture, it considers economic sectoral analysis with reference to work undertaken by German analysts of the restructuring of the East German economy after 1989, indicates some issues of representation of culture in Chinese tourism promotion and ends by reference to the models of tourism development offered by Weaver (2002).

DETERMINANTS OF DESTINATION AND CULTURAL REPRESENTATION

There is a wide-ranging literature pertaining to destination image, cultural representation, indigenous tourism and directed tourist gaze. It is not the intention of the authors to offer yet another review of this literature other than to pick a few themes. What is sought is some new form of thematic analysis that can be applied to the Chinese context, which, within the literature of mainland Chinese academics has tended, arguably, to the descriptive and/or been an extension of the western literature relating to the social impacts of tourism on local residents. Generally that literature has tended to be a series of case studies playing out increasingly familiar themes of perceived economic benefit being purchased at the cost of varying degrees of tourism intrusion caused by congestion and noise mitigated by spatial distance, zones and/or seasonality (for example see Lindberg, Dellaert and Rassing 1999; Williams and Lawson 2001). Such western studies have found that attitudes have been found to be determined by variables such as source of employment, years of residence in an area, degrees of contact with visitors and possibly income and age (for example see Long, Perdue and Allen 1990). In a few instances studies have sought a more significant theoretical underpinning by, for example, reference to theories of place attachment, shaping of the destination life-cycle with reference to community involvement and the process of glocalization as in the case of past work by the current authors, but in the final analysis the studies seem to conclude that any application of these variables may be time and place specific. It is suggested that similar themes would emerge in China, albeit with some differences relating to the relationship between local residents and local governments, as is suggested by Gu and Ryan (2008) and in other chapters of this volume, such as that of Zhang, An and Liu in Chapter 17.

With reference to cultural representation in tourism, Ryan and Aicken (2005) and Butler and Hinch (2007) provide comprehensive summaries of research relating to cultural representation of indigenous peoples within tourism. In these studies processes of commodification are discerned, the implications of this for minority peoples are indicated, for there are both advantages and disadvantages, and the social ramifications are illustrated. These include income creation, the creation of monetary values for culture, and related issues by increased participation in 'mainstream' economic, social and political life. Examples are provided of encroachment upon and utilization of the culture of minority peoples by non-members of those societies for monetary gain. Yamamura (2005) provides such evidence in the case of Dongba art in Lijang, where outsiders now sell local art-forms while Kolas and Thowsen (2005) explored links between the development of tourism in Shangri-La (Diqing Tibetan Autonomous Prefecture, Yunnan) and representations of place and notions of Tibetan identity. For her part, in this book Shi provides evidence of 'fake monks' selling incense in Wutaishan to unsuspecting visitors. In short, a range of negative and positive responses are raised when considering the relationship of culture, destinations, images and visitors.

In many cases this may be due to the patterns of variables that are being considered. For the most part they are based on empirical measures of intrusion (for example, perceived levels of congestion, changing land use patterns and crime statistics) or on psychological reactions to visitors and perceptions of social distance. A wider context of social change and political frameworks are rarely considered or incorporated into survey construction, and as pointed out by Pike (2002) with reference to studies of destination image, most have been quantitative in nature. The same might be true of the related studies of tourism impacts on local communities, although some qualitative work has been undertaken (for example, see the next section of this book).

In the period prior to the Beijing Olympic Games, tourism promotions undertaken by the China National Tourism Administration (CNTA), as exemplified by its video and DVD entitled *Welcome to China—2008 Beijing*, sought to establish a series of themes. These included China as a land of long-established civilization and images were created that referred to past architectural glories, and the past traditions of Buddhism and Taoism. Little reference was made to the period from 1919 to 1978. Truly it can be said that silences articulate a discourse as surely as anything that is stated. Second, China was to be seen as a land of diversity, and images were used that made reference to the different landscapes, from ice-capped mountains to deserts, from large lakes to green fields, and as part of this image creation, reference was made to the different sounds and sights of China through the music, dance performances and dress of China's minority peoples. Thirdly, China was a modern country, and thus the images of the past and its heritage, traditions and different cultures rooted in an agricultural and non-urban setting were counterpoised by images of modernity—of cities of gleaming sky-scrapers, modern hotels and shopping centres thronged with an affluent urban class purchasing internationally branded goods. Fourth, China was welcoming and friendly, and these images were reinforced by pictures of smiling people, children playing and groups all welcoming the overseas visitor to the Beijing Olympics. Any western stereotype born of past periods of the Cold War of a uniform society based on a militaristic readiness to counter western values was thus squarely banished, replaced by a portrayal of China at the commencement of the twenty-first century.

As in any good advertising, the images are based on truths, and again as in any advertising anywhere, they are selected truths because the advertiser wishes to present a positive image that is persuasive and, in this case, to encourage a potential tourist to visit China, or to address any perception that might perceive the remnants of a restrictive society.

SECTOR CONCEPTS, POLITICAL SCENARIOS AND CULTURAL REPRESENTATIONS

If there has been a *leitmotiv* in the book thus far it has been that tourism development in China is never far removed from considerations of national economic planning. China's policies increasingly lie at a nexus of being a command economy slowly being dismantled and seeking to utilize a private sector in the furtherance of goals that will economically benefit the majority of the Chinese population. It might be said that today the Chinese government increasingly rarely seeks to legitimize itself by reference to past ideologies of the nineteenth century but rather does so by reference to economic gain that enhances the standards of living for its people. In doing so it has created a series of transformative fields and thus the theories of Czada (1997) and Lemhmbruch (1998) possess interest. In this section of the book the work of Bähre (2007) is used to provide a framework of analysis.

Economic theory usually categorizes industries as falling into a three- or fourfold classification; that is: primary (the agricultural sector and extractive industries), secondary (the manufacturing industries), tertiary (the service sector—for example, retailing, insurance, banking and finance) and some writers argue for a separate quaternary sector comprising the public sector of governance and regulation given its emergence as a major employer in its own right even within 'capitalist' countries. Additional arguments are also advanced for a quinary sector based on knowledge creation industries and/or information technology—a contention with long antecedents as evidenced by the work of Porat (1977). Included in the knowledge creation industries might be the creative arts. Domberger and Jensen (1997) argue that under neo-liberalism, the quaternary sector has been subjected to such degrees of contracting out that it becomes a sector characterized by its paradigms. These might be said to be a drive for efficiency under conditions of price taking imposed by a process of tendering for contracts. However and whatever the classifications endorsed by different writers, what might be observed is that these sectors are first heterogeneous and, second, exist in states of flux and transformation. The very uncertainty about classifications is itself an indication of this state of flux. An additional observation made by Czada (1997) and Lehmbruch (1998) is that the sectors are not only distinguished by the products and services produced but are layered by the nature of ownership with varying degrees of state intervention. Accordingly a further threefold classification can be identified:

1. Sectors defined by the free market alone.
2. Sectors defined by competition but subject to state intervention.
3. Sectors dominated by policy domains with high levels of state intervention.

To further complicate matters Bähre (2007) identifies three classifications of high state intervention with reference to sector-specific property rights, these being:

1. Sectors with a proximity to the state where 'frameworks conditions or basic institutions, has long specified the concrete structure of sector-specific property rights in detail' (Bähre 2007: 36).
2. Sectors in high proximity to the market, yet still characterized by high degrees of state intervention.
3. Sectors comprised of state-owned enterprise alone.

To this Czada (1997) adds three more modes that revert back to the earlier discussion, these being the service sector, manufacturing and large-scale technical infrastructure sectors.

Much of this work arose from the German experience of the absorption of the former East Germany into the new German state after the fall of the Berlin Wall and the need to dismantle and reconstruct the economic system. In that sense the question arises as to whether the model can be applied to China and its withdrawal from statist intervention, and what implications it might have for the role of tourism.

Bähre (2007:37) poses a diagrammatic model which has here been adapted as shown in Figure 8.1.

Figure 8.1 can be applied to China. Compared to the German model the schema has been changed for some aspects of the Chinese state. First, to be more comprehensive the state has been divided into three to represent three powerful groups. There is first the Central State comprised of the State Council of which the Standing Committee members represent the core and where the Premier might be said to be the effective ruler of China. The People's Liberation Army (PLA) has been separately identified not as a military force but by reason of its ownership of many companies in China and in that sense it is an arm of economic policy albeit with, at times, differing objectives from normal economic concerns even though operating within the general economic sphere. For example, with reference to tourism, it owns one of Beijing's major hotels, The Palace, other properties, China United Airlines and has an interest in several airports. However, as noted by Cheung (2001) it divested itself of many of its businesses, although the process was hindered by a private sector refusal to purchase what were seen as poor performing concerns. One consequence has been that the PLA has become more entrepreneurial to turn these businesses around (Cheung 2001). A third important component is the level of provincial and municipal government that has significant powers. A continuous tension can be said to exist between the central government and the provinces where distance plays a role in providing degrees of local autonomy as described in different chapters, such as that of Arlt and Xu, in this book. It should be noted that just as the PLA may engage in commercial activities, so too may municipal and provincial governments (for example, ownership of hotels), or alternatively they may license commercial activities and derive revenue from the sale of permits. With the political and economic reform of China, market forces have certainly come to be stronger even while state-owned companies and other types of institutional tourism and hospitality organizations compete in the market. In short, privatization and internationalization have emerged as strong economic forces in China's tourism industry.

In addition to the pattern of potential tension and the implications these arrangements might have for businesses, another issue referred to by Ma, Si and Zhang in their contribution to this volume is that of differing ministries failing to co-ordinate when involved in the planning of heritage and tourism

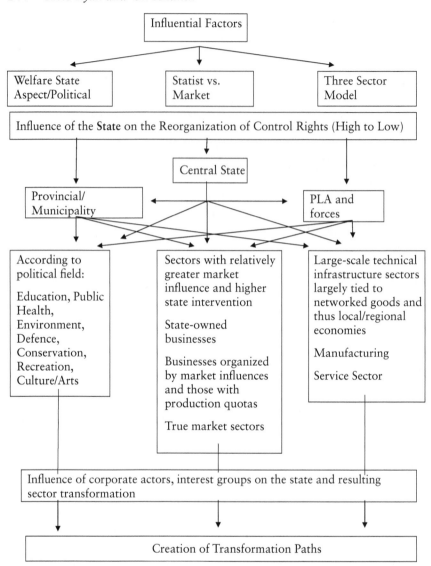

Figure 8.1 Schematic of transformative economic systems.

sites. Different governmental policies overseeing transportation, conservation, recreation and the arts, as well as tourism, require new forums for liaison and joint planning and given the local context this too begins to involve tourism in an empowerment of the local and regional as against central government. Indeed, in March 2008 these issues were recognized by the NPC and five new 'super ministries' were established to improve co-ordination, including one covering the environment. However, the empowerment of local people and institutions rests on several foundations, including freedom of movement of peoples. A system of internal permit requirements exists and there is a need to show personal identification documentation when, for example, booking a hotel. To some extent this constrains growth in tourism demand. However the *hukou* system of permits required of rural residents for movement into provinces other than the normal province of residence has been abolished in 11 provinces and is being dismantled in others while also being slowly eroded through the law not being enforced. Nonetheless Au, Nan and Zhang (2007) still maintain that its vestiges remain and fall primarily on female workers.

Another issue is the degree to which Chinese enterprise permits external organizations to establish a business in China. As in any country a series of stages related to tax registration, certification of business name, insurance coverage etc. are required and Table 8.1 provides a list of a number of required stages. In addition, specific regulations existed related to whether a business is a registered Wholly Foreign Owned Enterprise (WFOE), a Foreign-Invested Commercial Enterprise (FICE) or a Joint Venture (JV) company with a Chinese partner. However, in the JV classification, while this may be a preferred option for both Chinese and overseas parties, further constraints are imposed—for example, shares in a JV are non-negotiable and cannot be transferred without the permission of the Chinese government. Prior to December 2004 significant restrictions were imposed on overseas companies wishing to establish a business in China—for example there were minimum investment criteria such as the need to invest 50 million RMB for retail operations, and geographical restrictions also applied. With China's acceptance and endorsement of World Trade Organization membership the country moved to ease these restrictions, and from 1 June 2004 applications could be made for trading permission with, in the case of retailing, the level of required investment being reduced to 300,000 RMB (Dezan Shira and Associates 2005). At the same time FICE arrangements were now permitted to last for either 30 or 40 years dependent upon location. At the same time WFOEs were now permitted to fully operate in the Chinese domestic market, that is, they could sell to the Chinese population as distinct from only being able, under given conditions, to export from China, and it is this that accounts for the new interest in hotel development in China by large overseas groups, including the United Arab Emirates–based group, Jumeirah, as part of its ambitious global expansion.

For China's tourism industry these movements are very significant in terms of certain parts of its tourism portfolio, and for relaxing the degrees of statism that exist. For the hotel sector in particular, Gu and Ryan (2007) have argued that the next stage of hotel development in terms of movements away from what they termed a core product of comfortable beds and clean accommodations into a range of boutique accommodations, spa-based hotels, themed resort complexes etc., is unlikely to occur unless led by an overseas hotel sector catering to international demand, thereby providing

Table 8.1 Stages in Establishing a Business in China

No:	Procedure	Time to Complete	Cost to Complete
1	Obtain a notice of pre-approval of the company name	1 day	RMB 80
2	Open a preliminary bank account; deposit funds in the account and obtain the certificate of deposit	1 day	no charge
3	Obtain registration certification 'business licence of enterprise legal person' with SAIC or local equivalent	1 day (if done in person); 14 days if by mail	0.08% of registered capital (registration fee) + RMB 10 for copy of Business Licence + RMB 500 for public announcement (see comment)
4	Obtain the approval to make a company seal from the police department	1 day	no charge
5	Make a company seal	1 day	RMB 300
6	Obtain the organization code certificate issued by the Quality and Technology Supervision Bureau	5 days	RMB 148
7	Register with the local statistics bureau	1 day	RMB 50
8	Register for both state and local tax with the tax bureau	10 days	RMB 100
9	Open a formal bank account for the company and transfer the registered capital to the account	1 day	no charge
10	Apply for the authorization to print or purchase financial invoices/receipts	10 days	no charge
11	Purchase uniform invoices	1 day	RMB 1.05–1.67 per invoice book
12	File for recruitment registration with local career service centre	1 day	no charge
13	Register with Social Welfare Insurance Centre	1 day	no charge

models for what is currently perceived as an immature domestic market that is still, to a large degree, price rather than quality oriented. In terms of the schema identified in Figure 8.1, such policies move the Chinese tourism industry away from a state-dominated enterprise into a more market-oriented business organization. This is thought important, and it may mean that China might avoid issues associated with tourism monocultures arising from government fixation with tourism as a source of economic planning, as seemingly happened under the Communist Party–run local government in Italy in the 1970s, as described by Conti and Perelli (2007). Conti and Perelli argue that the Communist Party development of tourism in destinations like Rimini led to a model eventually characterized by 'stagnation and the need for a structural renewal based on significant economic investments' (Conti and Perelli 2007: 246), with all the associated dangers of an economy highly dependent on one single source of employment. They argue that a key factor in moving Rimini away from this over-dependence was the fading of support for the Communist Party in Italy after the fall of the Berlin Wall in 1989. Other variables have now also entered the discourse in Rimini, including the requirements to meet a Local Agenda 21 process that emphasizes longer term environmental sustainability.

China's whole tourism structure is thus in a state of transformation, but evidenced by following chapters, this state of flux eddies around seemingly fixed rocks—and often those rocks are to be found in areas where minority peoples are found. The reasons for this are not difficult to discern. There are, according to official policies, 56 different ethnic groups in China, of which the Han are the most numerous, accounting for about 92 per cent of the population. The remaining 55 groups total approximately 96.5 million people, and 15 groups each have more than 1 million people, these being the Zhuang, Hui, Uyghur, Yi, Miao, Manchu, Tibetan, Mongolian, Tujia, Bouyei, Korean, Dong, Yao, Bai and Hani. All have certain rights and privileges that include tax exemptions, government subsidies, exemption from the one-child policy and lower academic requirements for entry into universities and colleges. In part these concessions are a reflection of their often lower levels of income and geographical marginality as well as a respect for cultural difference. However, as already noted, cultural difference becomes a source of tourism product, a means of presenting difference and hence an attraction for those seeking, whether out of idle interest or deeper curiosity, something 'different' to gaze upon. While this is true for domestic tourism, as is evident at any site where people pay in order to dress in a different local costume and have their photographs taken as shown in Figure 8.2 (which illustrates a visitor to Stone Forest National Park dressed as local Sani person and taking her photograph using her cellphone), anecdotal evidence suggests that many overseas visitors have a deeper interest. For example, the authors found that at a Sani Art Gallery at the same park, the overwhelming majority of purchases (well over 90 per cent) were made by international visitors. Sharpley (2002: 312), drawing on a postmodern analysis, argues

Figure 8.2 Visitor to Stone Forest Park, Yunnan Province, dressed as local Sani person.

that within a western context, the relationship between tourism and culture has become a complex one and that when culture becomes a product 'commodities, whether goods or services, embrace a meaning beyond their economic exchange or use value'. The international visitor may therefore, like his or her Chinese counterpart, be a tourist and not a lay anthropologist, but the meaning of the postmodern gaze may be more complex than a gaze located within modernism, regardless of nationality of the gazer. On the other hand, as indicated by Li and Sofield in this book, the gazes are filtered through cultural meanings. Many implications arise from this, but possibly one of importance for future Chinese destination planning and research is a need to avoid a mistake arguably common in the West, and that is to decontextualize tourism planning from its wider social mores and political structures. Rather it must be recognized that the potential catalytic consequences of tourism represent challenges for the socio-economic and political status quo. To simply concentrate on economic and industry structures is itself an incomplete form of understanding the nature of tourism.

Yet, it has to be recognized that currently 'cultural difference sells', and thus is an important constituent of Chinese regional economic policy. As China increasingly becomes part of the wider world, it is bound by the conventions of that world, and one such convention is the United Nations Draft Declaration on the Rights of Indigenous Peoples ratified on 13 September 2007, albeit as a non-binding text. As noted by Ryan and Aicken (2005) and Higgins-Desbiolles (2007), this Declaration poses important issues for both the tourism industry and the political context within which it operates. Indigenous culture is for the enhancement of local indigenous peoples, and to be controlled by them and not exploited by outsiders. The Declaration supports claims for indigenous ownership of symbols, culture, land and environment, and it requires states to secure the assent of indigenous peoples to any development in their traditional lands and recognizes claims to self-determination. Set within the contexts of Figure 8.1 and China, it can be seen that cultural tourism based upon the minority peoples of China represents a further erosion of statism. Yet simultaneously it requires state support to achieve recognition of minority cultures and to offer access to the tourism markets and mainstream of economic life while offering diversity to the portfolio of China's tourism product. Adding to this is the point raised by Keitumetse (2007), namely, that different interpretations of indigenity exist, one of which is associated with anthropological primitivism. For the domestic Chinese market there is a danger that tourism product/representation is based upon themes of 'closeness to nature', 'a people free from worry', 'a people of song', 'a people of unchanging traditional ways'—all descriptions that cut minorities off from daily (urban) realities and present them as almost irrelevant subjects to be gazed upon. Representations of culture and heritage thus count, not only in terms of representing images thought attractive to tourists, but also as confirmations that fit within social and political policies. If people are irrelevant, then claims to ownership are frivolous! These issues of competing needs, of people to possess their culture, or tourism product that improves regional economies, of sensitivities as to how best incorporate people into social and political processes, all framed within needs to enhance the living standards of both Han and minority peoples is, it is suggested, only now beginning to be played out within the Chinese tourism industry.

The same issues can be noted with reference to other cultural and heritage aspects of China's tourism. The long classical traditions of Chinese culture from a past imperial age are obvious, frequent and common in terms of an architectural heritage of palaces and the houses of local aristocracies, and Confucian traditions of respect for the achievement of past masters have to a large extent permitted the safeguarding of that heritage with only a few exceptions such as in the worst excesses of the Cultural Revolution. Yet the presentation and interpretation of that heritage places them in a past that has seemingly little relevance to the current modern state that is emerging. Yang and Chen argue in this book that guides offer interpretations based on stories of a past land and time that constructively fantasize the past and thus form entertainments. They argue for a changing emphasis in interpretation and suggest increased content with reference to the science of the landscape. On the other hand there is a state-sanctioned tourism of the recent past, namely 'red tourism'. Red tourism is based on assets and resources that are the commemorative places and memorials that refer to the revolutionary period of China's history and the emergence and eventual success of the Chinese Communist Party. Gu, Ryan and Zhang (2007) describe such tourism with reference to Jinggangshan, noting also its role in reinforcing patriotic feelings along with infrastructure improvements and economic regeneration of rural areas.

Nonetheless, in spite of such examples as 'red tourism', it might be said that as tourism policies begin to evolve away from the past 'statism' of a command economy into one of directorship in partnership, evolving models of tourism destination management will occur. It might also be noted that these will increasingly incorporate the local at varying different levels, and thus there may be value in looking at models of tourism development.

DESTINATIONS AND TOURISM DEVELOPMENT MODELS

While, as noted in the opening chapters of this book, Chinese tourism policy is an extension of economic policy; what has not yet been fully discussed is a growing articulation of sustainability. In both the 10th (2000) and 11th (2005) Five-Year Plans announced by the CNTA, commitments are made to 'sustainable development strategy by carrying out green development, producing green products, popularizing green management and building green system' (CNTA 2000). Such pronouncements are not restricted to tourism alone, and a growing awareness of environment-led policies are beginning to emerge even while China still sees as its foremost priority the reduction of poverty for many of its people. In this respect it is of interest to note that Hunter (1997) and Bramwell and Sharman (2001) proposed a four-stage model of tourism sustainable development that can be applied to community- and heritage-based tourism. The stages are as follows and are commented upon by references to the earlier discussion.

Sustainable Tourism Through a Tourism Imperative

At this stage tourism is in an early stage of development and is embraced simply because it is more sustainable than other economic activities. Resource loss is thought acceptable so long as it is less than that which would otherwise occur and is not at such a level as to threaten the attraction of the site as a tourist destination.

At first sight this may appear wholly applicable to the Chinese situation, with the imperative being determined by socio-economic policies of rural poverty reduction. However, policies of economic growth in other arenas of economic life have created well-known problems of environmental degradation and, as evidenced in other chapters of this book, it is recognized that environmental degradation impedes long-term success in tourism, particularly when much of that tourism depends upon environments, heritage and culture unique to China. Tourism sustainability is thus dependent upon protection of those assets from wider industrial pollution, which itself is increasingly perceived as much a long-term threat to the health of the Chinese population as surely as is poverty itself.

Sustainable Development Through Product-Led Tourism

At this stage, tourism is more developed and the primary concern is to maintain existing products and develop new products for the well-being of the local community. Bramwell and Sharman (2001) note that there may be a wide range of concerns, but these are primarily reactive—that is, they emerge only when threats to tourism products arise.

Consequently Chinese tourism development, for the reasons given in the previous section, can be discerned as having moved into this category of development, and the following chapters provide many such examples, as at Baiyang Lake. The socio-economic imperatives remain and tourism is still often perceived as the best means of enhancing local community lives, but more attention is being paid to the development of specific tourism product.

Sustainable Development Through Environment-Led Tourism

Bramwell and Sharman (2001) note that while this form of development may be associated with tourism from the outset, it is more likely to be found in already developed tourist destinations seeking to reorientate their tourism along more sustainable directions. They argue that at this stage priority will be given to those forms of tourism dependent upon maintaining high-quality environments and establishing a linkage between tourism success and environmental quality.

It can be observed that this differs from the normal constructions of eco-tourism and its cultural/heritage connotations because this categorization permits the existence of well-established and possibly high-flow tourism. In some instances this form of tourism may be a supplementary form based on a wish to sustain zones of cultural or environmental heritage from the encroachment of yet more mass tourism developments that are perceived as having high carbon footprints. For example, in Dubai, developments such as the Dubai Desert Conservation Zone and the luxury resort of Al Maha combine luxury tourism with a desert reclamation and conservation scheme that covers 5 per cent of Dubai's land area while also being a counterpoise to the high-rise developments normally associated with Dubai (Ryan and Stewart 2008). In China, the movement towards the categorization of an increasing number of sites as UNESCO World Heritage Sites like that of Stone Forest National Park in Yunnan, the classification of many areas as natural reserves and the establishment of criteria for landscape classifications indicate a taxonomic approach to site designation based on cultural, environmental, heritage and aesthetic grounds (as separately discussed by Li and Sofield; Ma, Ryan and Bao; and Ryan, Gu and Meng in this book). However, a feeling remains (as indicated by Ma, Ryan and Bao; Zhong, Deng and Xiang and other contributors), that many of these categorizations of place still primarily serve a function as being markers of place to see, rather than specific actions of place protection. However, as always, exceptions and countervailing currents might be found, as in Ghirmire's (1997) analysis of Panda Reserves in China, where he argues that the focus of effort has been such that while environmental goals are being achieved, in the longer term the failure to obtain equal degrees of local social development may well undermine the longer term effort to sustain the panda's terrain. That this is part of a global and not solely Chinese issue is exemplified by Ghirmire and Pimbert's (1997: 17) observation that 'Declaring bio-diversity-rich areas as "internationally important" conservation sites is meaningless for local resources users as long as the issues that emerge out of such declarations have not been discussed and resolved to the satisfaction of local communities ... conservation benefits should be quantifiable—if possible immediate—with local people getting a fair share of the benefits accruing from the protected area'.

Sustainable Development Through Neotenous Tourism

This policy adopts active discouragement of tourism in order to protect resources. Within the context of developing countries this approach is problematical, and indeed, it is also problem-ridden in developed countries. Essentially it is based upon a number of assumptions that include:

1. The absence of any population that has been economically, socially and politically marginalized in the past, and which requires income generation to improve future life opportunities.
2. If such a population exists there are alternative means of offering people life-enhancing opportunities in ways less damaging to the environment.

As noted in Chapter 1, in the Chinese context where tourism is specifically used as a social and economic generator in a rural population still characterized by poverty, the presence of possible tourism resources near such populations represents significant assets. These are likely to be utilized, especially when such utilization is consistent with wider economic, political and social goals. This is discussed in different ways by contributors subsequent to this chapter, but one example that can be cited here is the increasing profile being attributed to China's minority ethnic groups. As already noted, the promotional video developed by CNTA for the Beijing Olympic Games provides many clips of China's minority groups to help represent the diversity of China's population and dispel any perception that others might have of China as some cultural or ethnic monolith. Visitors to those provinces of China where ethnic groups exist will find many examples where local communities are encouraged to provide representations of their culture through dance performances (for example, at the Stone National Forest in Yunnan where Sani people act as guides and also entertain guests as shown in Figure 8.3) or where religious beliefs are used as part of a government-supported programme of special events (as in the Wutaishan Buddhist Festival in Taihua Town, Shanxi Province; see Ryan and Gu forthcoming). In the

Figure 8.3 Sani guides in Yunnan Province, China.

following chapters Arlt and Xu and Ma, Si and Zhang also provide similar examples—and the latter, when discussing Qufu, suggest that promotion of the birthplace of Confucius has reached a point where one might even propose that, in marketing terms, a 'brand Confucius' might be said to exist.

Telfer (1996, 2002) proposes a detailed matrix of the relationship of tourism as an agent of development under four scenarios of modernization, dependency structuralism, economic neo-liberalism and alternative development. The first of these scenarios, modernization, relates to an emphasis upon the economic where tourism is perceived as a means of technology transfer, infrastructure improvement and a force for generating economic growth as discussed in Chapter 1. Dependency structuralism is closely related to this because, again, tourism is perceived as a means of tackling issues of poverty and poor economic prospects, but Telfer tends to locate this within newly independent countries only recently escaping from a colonial past. For example, he notes, 'The predominance of foreign ownership in the industry imposes structural dependency on developing countries . . . in a core-periphery relationship which prevents destinations from fully benefiting from tourism' (Telfer 2002: 54). Dependency can take several forms, including cultural dependency and a borrowing of models seen pertinent to international markets. China represents an interesting case from this specific perspective. China has its own successful chains such as Jinjiang Jianguo, but as Zhang, Pine and Lam (2005) point out several times in their book on the Chinese hotel industry, in 1978 China had only 137 star-graded hotels. By 2005, there were 11,828 hotels with 1,332,110 bedspaces (CNTA 2005). While the majority are under Chinese ownership, many remain in the hands of different local municipalities and not all are under private ownership. Indeed many of the larger international hotel companies operate in China with more luxury properties than many Chinese counterparts. These tend to operate in the more international-oriented destinations of the major cities, and are more profitable than the Chinese-owned and managed hotels (Zhang, Pine and Lam 2005). Hence, in this instance at least, while by no means a newly independent country, China displays some characteristics of structural dependency in this regard. Further examples of dependency also reside in the rural–urban divide mentioned in Chapter 1, where rural areas may be dependent upon urban centres for investment capital and markets and additionally lose control of their marketing to urban-based travel services (Ministry of Civil Administration, 1992).

The third of Telfer's scenarios is labelled neo-liberalism—a term relating to an economic philosophy regarding the free working of the market as the best means of establishing growth, and which led to the dismantling of government controls in order to 'free' markets. Internationally the policy was associated with IMF and World Bank policies that sought to create free trade and the removal of trade restrictions, thereby permitting inbound investment by major companies—as just noted in the case of the Chinese hotel industry. Again China represents an interesting case in its adoption of competitive entrepreneurship within the bounds of a command economy that still identifies key economic sectors, such as tourism, as needing government support. The final scenario is that of alternative development that includes community involvement and command of tourism entrepreneurship at the local level—a theme to which much of the remainder of this book is devoted.

Within these four scenarios Telfer (2002) lists 25 components divided into two broad categorizations, namely, 'scale and control of development' and 'local community and environmental linkages'. For each of these he lists possible positive and negative attributes. For example, he notes 'alternative tourism' is bottom-up in its planning, involves local ownership and promotes sustainability and self reliance. The potential disadvantages include tourism inexperience, loss of profit, vulnerability to market change and with reference to guidelines for sustainability, excessive regulation.

While all of these factors would not normally be considered as being immediately pertinent to issues of cultural representation and destinations, because of the economic imperatives listed in Chapter 1 and the planning paradigms discussed in Chapters 6 and 17, the parameters implied within the frameworks suggested by Hunter (1997), Telfer (1996, 2002) and Bramwell and Sharman (2001) apply to the Chinese context. These are evidenced by the following chapters. In Chapter 10 Arlt and Xu discuss the impacts of tourism in Gansu Province and its ready adoption of the legends of Shangri-la. They indicate that the tourism is resource-led. Yet they also make the telling point, citing Nyiri (2006), that commercialism is rampant in China, but it is an open, undisguised modernism with none or very little of the playful disguises of tradition that can be found in western countries. They argue that the tourist searching for an imagined 'authenticity' will find an evident packaging, and from this Arlt and Xu propose that the product should be based on the premodern if it is to succeed as a tourism product. Visitors to the cultural shows that abound in China such as at Shenzhen, Kunming and elsewhere will find examples of extravaganzas of large casts, laser lights and modern media. Yet, examples of traditional dances can be found, but these are generally located in staged performances working to the schedules of visitors, and often geared to the domestic tourism market.

Aiping, Si and Zhang also provide an example of resource-led and resource-manufactured tourism built upon a tradition—this being the birthplace of Confucius at Qufu. In many ways the representations and the use of staged festivals might also be interpreted in a wider social context, as indicated by Ryan and Gu (2008) in their paper on the Wutaishan Buddhist Festival, in that the Chinese Communist Party uses the culture and heritage of a distant past, but by creating a show located in the modern effectively, at least to western eyes, decontextualizes the event from the past. However, that is not necessarily the view of Chinese participants in these events, and a tension exists between respect for the heritage of the past, and a harmonious present where the past is made real for the present, and in doing so secures the goals of achieving economic growth in more rural or poorer provinces. Yang and Chen are also concerned with the representation of culture, but as noted earlier, this time in the way sites are interpreted by guides. Shi presents a slight change in emphasis, turning from the given interpretation of sites to the perceived and received experience of the site, in this case the Buddhist area of northern Shanxi Province. She reports in this case confirmations and disconfirmations of expectations with reference to visitors' experiences and interactions with those who actually represent, and those who purport to represent, the Buddhist faith. Again commercialism is present at the local level as people seek to make a living from the tourists. Tensions are noted with reference to tourist experiences as they have to cope with mixed degrees of a desired authenticity, an experience that is rooted in the Buddhist faith, and the commercialization of that faith including 'false' monks.

The chapter by Li and Sofield is important in that it directly addresses an issue mentioned by Ma, Ryan and Bao, and by Ryan, Gu and Meng, namely, the issues of cultural representation within the Chinese context and the designation of World Heritage Sites with reference to the norms of a positive rather than normative-based system of classification. They refer again to the role of culture in National Park planning and the concept that harmony between nature and man is one whereby the human exists and interprets the natural within a culture of story-telling, and not by reference to the scientific alone. Li and Sofield reverse the argument proposed by Yang and Chen, who argue that interpretation is too biased towards the fantasies of legends, implying that a balance has yet to be found between the filters of popular classical culture and the wish to preserve a natural setting unsullied by human presence.

In total, the chapters identify many themes. There is the mix between local and centrally directed tourism, and evidence that the statism that once characterized China's policies is being eroded even while, in conjunction with regional and municipal governments, central policy lies in the creation of destination plans. For the most part though, it would seem that these plans are resource led, and in that sense Chinese planning is at the first two stages of the model suggested by Telfer. However, as indicated in Chapter 15 and beyond, the local does possess a voice, and it is increasingly being heard. Issues of a mix between the cultural, the natural, and the commodification of cultures, reinstatement of local identities, economic regeneration and legitimization of difference can be found in these chapters. Together they provide a picture far removed from the homogeneity of a Maoist state that once dominated images of China.

9 Huangshan (Yellow Mountain), China

The Meaning of Harmonious Relationships

Fung Mei Sarah Li and Trevor H.B. Sofield

There is no mountain as beautiful as Mt. Huangshan,
No other mountain under heaven do I want to see after visiting Mt. Huangshan

(Xiu Xiake, Ming dynasty geographer, 1586–1641)

INTRODUCTION

In their chapter Ma, Ryan and Bao argue that transferring western, and especially North American values of 'wilderness' as the core function of a National Park is to undervalue the role of cultural understanding of the nature of human–nature relationships as practiced in China. Nowhere is this perhaps more evident than in the applications made by the Chinese government for World Heritage Status (WHS) with reference to a number of China's places of natural and cultural heritage. This chapter uses the example of Huangshan to scrutinize the (western) paradigms that govern WHS assessments and to counterpoise that with a Chinese world-view that sees cultural and natural heritage as a single unitary construct. Such differences inevitably cause tensions, and the practical outcomes are that heritage site management gets caught in a conundrum of applying western models in a Chinese context and value system. Problems thus arise due in part to the prescriptive authority of western paradigms that perhaps pay insufficient attention to culturally determined values of 'Others', a discrepancy that might be interpreted from the framework of Said's (1978) Orientalism versus Westernism. This chapter primarily concerns itself with this difference in world-views, and hence represents a divergence from some of the other chapters that have tended to emphasize the economic role involved in awarding WHS. This is not to deny the importance of the economic, but even the economic roles ascribed to tourism reflect a western empiricism passed down and then filtered through Maoism initiated by a Communist Party. Chinese value systems in assessing natural heritages are thus influenced by the remnants of such thinking and a renewed interest in more classical Chinese perspectives. It can be argued that within the professional cadre of the Chinese park management systems, these perspectives are of importance, because it is through these appeals to a domestic tourism market that the economic returns result.

HUANGSHAN

In the south of Anhui Province lies a small mountain range that has been extolled by Chinese poets, essayists and artists for more than 2,000 years. For Chinese, whether they reside in China itself or are part of the Chinese global diaspora, its fame is perhaps greater than that of any other range. This is Huangshan or Yellow Mountain, named after the mythical/historical Huang-di, Yellow Emperor, who ruled a small kingdom about 5,000 years ago. He is credited with being extremely wise, and giving the Chinese their distinctive title, the Yellow Race. According to legend he practiced alchemy, and produced an elixir for immortality from a purple fungus growing in Huangshan which he imbibed before ascending into heaven on a dragon.

More than 20,000 poems and 200 essays have been written about Huangshan over the centuries, many of them incorporated into Chinese 'common knowledge' and known to Chinese all over the world. Ancient—and new—calligraphy adorns many rocks and cliff faces (calligraphy is more than 'writing': it is a high art-form and it is believed that 'the inner man' is revealed in the way one composes calligraphy). Hundreds of paintings have also been created, and they capture for Chinese the quintessential mountain landscape—sharp peaks, deep gorges, swiftly flowing streams and waterfalls and contorted pines clinging to precipitous cliffs, with a pagoda or temple sited in the landscape according to *feng shui* principles. By the Yuan Dynasty (AD 1271–1368), 64 temples and many pavilions had been constructed around the 72 peaks of this 154 square kilometre area. More were subsequently built, e.g. Fahai Meditation Temple and Wonshu Temple in the seventeenth century, connected by steps cut into the mountain (Peoples Republic of China [PRC] 1989). Although only 20 temples now survive they form a focal point for many Chinese visitors to Huangshan.

The mountain range, isolated from other ranges by surrounding plains, encompasses several distinctive ecological niches and is rich in biodiversity with 1,450 native plant species, 28 significant endemic plants and 300 vertebrates (PRC 1989). It is not surprising then, to find that Huangshan was accorded WHS listing based on its outstanding natural and scenic qualities as well as its cultural features. What is perhaps surprising is that in its assessment of the Chinese government's nomination, the International Union for the Conservation of Nature (IUCN) (IUCN 1990: 11) declared that Huangshan's natural values 'are predominant over its cultural heritage'; and the International Council on Monuments and Sites (ICOMOS) (ICOMOS 1990) originally deferred its recommendation for cultural heritage listing, citing a lack of supporting evidence. The IUCN (1990: 12) also recommended that the Chinese authorities 'should be encouraged to reduce the human influence on the mountain', a startling comment to these two authors given that the mountain has for more than 2,000 years been a very rich cultural site and exemplifies the Chinese world-view of 'man and nature' as a single unitary construct (this phrase accurately reflects Mandarin usage and should not be interpreted as unthinking sexist language on the part of the authors). The division between cultural and natural heritage is a characteristic of a western, positivist, scientific approach—reinforced by UNESCO's classification system for WHS listings—in contrast to a more holistic Chinese world-view, and an examination of this binary classification system constitutes the focus of this chapter.

THE CULTURAL HERITAGE/NATURAL HERITAGE DIVIDE

The Chinese world-view is both anthropocentric (humans first) and anthropomorphic (attributing human characteristics to non-human features, animals, plants etc.). The Chinese word for 'nature'—*da-jiran*—may be translated literally as 'everything coming into being' and expresses the entirety of mountains, rivers, plants, animals, humans, all bound up in their five elements—metals, wood, water, fire and earth (Tellenbach and Kimura 1989). 'Man is based on earth, earth is based on heaven, heaven is based on the Way (*Tao*) and the Way is based on *da-jiran* (nature): all modalities of being are organically connected' (Tu 1989: 67). Under Confucian values scholars and mandarins were exhorted 'to seek ultimate wisdom in Nature' (Overmyer 1986). Confucian thought and Daoist philosophy encompassed the need for man and nature to bring opposing forces into a symbiotic relationship where 'harmony' rather than 'difference' or 'opposites' was dominant (Rawson and Legeza 1973). This is an anthropocentric perspective with a sociological definition in which, because nature is imperfect, man has a responsibility to improve on nature (Chan 1969; Elvin 1973). It is thus distinct from a western perspective that separates nature and civilization (humans), which views nature ('wilderness') ideally as free from artificiality and human intervention.

Mountains were particularly venerated and the complementary force-fields of man and nature came together most powerfully in the Daoist concept of *yin-yang*. Like a magnet with its different force-fields, both are needed for the magnet to function, and man is seen as indivisible from nature (Ropp 1992; Spence 1992). Under the religious belief system that evolved over centuries, there were nine revered sites of particular significance—five sacred *shan* or mountains and four rivers (Chinese History Museum 1994). It was a fundamental responsibility of Chinese emperors to visit these sacred mountains on a regular basis to propitiate the spirits, gods and ancestors. Failure to do so could place the entire prosperity and well-being of the empire at risk. Grand roads were constructed for the emperor to approach the sacred mountains (the imperial way). Steps, termed 'staircases to heaven' since the emperor was revered as the son of heaven, were carved into their slopes for his ascent to the summit. Pavilions, tea-houses and inns were erected at regular intervals (Sofield and Li 2003). While Huangshan is not one of the sacred mountains it is revered and manifests much of the pilgrimage development associated with China's five sacred mountains.

In examining how Chinese values about landscape and wilderness are translated into tourism attractions, the anthropocentric position encourages and facilitates programmes to alter the physical and biological environment in order to produce desired 'improvements' (Sofield and Li 1998). These may include landscaped gardens, artificial lakes and waterfalls, facilities for recreation and tourism, roads for ease of access, observation towers and so on. Increasing direct human use is the objective of management and the character of the 'wilderness' will be changed to reflect the desires of humans and contemporary standards of 'comfort in nature'. Styles of recreation and tourism will be tuned to the convenience of humans, so trails will be concreted, resorts and restaurants permitted inside reserves, cable cars approved and so forth. Huangshan exhibits all of these examples of 'man improving on nature'.

The biocentric approach that underlies the IUCN's approach to WHS assessment by contrast emphasizes the maintenance or enhancement of natural systems, if necessary at the expense of recreational and other human uses (Hendee and Stankey 1973, cited in Hendee *et al.* 1990). 'The goal of the biocentric philosophy is to permit natural ecological processes to operate as freely as possible, because *in the western system of values* wilderness for society ultimately depends on the retention of naturalness' (Hendee *et al.* 1990: 18; our qualification in italics). It requires controlling the flow of external, especially human-made, pressures on ecosystems by restricting excessive recreational or touristic use of the bio-geophysical resources. The recreational use of wilderness is tolerated with this position only to the degree that it does not change the energy balance inordinately. A biocentric philosophy requires recreational users to take wilderness on its own terms rather than manipulate it to serve human needs. Like the anthropocentric approach the biocentric approach also focuses on human benefits, but the important distinction between them is that biocentrically the benefits are viewed over a longer term and as being dependent upon retaining the naturalness of the wilderness ecosystems (Hendee *et al.* 1990).

HUANGSHAN AS AN EXEMPLAR OF THE INDIVISIBILITY OF 'MAN AND NATURE'

Given the binary natural/cultural heritage approach that is integral to WHS listing, the Chinese authorities had no alternative but to seek inscription for Huangshan based on separate assessments of its qualities even if their own world-view does not draw the same clear-cut distinctions between the two. Thus, a major component of the Chinese submission for WHS (PRC 1989) described its geological attributes and enumerated its flora and fauna, emphasizing the diversity and endemism of the site and identifying rare or endangered species in conformity with the western scientific paradigm in order for Huangshan to qualify for inscription under Natural Criteria (iii) and (iv). Its intrinsic cultural values were detailed separately under Cultural Criterion (ii) (World Heritage Centre 1990; see Appendix 9.1).

Despite the forced separation of geological and biological features from cultural features, aspects of cultural heritage and anthropomorphic descriptors were interspersed with the scientific terminology necessary to make the case for the former. For example, in describing the vegetation of Huangshan and its endemic pine species, *Pinus huangshanensis*, the submission by the Chinese authorities to IUCN noted, 'A number of legendary trees are celebrated on account of their age, grotesque shape, or precipitously perched position, and more than 100 bear special names' (PRC 1989), such as 'Two Lovers Embracing' (two pines with intertwined trunks) and 'Welcoming Guests Pine' (so named because its branches open out like the arms of a host gesturing to visitors to enter his/her house). Such anthropomorphizing is regarded as inappropriate in western scientific texts but acceptable in a Chinese context. While the Huangshan pine is of intrinsic botanical interest to western science, its significance for Chinese visitors to Huangshan lies in cultural attributes, many of the trees are so well known from literary references over the centuries that they form 'must-see' sights. The pine survives on precipitous cliff faces at high altitudes, its roots often twice the length of its trunk, buffeted by strong winds and heavy mantles of snow and ice in winter. It has thus been anthropomorphized as tenacious, strong, steadfast, determined, iron-willed and brave—all human attributes. In this context, one of China's most famous calligraphers, Ouyang Xun (AD 557–641), who lived in the Tang Dynasty period, has been categorized in a similar way. His writings and his character are both infused with sentiments such as strong, tightly structured and as steadfast as the lone pine tree clinging to a high mountain cliff buffeted by winds and storms (Fan Yunkuan 1996). Chinese visitors will immediately recognize the 'strong, steadfast' Ouyang Xun style of many of the engravings that may be found all over Huangshan. They will nod approval because of its appropriateness in such a setting, making the association between the calligrapher and the pine tree in a fusion of history, psychology, botany and literary art. They 'see' a profound cultural element of Huangshan in a botanical species that is invisible to non-Chinese (and was obviously invisible to the IUCN 1990 assessment panel). There is no interpretation of this phenomenon provided for visitors to Huangshan because the Ouyang Xun style is one of the three main Tang dynasty styles on which the art of Chinese calligraphy is based, and in learning how to write the characters of the Chinese language all literate Chinese are familiar with it, so it is part of Chinese 'common knowledge'. Yet for many western visitors to Huangshan such detail, full of meaning in a cultural context, would prove of real interest.

As with the section of the WHS submission on flora, so the section on geology was also moved by its Chinese authors beyond 'western' technicalities to incorporate cultural elements that might seem inappropriate in a purely scientific document. Thus in outlining the geology of Huangshan and noting the presence of numerous 'erratics' (a technical term for boulders which have weathered differentially and perch atop mother lodes), the Chinese submission to UNESCO added that: 'Many of these grotesquely shaped rocks are individually named, such as "Pig-headed monk eating water melon", such names having ancient literary, historical or mythical significance' (PRC 1989). Another isolated porphyry column with a pine tree emerging from its tip—one of the famous geological features of Huangshan—is named 'A Flower Growing from a Calligraphy Brush Tip as if Seen in a Dream'. For Chinese visitors it is a compulsory sight, a pilgrimage to a site associated with one of China's most famous poets, Li Bai (AD 701–762), whose personal calligraphy also adorns other sites around Huangshan. Again, in learning how to read and write Mandarin, all literate Chinese will be familiar with several poets' eulogies to Huangshan. Their poems, combining descriptions of the natural with social and human emotional responses, are part of a major Chinese literary and artistic movement called *shan shui* culture (literally, *mountain water*) that was firmly established between the eighth and eleventh centuries and continues to the present day. By incorporating values that imbue nature/natural scenery with a range of human social and cultural values, such as likening mountains to benevolence and waters to wisdom (*The Analects of Confucius*, fifth century BC) *shan shui* philosophy has had a profound influence on the aesthetics of natural landscapes.

Because the Chinese world-view privileges literary and cultural heritage before the sciences, Chinese tourists to National Parks like Huangshan will interpret their experience through the culture of *shan shui* rather than through western paradigms of biological and geological sciences, or 'wilderness' which in the ideal western construct has no visible presence of humans. Their appreciation of the landscape and their motivation for visiting has a somewhat distant relationship to IUCN precepts. Many Chinese visitors follow routes established by authoritative figures over the centuries, following a preordained sequence to certain peaks, temples, pavilions and scenic sights, counting them off in much the same fashion as dedicated birdwatchers ticking off sightings from their list of rare and endangered species. Each and every step along the way will have been eulogized by poems and essays that will be familiar to most Chinese visitors and often they will recite such texts as they view the different sites. They will walk along paved trails, steps carved into the rock, bridges erected across chasms, with pavilions and kiosks conveniently located every half kilometre or so. A common reaction to Huangshan's miles of paths and stairs by western tourists is a perception of visual pollution, of a geometric, reinforced-concrete invasion of nature. But semiotically the Chinese will understand that such paths and flights of stairs symbolize hundreds of years of visitation by their emperors to sacred mountains to worship their ancestors and gods, and that vertical flights of steps ascending to the summit signify 'stairways to heaven' that once only their emperors would have been privileged to climb.

MANAGEMENT

Management of Huangshan as a WHS inevitably involves tension between the conservation ethic espoused by the western scientific paradigm for protecting natural and cultural heritage which is embedded in WHS listing, and Chinese anthropocentrically oriented values which see no contradiction between major constructions for the comfort and convenience of visitors being located inside the boundaries of a designated WHS. Under the Anhui Province *Plan for the Places of Scenic and Historic Interest in Huangshan* (1987), the reserve is divided into six tourist zones and five protection zones, and while it proclaims that 'no construction will be permitted if it will impinge on the quality of the landscape', a western perspective and the Chinese understanding of 'quality landscape' differ significantly. For example, if the ideal shape is a spire, then in terms of the Daoist need for humans to improve on nature, constructing a pagoda on the summit of a rounded hill will enhance its beauty; if the pagoda functions to provide shelter or refreshments for travellers then it will bring humans and nature into harmony; but a western perspective might see the intrusion of a human-made construction dominating the landscape, creating visual pollution and perhaps destroying its intrinsic naturalness.

The anthropocentric Chinese approach to nature is evident in an alpine valley adjacent to the North Sea of Clouds which now has seven hotels (three-, four- and five-star ratings) and two hostels with a total of more than 3,000 beds to facilitate travellers being able to experience the dawn rising over the peaks. There are about 70 km of concrete paths and stairs to facilitate the flow of huge crowds around Huangshan. There are three cable cars, one of them at 7 km the longest in China. There are two sites with large restaurant complexes and about 30 refreshment kiosks and souvenir stalls dotted around the peaks. Two alpine streams have been dammed to provide water for these facilities. They are all functional elements that enhance the experience of a visit to Huangshan for Chinese travellers. The Bureau of the World Heritage Committee (WHC) might recommend that management 'not permit the development of new hotels in the vicinity of popular scenic spots' (Bureau of the World Heritage Committee 1998); but that recommendation is based on western perceptions of conservation and visual and environmental degradation and fails to take account of Chinese cultural values concerning the interaction of humans and nature. Where tourist facilities create environmental damage and water pollution through, for example, inadequate sewage treatment and waste disposal, then concerns are valid. But when assessments of what constitutes visual pollution or the quality of a landscape are based on subjective criteria—which are in fact culturally determined from both sides, whether western or Chinese—then it is difficult to argue that one approach is right and the other is wrong. Rather they are different. And in the context of a site like Huangshan insistence on a western-oriented definition of what is appropriate may in fact be a denial of the very essence—the 'Chineseness'—of the site.

Until recently there was little signage around the mountain other than directional signs devoid of interpretation. But in response to urgings from the Bureau of the WHC that the significance of Huangshan's natural heritage be communicated to visitors (e.g. Bureau of the World Heritage Committee 1998) there are now about 40 metal plaques located at key points around the reserve. However, as with the original submission seeking WHS listing, much of this signage amalgamates both western scientific technicalities and Chinese cultural values. As just one example, the interpretation plaque at Xihaimen (West Sea Gate) lookout has its first two sentences utilizing technical terms to describe the geological formations and processes, with the remainder of the information referencing a range of Chinese-specific myths, legends and classical tales. The Chinese characters are followed by this English language translation:

> Xihaimen is the most profound and beautiful part of the Huangshan Scenic Spot [Reserve]. A blaze of multifarious colours of medium-to-fine-grained porphyritic granite bodies, and densely distributed vertical and horizontal joints add much to the high and steep granite peaks as well as interesting and odd stones, from which countless fairy stories and sayings are handed down. The NW-SE-trending fault zone cuts into granite bodies, thus forming a quiet, deep and precipitous dreamland in the Xihai (West Sea) Canyon. Stone scenes gather together before Paiyunting (Clouds Overwhelming Temple), such as An Immortal Airing His Boots, Wu Song's Fighting Tiger, Memorial Archway Stone in the near [foreground], Immortals Walking on Stilts [a line of ascending pinnacles], Female Immortals Embroidering, Heaven Dog's Watching Moon [Rock], King Wen Pulling a Wagon, etc. in the far [distance].

As with most other signage in Huangshan, the information relies upon Chinese common knowledge to deliver understanding and Chinese visitors will automatically draw upon their knowledge of ancient poets, Confucian and Daoist philosophy and religion, imperial history and Chinese classical literature to recognize the significance and symbolism that is captured in the scenery before them. But this information requires very lengthy interpretation if it is to be comprehensible to non-Chinese visitors. For example, the very name of the lookout is puzzling: What is this West Sea several hundreds of kilometres inland, high up in a mountain range with no lake in sight? The answer lies in classical literature when an early visitor to Huangshan climbed to its highest peaks and looked down on alpine valleys filled with surging clouds. This scene has been immortalized in numerous poems about Huangshan's West (and North) Sea of Clouds.

Other references similarly rely upon Chinese common knowledge to make sense. 'Immortals' in Chinese culture are integral to Daoist belief. They are not gods in the western sense of that word although they may be worshipped and shrines built for them. Nor are they angels although they are celestial beings; they are mischievous, fun-loving, carousing creatures with superhuman strengths and skills, constantly playing tricks on each other. They dwell in mountains, and caves (the Daoist 'passage-way to Heaven') are often their abode. Evidence of the presence of immortals in mountains thus abounds and Huangshan is no exception as this example demonstrates.

The reference to Emperor Wen also links Immortals to Confucian philosophy. In the famous *Analects* Confucius discussed the meaning of an ancient story in which Wen pulled the cart of an Immortal for 800 steps before stopping, exhausted. As a result the Immortal blessed his descendents with 800 years of unbroken rule. The reference to Wu Song fighting a tiger is taken from two of China's most famous classics, *Outlaws of the Marsh* (Shi Nai'an and Luo Guanzhong, circa AD 1350) and *The Plum in the Golden Vase* (anonymous, circa AD 1618). Every educated 10-year-old Chinese child knows that the character Wu Song personifies manly strength because he killed an attacking tiger with his bare hands.

During a field trip to Huangshan in 2004 to prepare a new tourism master plan for the region it was noticeable that perhaps less than 5 per cent of the Chinese visitors to the mountains actually stopped to read the newly installed plaques. Questioning revealed that, on the one hand, they were on a pilgrimage to validate their knowledge of ancient cultural heritage (Petersen 1995) and were not interested in the scientific information ('We are here on holiday, not to go back to school'); and on the other hand, the references to Chinese culture were superfluous. These observations reinforce the conclusion that for Chinese, Huangshan is a cultural landscape before it is a natural landscape even if WHS listing favours the biological over the cultural. They are there to view the beauty of the physical features and forests but they 'gaze' at them through cultural lenses (Urry 2002), denying the validity of the IUCN's attempt at a scientific imprimatur.

Calligraphy may be considered a signifier of China's unbroken 4,000-year-old civilization for Chinese visitors to Huangshan. It is common all over the mountains, carved into the living rock, often recently highlighted in red, yellow or blue paint, and enhances their appreciation of a site. This is in contrast to a western perception of 'seeing' graffiti that degrades the intrinsic values of mountain wilderness and imprints the dominance of *Homo sapiens* over nature. Most western visitors will fail to understand the deep significance of calligraphy as high art and as a gift from the gods with semi-sacred connotations, or the qualities inherent in a particular style of calligraphy, the historic–cultural significance of the calligrapher and a particular text, or the fact that the colours themselves have deep symbolism for their Chinese viewers—red for happiness, blue for longevity, yellow for prosperity (Sofield and Li 1998). On occasion, management may newly inscribe a bare site with calligraphy as a deliberate act to improve it and Chinese visitors will see this as an appropriate enrichment. In other instances, consistent with the ancient Chinese tradition of authoritative figures inscribing their thoughts for future generations, an important person may be invited to display their calligraphy skills (e.g. Mao Zedong at Huangshan, 1958). ICOMOS may recognize ancient inscriptions as significant because it places a value on antiquity, but it will frown on Dulux gloss paint being used to highlight such ancient texts and will want to prevent new additions on grounds of destroying the integrity of the historic–cultural fabric of a site. But for Chinese, a newly engraved inscription may have an authenticity similar to a much older inscription because they 'see' the continuity in an age-old process that should not be 'museumized' according to some western notion of separating the past from the present. Authenticity is culturally defined, not a concept that can be scientifically and objectively measured and universally applied. It is one further aspect of dissonance in the management of Huangshan that resists the western paradigms governing WHS listing.

CONCLUSIONS

Huangshan illustrates the limitations of an approach to WHS listing that is fragmented and divided along western-oriented constructs. It is suggested that there would be advantages in an increased capacity for incorporating non-western cultural values in considerations of WHS management regimes. However this is difficult: The IUCN's experts are enjoined to focus only on natural heritage to the exclusion of other factors, and their scientific training, as per Foucault's (1980) treatise on power, dictates that they will privilege their knowledge base over other knowledge bases. ICOMOS similarly has western precepts guiding its assessments. But without a more open, more integrated multidisciplinary approach to consideration of the merits of a place for WHS listing, we are left with a variation of Said's (1978) engagement of 'Orientalism' versus 'Westernism'—the assumption of an unquestioned 'western/modern' authority accompanying the ownership of the concept of WHS listing that subordinates Others' values to its own.

APPENDIX 9.1

World Heritage Centre (1990) 'Operational Guidelines for the Implementation of the World Heritage Convention'. Huangshan was inscribed on the

World Heritage listing under Natural Criteria (iii) and (iv) and Cultural Criterion (ii) as follows:

Natural Criteria

(iii). contain superlative natural phenomena or areas of exceptional natural beauty and aesthetic importance; and

(iv). contain the most important and significant natural habitats for in-situ conservation of biological diversity, including those containing threatened species of outstanding universal value from the point of view of science or conservation.

Cultural Criterion

(ii) exhibit an important interchange of human values, over a span of time or within a cultural area of the world, on developments in architecture or technology, monumental arts, town-planning or landscape design.

10 Tourism Development and Cultural Interpretation in Ganzi, China

Wolfgang Georg Arlt and Xu Honggang

INTRODUCTION

As noted in the introduction to this section of the book, and in the chapter by Ma, Si and Zhang in the case of Qufu, cultural heritage tourism almost inevitably produces the contradictory phenomena of, on the one hand the (re-)strengthening of the visited local cultures, albeit in commodified form, and on the other hand the enablement of local inhabitants through contacts with the outside world and increased income to leave their traditional lifestyle behind in favour of a more modern way of life. It would perhaps be romantic to believe that there would be any other outcome, especially perhaps in a country like China where millions exist on incomes of as little as 100 RMB a month in rural areas, but all wish to achieve better life opportunities, if not for themselves, then for their children. Such opportunities often mean the adoption of value systems consistent with the dominant means of doing business, of accepting the modern technologies of the world, and also of acquiring the work schedules and time management of potential business partners—all of which might be foreign to cultures still tied in many ways to premodern modes of thought.

Substantial research has been conducted in the area of tourism and its cultural impacts. Jafari (1990, 2005) summarized the research and identified five 'platforms': the Advocacy Platform, Cautionary Platform, Adaptancy Platform, Knowledge Base Platform and Dissemination Platform. In the knowledge platform he basically calls for objective and scientific study of the place. Sofield (2003) also stressed that it is far too simplistic to utilize value judgements as to whether tourism development is 'good' or 'bad', or that tourism development is 'wrong' and should be avoided; rather, it is necessary to subject a specific situation to detailed research and examine conceptual issues such as conflict, coexistence and symbiosis to reach an understanding of tourism development.

This study attempts to use Ganzi as a case study to understand the cultural context on which its tourism is based and the interaction between tourism and local culture. Based on the involvement of the authors in the research for the Greater Shangri-La Region Tourism Development Plan, the paper attempts to understand how the *Shangri-La* is interpreted differently by different stakeholders. Due to the special historical circumstances of Chinese Tibet, these phenomena can be observed in a specific form with contradictory interpretations produced by the different stakeholders involved. Without a clear understanding of the cultural background and consequences of the tourism development, it is difficult to formulate proactive policies to help locals find a balanced way towards both cultural protection and development within a context of inward Han migration. However, this is essential given the mantra that successful tourism development requires local community involvement in order to ensure the protection of local resources, natural, social and cultural, for which local communities are guardians, interpreters, and gate keepers. Ryan, Gu and Meng examine these issues more closely in the introduction of the next section of the book, and thus these issues will not be duplicated here, other than to note that given China's need for development in rural areas, any tourism development that fails to provide economic benefit for local communities is not likely to win their support or participation.

THE CONTEXT OF THE STUDY

Ganzi is the border area between Sichuan, Yunnan and Tibet, and is an autonomous province comprising 16 counties with a total population of 880,000. With a population density of 5.8 people per square kilometre it is one of the least densely populated areas of China. It is one of the areas where the Kangba Tibetan people are located, and in total they accounted for over 70 per cent of the province's population in 2000, thereby outnumbering the Han and the Yi, who form the second and third largest ethnic groupings. Other ethnic groups present in the province include the Qiang, Hui, Naxi and Bai peoples. Recent years have seen an influx of Han migrants as in Tibet itself. Its proximity to Tibet means that close relationships exist with that region—indeed for two years a Tibetan invasion in 1930 meant the area was under Tibetan rule. Various researchers, including Lin Junhua (2004), have pointed out that this cultural diversity is embedded in the diversified natural landscape and is one of the major characteristics of this area. It is, in fact, the unique cultural landscape that attracts the tourists. Being Tibetan the region is dotted with temples, one of the oldest being Garze Monastery, over 540 years old, but now partly rebuilt in a mix of Han and Tibetan styles.

A feature of this part of China are the wooden buildings and their associated architecture, and high levels of carpentry skill are found. Most middle class residents will live in quite spacious wooden houses, which will also be beautifully painted. The traditional skills and arts are still widely practiced, and a particular form of art, such as thanka, is still present. Thanka is a form of painting based on religious themes, usually with a specific centre-piece surrounded by replicated patterns. However, because of a combination of factors that include an expanding population, the costs of traditional building, the time required, the need for high-rise office and apartments in built-up towns where land is more expensive, concrete and steel buildings are beginning to make their presence felt. Another reason is that these forms of building are more fire resistant (Logan 1998).

In spite of recent growth, and due to its poor accessibility, this area lags behind the development of many other regions in China. Traditionally, the region relied on agriculture and forestry. After the implementation of the Natural Forestry Protection Project, the region's economic growth slowed due to a loss of employment in the forestry industry necessitating alternative sources of employment and income to be found. Thus, witnessing the rapid development of tourism inside China in recent years, Ganzi now also attempts to grasp the opportunity to use tourism development as a tool for regional development. In addition, the Chinese central government provides support in this regard by providing infrastructure and technical support. The Chinese National Tourism Authority recently initiated the Greater Shangri-La Region Tourism Development Plan, of which Ganzi is a part of the core zone. As a result, the enthusiasm for tourism development is very high across the region.

It can thus be argued that the region, in the period immediate to 2008, was characterized by the following features:

1. The rapid development of tourism in the region.
   In the seven years from 1999 to 2005, as shown in Table 10.1, the numbers of tourists have climbed from 6,885 to 2.25 million, and tourism's contribution to regional GDP has grown from almost nothing to over 30 per cent.
2. Resource-based tourism development.
   Like many tourism destinations in less developed regions, tourism development in Ganzi is based on its own existing resources. When tourism began, the local people found that their cultures and cultural landscape, which formerly were not well known and appreciated, suddenly could be turned into attractions and bring economic benefits. These hidden tourism resources are in the process of being 'discovered' and being turned into commodities. Yet, considering the special social, economic and physical environmental contexts, this commodification is facing challenges. In the short run, a misunderstanding of the market can mean this commodification fails and brings no economic return. In the long run, when proactive actions and policies have not been formulated, the local communities may lose control of the way local cultures develop and present themselves.
3. Such rapid expansion of tourism also poses questions of who authorizes and controls the process of cultural presentation and interpretation. It has been found in other similar cases that outside business

Table 10.1 Tourism Development in Ganzi

Year	Domestic Tourist Income (10,000RMB)	Domestic Tourists (10,000)	Foreign Currency Earnings US$10K	Overseas Tourists 10,000	Total Tourism Income 100M RMB	Tourist Income as % of GDP
1999	15.1	0.6885	12.38	0.1200	0.011728	—
2000	6430.4	30.1600	21.33	0.1200	0.660637	2.7
2001	36716.0	66.7470	57.04	0.3600	3.118558	13.49
2002	22120.0	31.6000	105.9	0.6700	2.299368	7.37
2003	49197.5	72.5000	508.62	1.8300	5.339362	15.16
2004	116545.1	179.8611	1532.19	4.8081	12.918574	30.69
2005	146491.5	225.3700	1709.00	5.9000	15.700000	31.13

Data source: Ganzi Tourism Bureau.
*Exchange rate: 2005, 1 US-Dollar = 8.10RMB, other years: 1 US-Dollar = 8.25RMB.

interests with capital and business connections to tour operators are often better able to capitalize on new tourist interest than local residents with little connection to the tourist generating zones—thereby meaning that their share of tourist expenditure is less than would otherwise be the case (Ryan and Aicken 2005).

THE STAKEHOLDERS' PERCEPTIONS AND INTERESTS

Culture is a complex and dynamic system. Stakeholder analysis is an appropriate methodology to assist in understanding a complex situation and it was therefore considered relevant in preparing the tourism plan. The first stage was to identify the stakeholders. In Ganzi tourism development, local communities, governments and related institutions, and domestic and international tourists, and their interpretations of the cultural resources, are reshaping Kangba.

Western Tourists

A western view of Shangri-La was developed based on an idealized pre-1950 situation. 'Shangri-la is a distillation of a borrowed piece of Tibetan mythology overlaid with a Western dream that was two centuries in the making. It is the mystical embodiment of the idea of a sanctuary—a place where all civilized yearnings are satisfied' (Schell 2000: 245). It represents Tibetan religion—or rather an imagined peaceful and sanitized form of

Lamaism—and also an isolated world. The western tourists are attracted by this image. As Ma (2006) described:

> The need to believe in an earthly paradise, a hidden utopia where men live in peace and harmony, seems to run deep among those who are discontented with the modern world. Westerners idealise Tibetans as gentle, godly people untainted by base desires and greed.

It has to be understood that when international tourists come 'questing for authenticity', their quest is not for the 'real' authenticity of the situation of today, but for the 'perceived' authenticity, the image the tourists have formed before arriving (Dodin and Raether 2001).

Therefore, westerners have difficulties in accepting present Ganzi development that co-operates with the cultural influences of Han China as well as other influences of the modern world, even though many of the changes are chosen by the local Tibetan people, especially that modernity driven by tourism development. Consequently many western tourists are heavily critical of what they see. What attracts the western tourists could be *stupas* or *chorten*, the experience of Tibetan street-life in the small town, and not the new bigger temples made of concrete and steel, and comprising hybrid styles. Yet it is these new temples that bear evidence of the continuing importance of traditional religious beliefs and values.

Although the western tourists, sometimes in combination with domestic intellectuals, are still in a small number, their impacts are significant. They act as explorers of this area; their arrival often implies a place worth visiting and thereby signals its tourism development potential. Their arrival and evaluations become the reference for tourism development on the part of local people. As a result, an increasing number of places are becoming involved in tourism development and the desire for using tourism for wider economic development—e.g. by building a need for more services, schools etc. (Kolas and Thowsen 2005).

Their persistent search for their perceptions of the 'real' and 'pure' Tibetan culture increases tensions rather than building a cultural understanding between hosts and guests. For instance, in order to maintain 'authentic' and 'pure' images, the local people are advised not to decorate the temples because the tourists prefer an old preserved temple. However, the local people always donate money for bigger, new and magnificent temples. Tourists also seem to want fulfilment of a stereotypical 'nice, meditative and simple people' and thus are sometimes unable to cope with the complexities that they find and with which local residents deal on a daily basis.

Domestic Tourists

Although place, people and culture cannot be separated for tourists to experience Shangri-La, a cross-cultural understanding of place and culture is not a priority for domestic (Han Chinese) tourists, many of whom may come from the larger Chinese cities, many of which suffer from pollution and consequently from bad air quality. Also there is, on their part, generally a lack of knowledge of local culture. Shangri-La, for the domestic mass tourist market, is a place of natural beauty, photo opportunities, exotic settings and amusement. Consequently many of the mass tourists are more interested in the magnificent nature of the region. Their experiences of the place often concentrate on a few nature spots rather than a comprehensive experience of place, people and culture. Their images are not really associated with visiting the holy places of Tibetan people, but rather with the mountains as symbols of nature and wilderness.

Unlike their western counterparts, domestic tourists not only tolerate new modes of cultural production but even appreciate these new and modern changes. Many new cultural products are developed to meet the domestic tourists' needs. Singing and dancing are not only the dominant component of home-stay tourism products, farm tourism etc., they are also bundled with the catering and other tourism activities. But as a matter of fact, many of the songs are not Tibetan songs; they are more Chinese folk songs. As Lu *et al.* (2006) also pointed out, the most popular song for domestic tourists, *Kangding Love Song*, which was used to help brand and promote the area, was written in Chinese, its story is also about Chinese lovers, it has little relationship with the Kangba culture and can hardly be said to represent the whole area. One source indicates the song originated from Sichuan.

Searching for the 'out of the ordinary' is also a motive stimulating many Chinese tourists to this region. Consequently many visit Lugu Lake because the local community of Mosuo still retain a matriarchal system and live in extended multigenerational groups of some 20 to 30 people descended from a common grandmother. In addition, the Mosuo of all ages engage in naked bathing at the Waru hot spring, seeing the body as simply a part of the normal pattern of things. For many Chinese such visits permit a view of fascinating customs found in a beautiful part of the country; thereby combining scenic valleys, a cultural difference that intrigues and access to holy places for those so inclined. It can also be observed that providing services to the domestic tourists is currently relatively easy owing to their proclivity to adhere to well-organized tours and groups, while being uncritical of current modernization.

Local Communities

For the local people, Shangri-La tourism offers development opportunities. An increasing number of people are becoming aware of the market value of the local culture and landscape and thus the desire to turn the resources into commodities is high. Consequently with the growing flow of tourists and the resulting increase of income, more and more local people can afford to buy outside products and live modern lives.

The local communities have very few development opportunities in this region and tourism, a low barrier to entry-level industry, is particularly welcomed by the local communities. Not only material goods, but heritage and the living culture are explored and packaged into commodities, and this includes the matriarchal system of the Mosuo people. In attempts to meet tourist demand, and perhaps gullibility, there is a new wave of production of local history and culture, often including factual misconstruction, to gain a unique market position while creating manufactured authenticities of Shangri-La in the market. Given the Chinese love of musical spectacle, local dances have been turned into dance shows. Myths about Kangba people have been created and spread out. These myths claim that Kangba people are usually tall and handsome. For example, they reputably share a common ancestry with the Germans! One village has already perceived the business opportunity of the Kangba man and changed the village name into Kangba Man village. Having perceived that many tourists came to this region to search for exotic and strange things, local people began to use the existence of the polyandry system to attract tourists. It seems there is no restriction on what can be turned into a commodity. One consequence is competition for ownership with, for example, competing claims as to being the origin of 'authentic' Xuanzi, a kind of local music, or for being the 'authentic' holy mountain.

Imported culture is readily accepted if it is thought to have market value. For instance, the local communities are quite ready to accept the imported brand *Shangri-La*. Although the word *Shangri-La* is James Hilton's corruption of the word *Shambala*, which refers rather to a state of enlightenment than to a physical place, and although local communities did not ask or seek for its replacement with *Shangri-La*, the name is accepted. This implies, therefore, a right to make corrections, and in doing so thereby draw comparisons with the pronunciation of specific place names with the word *Shangri-La* to claim its authenticity and originality, and thereby create another story for the tourists.

When the household income grows, the desire to build new functional houses is strong. With traditional materials for housing building such as wood and stone becoming more difficult to access, especially in the larger towns, new houses are built using new materials and in new styles. This not only affects individual houses, but increasingly the layout of village and towns. How to represent their culture in these new environments is a question not easily answered. Often imitation and simple solutions are provided. Similar cases are found in the east of China and in Chinese big cities where the new Central Business District may be just an attempt to copy the Central Business District in New York.

The Governments

The awareness of the importance of this region for its ecological significance has been increasing at the level of national government. In the 11th

Five-Year Plan, the area is found under the category of 'restrictive' development. Therefore ecotourism, not mass tourism, is proposed as an alternative development tool. However, since no compensation policies have been implemented yet, it seems that such concepts are not wholly welcomed by the locals who have a strong desire for development and modernity.

The local governments in the less developed regions are under pressure to play multiple roles. They are expected not only to be entrepreneurs, but also protectors of the natural and cultural resources, facilitators who should provide the necessary technological advice for the local people and outside investors, and financiers to obtain enough financial resources for running the local public sector. The local governments always attempt to find a balance among these objectives and roles. Within all these multiple roles and objectives, development, especially growth, is the main goal for local government. When they successfully lead their local area into greater modernity with visible effects of more modern buildings and higher local incomes, their political goals are met. When tourism is seen as an enabler of development, it is linked with an increase in capital investment and the arrival of mass tourism. In consequence, major infrastructure development in terms of roads, hotels and public buildings like museums are built to facilitate outsider investment and to aid the movement of the large numbers of visitors.

Because the natural reserves and national forestry parks are well-established products in the tourism market, significant efforts are made to turn particular landscapes into parks for which the main purpose is the promotion of natural beauty, such as, Hailuo Guo National Park and Daocheng National Nature Reserve. At one level, the award or classification of 'Nature Reserve' does not really restrict major development for mass tourism, and on the other hand, the personnel recruited are primarily educated in the natural sciences and so lack an understanding of the cultural issues. As a result, contentious issues arise such as the use of religious symbols and artefacts as tourism decorations. Over time, as the Beijing government has become increasingly concerned with environmental issues and the need to protect and conserve natural environments, the regulatory hand over these parks has become stricter, to the point now where local governments are expressing the view that they do not want additional natural reserves because they become a barrier to development, while some voice the view that some areas should be declassified on the same grounds.

As for entrepreneurship, in a region in which development is primarily based on a local resource and where managerial expertise and the entrepreneurial base is not sufficiently present, the identification and evaluation of the potential tourism resources and their transformation into economic capital are very much the responsibility of local governments. The local governments are not solely engaged in exploitation of the tourism resources, but also compete for control of these resources. One sign of this is the competing brands and promotional messages associated with sources of 'authentic' Kangba culture and the *Shangri-La* brand.

At the level of local village administration a tension can arise as it is a function of local governments to both protect and promote local culture. They are monitored by higher level government agencies and they will be criticized and perhaps lose funding if the local resources are not effectively protected and utilized. Cultural tourism is thus often regarded as a link between cultural protection and turning these culture capitals into economic gain through cultural production. Easy and quick solutions are often implemented in cultural tourism development. For instance, a quick solution to the cultural issue is the building of museums which can be a tourism attraction and serve as a source of knowledge acquisition and cultural protection. As a result, museums with poor displays are often promoted as tourist attractions. Dancing and folk performances are two products strongly promoted to meet the demand for entertainment from the mass tourist. However a combination of factors that include fitting performances into desired time-frames that suit tourist productions, a love of spectacle and drama have meant a simplification of local dance performances (McKay 2001).

A lack of cultural understanding and expertise amongst governmental advisory services is also problematic, especially when a need for quick decision-making arises. Kangba culture is one of the major images that the county local governments want to establish. Conceptually it is understood that preserving Kangba culture in the initial development phase is good for longer term sustainable development and for the long-run competitiveness of the region. Yet, how to proactively protect and develop Kangba culture is a challenge. Debates express concerns and tensions over the physical appearances of cultural forms and how to manage them, and how to plan and design a Kangba township and city. When the pressure for significant, quick change is great and the stock of available advisory supports is either not available or consulted, many of the new cultural productions lack a detailed understanding of local culture and become theatrical performances with only a veneer of local culture.

The Tourism Business and the Media

The tourism industry and media manipulate local images. The development of branding and image generation owes much to the media. *Shangri-La* is an image of a 'placeless utopia' (Bishop 1989), based on a fictitious description of a paradise. 'When Westerners think of Tibet today, the name conjures up a series of images around terms such as "Dalai Lama", "Roof of the World", and "Shangri-la"' (McKay 2001: 67). These images were further shaped in the late 1990s by a number of very successful Hollywood films, the most important being *Little Buddha*, *Seven Years in Tibet* and *Kundun*. Guidebooks, such as *The Lonely Planet Guide to China*, and different web sites are probably the most popular source for information about Ganzi cultural heritage, and are widely read by international tourists.

Inside China, the *Chinese National Geographic* magazine plays an important role in the shaping of regional images for the past few years. The most popular sites in Ganzi are Hailougou and Shanshenshan in Daocheng, the two nature reserves. The popularity of these two sites is largely attributed to the *Chinese National Geographic*. In the October 2005 issue, the two mountains are listed among the top ten beautiful mountains in China. The criteria used in these evaluations were (a) unique appearance (0–40); (b) high scientific value (0–20); (c) relative height above 1,000 metres (0–20); (d) high cultural value (1–10) and; (e) degree of development (0–10). It can be seen that natural beauty plays the key role.

Cultural Elites and Academicians

There are cautionary views expressed by various local culture elites and academicians. When the major theme is development, the voice of environmental protectionism is seldom heard. In addition, there are no structured means to make the voices from scattered academicians and people interested in the protection of local culture heard, and therefore their impacts are limited. Due to a lack of support for research into local traditions, architecture and cultural practices, the environmental and heritage lobbies also have to face the fact that there is little they can do to actively protect the local culture while allowing the processes of modernization to meet the local demand. Perhaps only a long-term commitment to research can help them to better state the nature of the problems and win some support.

DISCUSSION

Tourism in Ganzi is thus shown to bring both opportunities and threats to the province's local cultures. Through this chapter's stakeholder analysis, it can be seen that different stakeholders have quite diverse views towards the interpretation of local heritage on which they reshape the local culture and landscape, although it can be argued that shared visions of the local cultures would avoid cultural conflicts. However, since economic needs dominate, tourism has become a dominating influence; or rather the perceived needs of what might still be regarded as an inexperienced domestic tourism market have led to quick solutions based upon popular patterns of demand. The authors, however, tend to a view that strategies should be formulated to build shared visions at different levels and for different time phases, both to anticipate future changes in demand as the market matures, but also, more importantly, to better safeguard the traditions of these long-established cultures.

One possible way forward lies in segmentation of the market. Yet, market segmentation only considers the sustainability of tourism *per se* and indicates the power of tourists to shape local culture. Within the short term, without proper preparation, the cultural resources are explored, reshaped and commodified with reference to current demand patterns. Management tools are needed to reach a cross-cultural understanding, between hosts and guests, of the local cultures through educating and managing tourists. Interpretation centres should be set up to reduce the gap between the perceived and actual images, thereby also reducing pressures on locals to adjust themselves to meet the tourism demand. Such centres could also help to show that local cultures are valued for what they actually are, and that there is value retaining a heritage rather than simply shaping tourism performances. Local people need this kind of support to find a way to develop their culture consistent with their own values, and to find a position in the global system, while avoiding being completely controlled by the tourism market.

In the short run, the tourism marketing of the Ganzi region as part of the Greater Shangri-La region has to be segmented into differentiated activities for the different markets. Image plays a central and critical role in the decision for or against travelling to a certain destination. Yet images also shape the visit to the destination itself as they influence the things and experiences sought, and are used as the yardstick to measure the authenticity or fakeness of such material objects and social customs. The basic requirement for opening a destination to a market is to have a detailed understanding of the images prevailing in the source markets. For such images it is not important if they are correct or incorrect from the point of view of the inhabitants and leaders of the destination itself. Images are formed from many different sources of information, and with only limited number resources the peoples at the destination itself may only have limited means by which to influence the pre-visit image. What is required for better tourism planning is an understanding of the pre-visit images so as to better meet visitor demands in ways consistent with the values of the host communities.

For western tourists the mystical and *religious* aspects of a region untouched by modernity as implemented by the use of the *Shangri-La* image can be placed in the foreground, with sub-images of Ganzi as the region of the specific Kangba culture and as a place for adventure outdoor activities. Yet such tourists should be aware that commercialization is not only rapid and uniform (in China) but also strikingly overt. Unlike tourism development in the West, no attempts are made to disguise commercialization through clever design, use of materials or sales behaviour. Ticket gates and service stations abound. Western tourists desirous of an 'authentic' or 'picturesque' photo shot purified of blatant signs of tourism activity look around in desperation. The result of such uniform and self-conscious commercialization is the erasure of differences between tourist practices that, in the West, are perceived as distinct, such as ecotourism versus sightseeing (Nyiri 2006: 54).

For the western tourist, the tourist products should be packaged in ways perceived as being premodern. Relying on Tibetan religion and culture as the lead brand, products exploiting the expectation of a mystical, unexplored, virgin region, and offering one of the last pockets of unspoilt areas that western travellers can find, should be developed and marketed. Emphasis can be laid on the point that within the marketing of China, it is still an area not yet polluted by Chinese domestic tourism developments. For Chinese domestic and overseas tourists the combination of traditional (commodified) *culture* and modernity, also in the form of easy accessibility and comfortable accommodation, will be the more important selling points.

These two approaches are obviously contradictory if a temporal and/or spatial separation of specific destinations for different source markets cannot be achieved. What is needed is an awareness that it is relatively easy to provide tourist products to the Chinese tourists who, as noted earlier, tend to be uncritical of product. However, one side-effect of this is that it tends to reduce the local incentive to develop product for a more demanding international tourist, and this perhaps might be prejudicial to longer term success in both the tourism market generally and the more specific heritage/cultural tourism segment. It is suggested that the following may help to overcome the difficulties identified earlier.

Proper Use of Interpretation Management

The proper use of interpretation centres and mechanisms is an opportunity to provide the tourists with cultural insights. Current interpretation has its limitations; it is a form of tourism consumption that has been constructed and packaged (Carr 2004), but still interpretation can be an effective means of enhancing cross-cultural understanding. Yet, interpretation without careful design can further reinforce stereotypes and develop a direction of causation from the tourist to the host culture rather than building a bridge by which tourists can better understand the local culture. For instance, the ways currently adopted to interpret the Holy Mountains, the popular tourist sites and also the more important cultural sites for the locals, are often problematically designed. How can this be overcome? It has been noted that many of the Holy Mountains have already been categorized as nature reserves or forest parks for which scientific interpretations are required. However, scientific interpretation also indicates the import of western scientific paradigms and their interpretation of the cultural landscape. This scientific interpretation divides the cultural landscape from the local communities. The interpretation demonstrates the value of the natural system but not a deep understanding of the cultural landscape. Few Chinese tourists really stop in front of the signs and read them. Li (2006) has made a similar observation. Chinese domestic tourists like stories of fantasy, mystery and poetry when they visit the tourism sites. For many mountains located in the centre of Han culture, the cultural landscape is always shaped by often well-known poems. Reciting the poems and experiencing the feeling of the poets are one of the more fulfilling travel experiences for Chinese tourists (Li 2006). Therefore, restating the poems is a way of interpreting the cultural landscape in these areas. Yet it has to be noted that most poems are reflections of the poets' feelings of place, and are not really a presentation or understanding of the cultural landscape from the perspective of local folk traditions.

Many popular mountains in Tibet are regarded as holy mountains and are worshipped as a goddess. As a result, walking around the mountain is more important than climbing over it. The believers symbolically pass hell, the earthly world and the heavens while they walk. Therefore, worship and respect are the basic attitudes that locals have for the mountains.

Cultural Protection Plans

Historically, this is a region where local cultures have always undergone change. Tibetan, Yi, Han, Naxi, Qiang, Muslin and other ethnic groups have been interacting and influencing each other for a long time. For instance, Han culture is well preserved in Luding but the ethnic groups are highly mixed around Kangding, the central town of Ganzi. Han eat food similar to that eaten by Tibetans. In the south and north part of Ganzi where Tibetans are the majority, Han people have life-styles similar to the Tibetan, with similar housing and cuisine. However, unlike the quick tourism development, this cultural simulation and process usually takes a long time. Overall, tourism development brings opportunities for locals and outsiders to re-evaluate and understand the value of local cultures. However, the revaluation of culture is too dominated by its economic potentials and is realized through an intense commodification process.

It is true that tourism helps to show the value of local culture. Yet, when the value of culture is mainly only appreciated and presented through market forces, distortions occur in cultural production. And when the cultural value is only understood through tourism, the cross-cultural communication becomes shallow. In the long run, the locals may lose their own identity with places, people and traditional culture. Non-market forces to protect and develop the cultures should be supported and given a proper role in the planning process, not simply for aesthetic or even romantic reasons, but rather for the reason that if every tourist destination in China reacts the same way, then a homogenous tourist culture will emerge, and why then should a domestic visitor travel to any destination if they all offer the same product? Protection of the local does provide points of difference, they do provide the potential for 'unique selling propositions'—and thus even from the 'hard-headed' perspective of future tourism development there is value in protecting the traditional for longer term financial sustainability of tourism product.

APPENDIX 10.1

The Kangding Love Song

Tr. Ed Peaslee 1999

Pao ma liuliude shan shang	On that liuliude mountain top
Yi duo liuliude yun yo	Is a liuliude cloud, yo
Duanduan liuliude zhao zai	Looking liuliu down on
Kangding liuliude cheng yo.	Kangding liuliude city, yo.
Yueliang wan wan	Crescent moon, crescent moon
Kangding liuliude cheng yo.	Kangding liuliude city, yo.
Li jia liuliude dajie	Li family's liuliude daughter,
Ren cai liuliude hao yo.	Is very liuliude pretty.
Zhang jia liuliude dage	Zhang family's liuliude son,
Kan shang liuliude ta yo.	Loves that liuliude girl, yo
Yueliang wan wan.	Crescent moon, crescent moon.
Kan shang liuliude ta yo.	Loves that liuliude girl, yo.
Yi lai liuliude kan shang,	First he liuliude loves her
Rencai liuliude hao yo.	For her liuliude beauty, yo
Er lai liuliude kan shang	Next he liuliude loves her
Hui dang liuliude jia yo.	Caring for the liuliude home, yo.
Yueliang wan wan.	Crescent moon, crescent moon.
Hui dang liuliude jia yo	Caring for the liuliude home, yo.
Shi jian liuliude nüzi	All the world's liuliude girls,
Ren wo liuliude ai yo	Let me liuliude love them, yo.
Shijian liuliude nanzi	All the world's liuliude boys,
Ren ni liuliude qiu yo	Let them love liuliude whom they please, yo
Yueliang wan wan	Crescent moon, crescent moon
Ren ni liuliude qiu yo.	Let them love liuliude whom they please, yo

11 The Evolution of Cultural Tourism

The Example of Qufu, the Birthplace of Confucius

Ma Aiping, Si Lina and Zhang Hongfei

INTRODUCTION

The purpose of this chapter is to provide an insight into models of Chinese cultural tourism development by reference to Qufu. After providing a description of Qufu, the text describes the emergence of heritage tourism in Qufu with reference to resource utilization and the characteristics of each stage of development. The authors argue that each stage reflects differing resources, the unique nature of the culture being presented and the commodification of that culture as, in marketing terms, a 'brand Confucius'.

CONTEXT OF THE STUDY

There has been a growing commercial and academic interest in cultural tourism for many years. Currently, although estimates vary as to the degree of interest that exists in the marketplace and the actual nature of tourist motivation (e.g. the study of cultural tourism in Hong Kong by McKercher and Du Cros in 2005, and Ryan's 2002 studies of degrees of interest in indigenous tourism) it can be stated that heritage and cultural tourism forms a distinct product within the total portfolio of tourism attractions on offer in the marketplace. It is also a market sector that has attracted academic interest in terms of issues of protection versus commercialism. Similar debates and developments over this matter have arisen in China as in other parts of the world. This chapter examines tourism in Qufu, the birthplace of Confucius. Qufu itself is a medium-sized town having a population of approximately 640,000. It is located in Shandong Province, and served as the capital of Lu during what is known as the Spring and Autumn period of the first half of the Eastern Zhou dynasty; a period that approximately lasted from 730 BC to 480 BC. It has developed tourism based on its historic and cultural assets. While the chapter discusses issues specific to Qufu, the authors would argue that the debate would be symptomatic of many other similar locations in China.

Qufu is on the list of China's historical and cultural cities released by the State Council, and was the first location to be categorized as an AAAA tourist scenic spot in China. In 1994, the Temple of Confucius, the Mansion and the Confucius family cemetery were simultaneously listed as world cultural and heritage sites by UNESCO during its 18th session. The Temple of Confucius was based upon Confucius' original home, and was a place of significance to his followers soon after his death. It is noted that in 205 BC, Emperor Gao of the Han Dynasty was the first Emperor to offer sacrifices dedicated to the memory of Confucius. The actual complex is large, exceeded in size by only the Forbidden City in Beijing. It covers 16,000 square metres, possesses 460 rooms with many of the current buildings dating from the fifteenth century, it having previously been destroyed by fire in AD 1214 and again in AD 1499. It last suffered damage on any scale in November 1966 when students and staff from Beijing Normal University destroyed large numbers of artefacts during the excesses of the Cultural Revolution. Today the Cemetery has 3,600 tombstones set among 10,000 trees, which gives the area a wooded feel. The oldest graves date back to the Zhou dynasty, and the area covers about 2.6 square kilometres. East of the cemetery lies the Kong family mansion, where over centuries lived the descendants of Confucius who cared for the temple and cemetery until the third decade of the twentieth century. The present buildings primarily date from 1887, again due to previous fires that had destroyed many of the original buildings.

In recognition of its specific and unique heritage, the Confucius Research Institute of China was established in Qufu in 1984, from which initiative has sprung an international network of universities and other institutions interested in the promotion of Chinese culture. From the perspective of tourism, the authors would suggest that as a heritage centre and tourist attraction, Qufu has passed through three stages of development. Initially, prior to 2000, tourism in the city was based upon the three sites of the temple, mansion and family cemetery. By the end of the twentieth century a more complex tourism product was emerging based on a signage of place as not simply a location of cultural significance within China, but one possessing a much wider touristic importance within the remit of policies of economic regeneration and infrastructure improvement. Publicity at that time was oriented towards a message of 'three Confucian spots, four mountains and two cities'. Since 2002 those products have been involved in subsequent enhancements of physical infrastructure, interpretation, image generation and promotion. At the same time new products have been created, including the introduction of experiential and participative products. Examples of this include the daily Confucius' Dream performance, the re-enactment of opening the city gate of the Ming dynasty twice a week and various forms of worship eight times a month. Additionally there are two significant special events—an International Confucius Cultural Festival and an International Confucius Study Tour Festival that have been created. The cultural festival has emerged to not only note Confucian philosophies but to become a restatement of wider Chinese culture and Chinese identity. In many instances these developments illustrate the trends discussed by academic researchers like Getz (1995), who noted the commercial, social and political imperatives that lie behind the development of festivals and special events. In short, festivals have economic benefits, are statements of local identity and moments of civic pride and an opportunity for local politicians and worthies to play a more public role that tends to be less controversial than at other times. From a wider socio-political context the emergence of the festivals serves not only an economic role but also characterizes a greater ease with a pre-Communist past while reinforcing notions of being Chinese on the part of the government.

Consequently, in Qufu, it is the local government that is responsible for event planning, industry co-ordination, tourism promotion and planning, and heritage protection. Festival performances of a theatrical nature are a highlight of these programmes, and the revenue derived from these performances with those gained from visitation to other locations and scenic sports are distributed to further develop each of the attractions.

LITERATURE REVIEW

Sites of historic importance and cultural and architectural value use their heritage as the core of their tourist product. However, from this core a number of ancillary products and enhancements can emerge, and tourism itself can become a significant local industry. This is particularly the case when local government takes an initiative for the reasons outlined earlier. However, as noted by Ashworth *et al.* (2000), such policies can quickly lead to tensions between the need to protect and the need to exploit. Issues arise as to what is sustainable, what are the goals of developmental policies, what are the needs of local communities and what are the responsibilities of local communities for wider tourism and economic development. Individual community assets may become part of regional portfolios of product that can benefit those outside the immediate historic zone. Issues thus arise as to not only the role of protection and exploitation within a historic zone, but between local and regional policies, and between commercial and non-commercial interests. Amidst these concerns, threats can arise to the original nature of the historic city. In China much of the responsibility for assessing, monitoring and solving the problems caused by these issues lies with the public sector.

In their seminal work, McKercher and Du Cros (2005) classified the concepts of cultural tourism into four: tourism product, motivation, experience and operation. The motives for cultural tourism tend to be associated with, on the one hand, a sense of curiosity about history, human works of art, performance and senses of identity, and on the other, simply a wish to see something new, to cease being bored or to share a social occasion with other people. The resources that are being used can include archæological sites, museums, castles, palaces, historical architectures, visual arts, music and dancing, primitive cultures, subcultures, churches and other things that convey local, regional and/or national cultures. In short, cultural tourism emphasizes specific types of resources and tourist experiences that can be engendered within and by it. Therefore, the basic attributes of cultural tourism can be understood from the dimensions of time, forms, motives and market demand. The major products of cultural tourism can be divided into six areas: heritage sites, museums, 'special performances/identities', theme parks, historical and cultural activities and artistic activities, all of which have a more specific nature compared with other types of tourism products (Xuzhihui, Dingdengshan and Xiangdong 2006).

Nuryanti (1996) argued that cultural tourism not only revealed tensions between tradition and modernity, and the unique and the universal, but it raised specific issues and questions for developing countries. These issues were about restricted resources and the very nature of development as countries sought to emulate the success of developed nations, but struggled to find places in a postmodern world for their own value systems. Traditionally, in the tourism literature, one means of solving these tensions has been calls for community participation as one of the key factors for success in destination and local community/government planning. Jafari (1992) pointed out that tourism should provide communities the opportunity to develop their own culture. A continuous refrain in the academic tourism literature has been that tourism has significant impacts on local people resident in historical and cultural cities (Borg, Costa and Gotti 1996), and that local communities should be encouraged to participate in the development of heritage exploitation programs through related policies (Caffyn and Lutz 1998). It is argued that, in the development and planning of cultural heritage, the community's wishes and ideas should be considered beside the opinions of experts (Teo and Huang 1995). These discussions are echoed in the Chinese tourism literature, and this book contains further examples of that. Zhang, An and Liu (in this book) discuss these issues with reference to how community participation aids community development, while Zhou and Ma provide evidence of approaches adopted towards heritage tourism by two rural communities located at the Baiyang Lakes.

Yan Lili (2006) proposes a model of tourism which includes temporal and spatial zoning where tourists use zones of culturally engaged areas of a community for shopping, entertainment, dining, sightseeing and other leisure activities. This model applies in particular to Qufu and its world heritage status, and is consistent with trends in the development of tourism; that is, tourism evolves from short periods of time for sightseeing to a process of deeper involvement and stays of longer duration at the destination. Yan Lili further proceeds to describe four more developmental models, namely:

1. a phased development model of small-sized and independent scenic spots with weak ecosystems and/or cultural resource;

2. a synchronized development model combining physical and immaterial heritages;
3. a regional model comprising core heritage zone, core environmental zone, 'cushioning' zones and marginal areas, each subject to different developmental and managerial policies; and
4. a dynamic cycles model which comprises periods of use and re-use.

With reference to such models, Wang Xingbin (2002) points out the legal constraints imposed on cultural heritage that are operationalized through different levels of executive administrations even while seeking economic and social optimization through market mechanisms. Xu Zhuhui (2006) has taken both of the earlier perspectives to develop an alternative classification of cultural tourism development, noting six models. These are:

1. The integrative model that takes cultural tourism resources from different regions to artificially create a cultural landscape, which is then packaged and promoted as a heritage tourism product.
2. Historical recovery model—which seeks to re-establish a lost heritage while drawing on past records to replicate a heritage and restore historic appearance, even in the absence of the original.
3. Concentration model—this is where local government or investors build a theme park of local cultural tourism in the most appropriate site to present the essence of an area's culture and heritage, and presents this as a cultural tourism product.
4. Direct usage model—this is where current cultural resources become tourism products and are used by tourists within the current and original milieu.
5. Theme development model—which is where culture and heritage is grafted onto another tourism product to generate difference in the marketplace.
6. Short-term performance model—where specific cultural resources are used for touristic promotional purposes for short periods of time. The usual examples of this are traditional ethnic folk festivals.

Zhou and Ma provide examples of classifications 3. and 4. with reference to the villages of Wangjiazhai and Dongtianzhuang, respectively, while 5. is suggested by Ryan (2002) as a means of development for indigenous peoples to avoid an oversupply of cultural products based on performance arts, thereby creating a replication of products that can simply threaten the financial viability of new and existing tourism enterprises. For his part, Liu Yuqing (2005) notes the development model of relationships within the 'heritage tourism—the heritage industry—culture industry'. He argues for a solution to the problems of the protection and development of heritage resources from the concept of 'cultural productivity'. By comparing the ancient town of Lijiang with Hongcun and Xidi villages in China, Lu and

Huang (2006) discuss two different models: cultural leisure tourism and cultural sightseeing tourism arising from the different cultural backgrounds of destinations, emphasizing that local culture is an important determinant of whatever policy or model is to be adopted. In turn, Tao Wei (2004) points out that the sustainability of tourism attractions depends on whether the culture is strong or weak. Li Xinhua (2006) argues that whichever model is adopted, the primary nature of the product is that common to all tourism, namely, the creation of an experience different than that of daily life. However, the essence of cultural tourism is the development of an aesthetic feeling generating pleasure, freedom and learning from a past—the features of which add to our understanding of self and the present.

QUFU—A CASE OF HERITAGE TOURISM

Resource Base

Past studies have been undertaken in Qufu. For example Zhou Changchun (2003) classified the resources of Qufu and analysed the advantages and disadvantages of these resources. Li Lunliang (2006) examined the relationship between the tourism industry and other industries. Wu Bihu (2003) has pointed out that Shandong has many tourism sites with special attractions and that Qufu is famous in China, East Asia and even in much of the rest of the world. Tang Shunying (2004) discussed the status of study tours, both at home and abroad and examined the developing role of study tours in Qufu. He argued that such tours should be based on participation, improving understanding and comprehension, content innovation and be multifunctional if such tours are to be successful.

This study commenced on the principle of classifying tourism resources, thereby replicating but updating the work of Zhou Changchun (2003). These are shown in Table 11.1. It can be noted that in addition to the historic resources described, there have been added the use of natural resources such as Mountain Shimen National Forest Park, Nishan National Park and Mountain Jiuxian, which have all become important components of Qufu tourism development. In recent years, Qufu has also undertaken a number of large-scale man-made tourist projects, such as the Confucius Six Arts City, the Steles of Analect, the aforementioned Confucius Institute and the Xingtan Theatre.

Table 11.1 shows that the tourism resources in Qufu have the following features:

1. The main resources relate to geography and heritage. The Confucius Temple, the Confucian mansion and Confucian family cemetery are the main Qufu tourism resources, and have all been listed as possessing world cultural heritage importance. This designation recognizes the far-reaching influence of Confucian culture in Asia. For about

Table 11.1 Classification of Tourism Resources in Qufu

Main Category	Sub-Category	Tourism Resources
Land forms	Mountains and Caves	Mountain Nishan; Mountain Shimen; Mountain Jiuxian; Mountain Fang; Mountain Shouyangl; Mountain Jiuyang; Mountain Maan; Confucius' birth cave
Water-based	Lakes and Rivers	Saint lake; Si river
Biological	Forests and Woods	Confucian family cemetery; Nishan National Park; a tree planted by Confucius; a tree planted by Zigong; Cypress of Han dynasty; Ginkgo
Heritage sites (ruins)	Palaces / Tombs / Plebian	Nanxingfu sites; Xixiahou sites; Shaohao tomb; Guozhuang sites; LG old city sites; Dongdacheng sites; Jingling palace sites; Chunqun school sites; the twelve palaces sites; Wuyun altar; Qufu city wall of Ming dynasty; the old banchi sites; Lingguang palace sites

(continued)

2,000 years Confucian culture has been the guiding ideology for China's society, and this philosophy of life has also had a profound impact on the national character of countries such as Japan, Korea and other south-east Asian countries. Its influence still remains strong. Indeed, and perhaps partly because of China's economic development, the influence of Confucian thought is being increasingly felt in Europe and America, and many foreigners are now studying Chinese Confucian culture. The promotion of the Confucius Institute by the Chinese Government is itself recognition of the heritage and contemporary significance of Confucian culture in an international arena, while also being a means of promoting Chinese culture more generally. Given these influences, Qufu as the birthplace of Confucius is important culturally and politically and thus attracts resource support as well as increasing numbers of tourists.

2. A large proportion of the historic resources have not been ruined but maintained over the centuries by adherents of Confucius, including past family members. Therefore Qufu tourism resources are not only numerous but also are of a high quality in terms of authenticity and architectural richness. Classifications indicate a total of 94 types of tourism resources in the city, nearly all of which are in the forefront of size and number of tourism and cultural resources in the province. From a Chinese perspective, Qufu, Beijing and Xi'an are China's

Table 11.1 (*Continued*)

Main Category	Sub-Category	Tourism Resources
Built Heritage (standing)	Palace and Temples	Shimen temple; Mosque; Chinese fable palace
	Art Work	Stone inscription of Han dynasty; Statue of Tang dynasty
	Traditional Architecture	Confucius Temple, The Confucian mansion; Yen Temple; Zhougong Temple; Zhusi school; Mencius former residence; Ancient county government; Ancient test place; Ni hill college; Shimen college; Drum tower
	Tombs	Confucian family cemetery; Liang tomb; Mencius' mother tomb; Yen tomb; Han dynasty tombs group; Shaohao tomb
	City Wall	The old Lu city
	Transportation	Si river bridge
	Water Resources (built)	Mountain Nishan reservoir
	Education and Cultural Sites	Confucius college; the steles of Analect
	Manufacture Sites	Kongfujia alcohol factory
	Entertainment Places	Six Arts City; amusement park of Kong
	Shopping Places	Wumaci street
Tourism commodity	Retail	Kongfu jade; china; Bronze; Cloisonné; Records of Confucian mansion; Garments/ clothing displays at Confucian mansion; Mountain Nishan inkstone; Seal; Qufu rice
	Productive	
	Agricultural	
	Museums	
Human activities	Human Records	Confucius
	Folk Custom	Wedding of Confucian mansion; Qufu folk custom; Traditional music
	Modern Activities	International Confucius Cultural Festival

Derived from Zhou Changchun, Fujian Geography, June 2003.

three major cities for cultural tourism. Of these, Beijing and Xi'an are ancient Chinese political centres and represent 'old' China's political, material and to some extent spiritual culture, but through its association with Confucius, Qufu can claim a special role as representing the cultural spirit of China.

3. Notwithstanding the creation of new resources and the use of natural assets, the majority of tourist resources are primarily historic and cultural. This possibly represents an over-dependency upon one type of tourist resource, and the comparative lack of leisure-based or ecotourism resources may mean the town is not achieving its full potential.

4. Consequently a lack of potential tourism resources perhaps limits the future development of Qufu's tourism. It certainly requires consideration of key questions for future development. Does Qufu seek to diversify its product and appeal by complementing its heritage assets, or should the town primarily seek to become solely a heritage centre?

The Status of Tourism Product

Even in the few years since Zhou Changchun's (2003) study, past trends have intensified and been reinforced. Tourism product continues to evolve. One example has been the development of different tourist trails and activities based on Qufu's heritage. For example, there is the 'Confucius Home study tour' that follows the track that Confucius travelled around the six countries. There is pilgrimage tourism, the Confucian Temple gourmet tours, the calligraphy and stele tour and the classic wedding tour. Meanwhile, as noted previously, Qufu has adopted cultural interpretation performances based on ancient ceremony, but with edited story-lines and time-frames to meet the needs of modern tourists and their schedules. A number of these have been initiated, including a daily performance (Confucius' Dream), an open-city performance held twice a week, eight moments of worship per month (the Confucius performance) and two major festivals held annually—the International Confucius Cultural Festival and the Study Tour Festival. In short, the heritage of bricks and mortar, of palaces and homes, is now being supplemented by artistic and religious performances of colour, spectacle and emotional appeal.

If the objective of this resource enhancement was the attracting of more tourists, then the policy has worked. As shown in Table 11.2, the numbers of tourists and their expenditures have increased dramatically within a three-year period from 2003 to 2006. The number of domestic visitors has increased by half to 5,500,000, and the number of overseas visitors has more than doubled to 220,000. The total direct income derived from tourism has now increased to 2,300 million RMB. Tourism is thus a major industry in Qufu, accounting now for 15 per cent of the town's gross domestic product. The role of the public sector in enhancing tourism product has attracted a vibrant and growing private sector in terms of travel agencies,

which co-operated with the Shenzhen Overseas Chinese Town Group and other companies to establish the Confucius International Tourism Co. Ltd. It is this company that created a series of cultural performances, including the very successful Confucius' Dream in 2001. In order to co-ordinate the relationship between tourism exploitation and heritage conservation, in 2004, the government of Qufu established a Heritage Management Committee. The authors would argue that there is, to some extent, a model of separation of four rights operating in Qufu where co-ordination is achieved by the Heritage Management Committee, private enterprise is through the Confucius Tourism Group and the two appropriate bureaux undertake protection and tourism promotion.

The second issue is exploration and usage of the sites. Because heritage sites are often very fragile, any reckless exploitation of the relics will cause inevitable disaster. There is, therefore, a need to maintain a balance between the protection and usage of cultural heritage that requires both a macro and micro approach to tourism planning (Ryan 2003). The purpose is twofold: (a) development of the local economy and (b) the maintenance of the authenticity of local cultural heritage. Yan Lili (2006) identified five exploration models seeking this balance, as mentioned earlier. It is the view of the authors that Qufu is seeking to implement two of the models, namely (a) the combined model, that is combining material and non-material heritage, and (b) the dynamic model of use and re-use periods. Prior to attempts to generate a more controlled approach to visitor management, tourists tended to visit just three places: the Confucius Temple, the Confucian mansion and Confucian family cemetery. As has been noted, this has now been significantly supplemented by new events and new infrastructure such as the cultural performances and the Confucius Six Arts City and Confucius Research Institute. This represents movement into a combined model. Additionally, Li Xinhua (2006) has suggested another approach based on experiencing city landscape and its ambience. This approach was based on a study of the old and new city layouts of Pingyao in Shanxi Province. The experience exploration model is one that creates aesthetic experiences, pleasure and multiple learning opportunities. This, as again already noted, has led to the development of various trails in Qufu, and is also an attempt to cope with an emergent problem of shorter stays. Over the five-year period the tourist average stay has dropped from 1.8 days to 0.4 days and it is argued that by extending a stay through the use of the experience exploration model, not only will valuable economic benefits accrue from tourism, but the tourist experience can be enhanced to better understand the Confucian culture. Partly, however, the falling mean duration of stay of visitors may be caused by the previously mentioned shortage of accommodation, thereby forcing larger numbers of visitors to participate in day tours. In another product initiative, Qufu launched the Confucius Study Tour Festival. Based on Confucian family traditions, the festival was successfully launched in 2006 on the theme of the rites associated with 'the ceremony of growing-up for teenagers'. The purpose of the festival was a restatement of Confucian traditions of respect for parents, teachers and the need for appropriate etiquette in relationships. It can therefore be argued that, in the terminology offered by Yan Lili (2006), these efforts are consistent with a dynamic model of spreading tourism both spatially and temporally. As a note, one can conclude from this discussion that any site or destination of cultural and heritage tourism will have its unique characteristics, and thus in terms of fitting places to models, or models to place, specific details will never dovetail exactly into any generalized model, and thus to some extent any application of place to model generates a hybrid mix of practice, principle and problem solution specific to a given place and time.

The Role of Government

In describing these developments, and remembering the Chinese nature of tourism planning, it is evident that government has a specific function and role. This is as true of Qufu as elsewhere in China. It is also true to note that tourism in Qufu has really only been developed since the commencement of the open-door policy in China from 1978. In 1987, considering the advantages, features and character of tourism resources in Qufu, the government proposed a plan to develop tourism and established the idea of a tourism-driven local economy. It thus established the Tourism Bureau of Qufu. Not that this was, strictly speaking, the first tourism initiative in Qufu. The town had established its own international travel agency branch in 1963 and, unusually for that time, most of its business was non-governmental in nature—a fact which itself reflects the importance of the Confucian tradition in Chinese culture. Furthermore, the Quelibinshe Hotel was constructed in Qufu, and this represented one of the first three-star international standard hotels built in China. Nonetheless, government has played a vital role in the development of Qufu tourism. Today much of the tourism business in Qufu is led by the government, such as tourism promotion and planning, industry management and conservation of cultural relics. A Heritage Management Committee co-ordinates specific arrangements and overall planning.

Prior to the 1990s, the only tourism experiential product in Qufu was the Confucius Cultural Festival, which mainly focused on historical records and events in order to commemorate Confucius. After the 1990s, seeking to meet the needs of the tourism market, Qufu gradually introduced the new initiatives indicated earlier and others. In addition to the Spring Festival etc. there is the Confucius family gourmet tour, the calligraphic inscriptions tour in Confucius' hometown and the Seeking Source tour. In short, imaginative and even slightly quirky products have been introduced. What can be observed has been a series of product developments and initiatives that effectively represent a commercialization of what might be termed 'the brand Confucius'. For example, there is Confucius Road, which the provincial Tourism Bureau has used as the centre of a promotional marketing campaign in Europe. The government has also sought to capitalize and expand upon what were initially smaller scale initiatives. For example, the Confucius' Dream Performance was originally based on a show with 'old style' music launched in the early 1990s by the Quelibinshe Hotel. Every attempt was made to internationalize the Confucius Cultural Festival with specific attempts leading to major expenditure on the anniversary memorial ceremony of Confucius' 2,555th birthday in 2004, which gained significant attention at both home and abroad, and which contributed to the growth in visitor numbers shown earlier. The International Confucius Cultural Festival is now held by the National Tourism Administration and Shandong Province each year, thereby representing higher levels of government participation above that of the Qufu municipal government.

This continuous expansion of product has required rethinking carrying capacity, and at present the government is unwilling to consider constraining demand, thereby meaning that its efforts have been oriented towards promotion and increasing the range of products. Spatially, tourism is being encouraged to move beyond the town itself, and in recent years, the government has put forward a blueprint of how to develop the potential value of other tourism resources under the theme of 'Three Confucian sites, Four mountains, and Two cities'. As a result of this idea the Mount Shimen National Park is now being exploited as a tourist asset, as is Mount Nishan, which was the birthplace of Confucius. To create new product the reconstruction of the wall of the old city of the Ming dynasty has been completed. There is also the concept of the 'museum city' and a number of large-scale projects have been completed, and are further being developed, such as the Confucius Six Arts City, the Landmark Court Analects of Confucius and, of course, the Research Institute. Qufu has also co-operated with the cities of Jinan and Taian, and together they have developed a new tourism product based on landscape and the 'saint's tourist routes'. This route is being currently promoted by Shandong Provincial Tourism Bureau in its European marketing as a Chinese equivalent of the pilgrimage routes associated with places like Santiago de Compostuela. It has become one of the two prime tourist routes in Shandong. Based on this success, other themed routes are being developed. For example, Qufu now co-operates with Qingdao and Zibo to launch the Qilu cultural tourism routes. Rizhao, Linyi and Qufu jointly launched the 'Emotions of the Saints Tours', and is also promoting the 'Tai Mountain—Han Dynasty—Confucius' tourist route with Tai Mountain and Xuzhou. Each of these is gaining success, and generating well-known profiles that enable enterprises along the routes to be profitable.

CONCLUSIONS

Following the work of Butler (1980) and his concept of the destination life-cycle, destinations take on different characteristics as they respond to changing demand, competition from other destinations and the initiatives undertaken by planning authorities. The theory implies that this is an inevitable process, even if the duration of stages and the physical attributes adopted by destinations as they evolve through the cycle are neither certain nor preordained. In the case of Chinese destinations based upon heritage and culture, the authors would claim that the example of Qufu permits some suggestions as to destination development pertinent to China.

First, there is a need to undertake market research to better understand market segmentation and the psychographics and motives of tourists who choose to visit such destinations. The academic literature suggests that different types of tourists possess different expectations, perceptions and needs. Equally, different nationalities may wish for different modes of presentation, and have different understandings of what comprises 'authenticity'. Within the western academic literature one might argue that western understandings of genuine experiential products include: local tours of homes, folk custom and customary activities, temple fairs and local festivals; and partially authentic products would include performances based on traditional art-forms but related to the time constraints of modern scheduling, replications of older art-forms; new art-forms informed by older traditions and lectures, talks and art competitions pertaining to local civilizations. It can be argued that the development of product in Qufu represents an understanding of this approach, but in a rush to meet the needs of growing demand (that in part results from new promotion and product); what is being produced is a hybrid product confusing old and new under the labelling induced by what was termed earlier as 'a brand Confucius'. The name of the philosopher is being used to legitimize product and performance, and, as in the reconstruction of 'authentic' ancient town walls, one is reminded of other reconstructions in the world, such as the ability to visit the authentically recreated tomb of Tutankhamen in the Luxor casino in Las Vegas. The product development of Qufu appears to be supply led rather than demand led—but such development can be justified by simply pointing to the growing numbers of visitors and the economic gains generated. That this is the case is not surprising given a local government objective of developing a tourist-led economic resurgence of the local economy. Quite where this leaves a concept of Confucian thought is unclear, and there is some danger, at least to western eyes, of creating a theme that becomes divorced from the original philosophies.

The role of government is another key factor. Local government has been motivated by economic need. Yet national government also has a role, and has seen the Confucian tradition and the birthplace of Confucius as important in the development and promotion of traditional Chinese culture that serves many purposes. Confucian thought reinforces respect for authority, even while reinforcing the responsibilities of those in authority to provide for people. It helps creates a difference for a post-1987 regime but reinforces a sense of Chinese identity and pride that is important in a period of transition. It allows the promotion of Chinese culture to an international audience in an uncontroversial manner. In doing these things though, the commercial necessity of engaging in marketplace transactions has meant the government has to liaise with private enterprise organizations, and overseas-based organizations of both public and private sectors. In addition to the hybridity of tourism product can be added the hybrid and complex nature of private–public/domestic–international relationships, co-operations and relationships. Qufu becomes symbolic of drastic change made manageable.

Finally, but not least, there remains a need for compliance to historical 'truths'—although the construction of these are also problematic given the concerns voiced in the previous paragraph. The writings of Confucius are part of an international historic record, and thus have an existence independent of what, for the purposes of this chapter, have been termed 'the brand Confucius'. Tourism product, even if of a commercial branded product, needs to adhere to the marketing cliché that at the heart of the brand there must be delivery of a promise. The promise of Qufu is an understanding of Confucius, his thoughts, writings and relevance to a modern and postmodern world. The issue is thus to carefully delineate between different positioning of the 'brand Confucius'—between that which strictly adheres to the culture, writings and philosophies of Confucius, that which accepts replication and enhancements of modern forms of expression that pay tribute to the original, that which creates pastiche for the relaxation, entertainment and enjoyment of visitors, and that which simply uses an appeal to things Confucian to develop tourism product in a wider region, thereby spatially distributing both tourists and economic benefits. The models of destination development indicate means by which this can be done, and one can take hope that the new policies being adopted represent responses to these issues.

ACKNOWLEDGEMENTS

The authors would wish to acknowledge the help received from Hui Puke, Li Miao, Wu Jia, Hang Xuemei and Ding Qi who all contributed to the research in various ways. They would also like to express their appreciation of the help offered by Professors Chris Ryan and Gu Huimin in the preparation of this chapter.

12 Evaluation of Visitor Experience at Chinese Buddhist Sites
The Case of Wutai Mountain

Fangfang Shi

INTRODUCTION

Cultural tourism is of growing importance in China, and from an international perspective, lies in the interest of the classical Han majority culture. However, for both the domestic and international market the period from 1995 has seen an increased interest and governmental support for regions associated with China's minority peoples (e.g. as with the Sami, Bai and other peoples of Yunnan province), and in the culture associated with Buddhist and Taoist religions. This chapter studies how visitors evaluate their experience at Chinese Buddhist sites using the case of Wutai Mountain. Due to the long history of Buddhism in China, many Chinese Buddhist sites are of significant historical and cultural importance and have been designated as locations of cultural heritage at municipal, provincial and national levels. With the recognition of their potential to bring tourism income, especially in previously marginalized regions, a large number of Buddhist sites are now promoted as visitor attractions, hosting various religious and cultural events and collaborating with tourism intermediaries nationwide. Examples of such sites include the Big Pagoda in Shaanxi, the While Horse Temple in Henan, the Yonghe Palace in Beijing, the Mogao Grottos in Gansu, the Potala Palace in Tibet and Wutai Mountain in Shanxi. While many efforts have been taken from the perspective of those responsible for the promotion of these attractions, there has been a lack of study of visitors' perceptions as to what they think are important about their visits to Buddhist sites. Appreciation of the needs and wants of the end-users of these sites is crucial for service design, quality management and the sustainable development of these sites as religious centres as well as tourist attractions. This chapter investigates the visitor experience from the consumers' stance and presents some data derived from field-work conducted at Wutai Mountain (Shi 2007).

TOURISM DEVELOPMENT OF RELIGIOUS SITES
AND THE MANAGEMENT DILEMMA

Religious sites are spread worldwide and attract a large number of visitors every year (Gladstone 2005). In addition to pilgrims, whose number is estimated at about 250 million each year (Jackowski 2000), there are other tourists who visit these sites for various purposes (e.g. Peretz 1988; Bate 1993; Fisher and Sharone 1994; Laushway 2000; Collins-Kreiner and Kliot 2000). With this substantial volume of visitors, the economic potential of religious tourism has been recognized and exploited by parties with diverse interests, including governments, entrepreneurs, retailers and religious institutions.

Governments, with a mandate to increase tourism revenues, are keen to utilize the physical and socio-cultural resources of religious sites and turn them into tourist attractions (Dietvorst and Ashworth 1995). Moreover, joint efforts have been taken by governments and religious organizations to facilitate and promote tourism development at pilgrimage destinations for various motives, including economic development and a wish to ease access by the faithful (Tilson 2001). Such examples may be found in many countries. These include:

1. In Jerusalem, the Dome of the Rock, the Islamic shrine that marks the spot from which Muhammad ascended in a vision to view paradise and which was regilded with 24-carat gold in 1994 with sponsorship from Jordan's King Hussein (Ullian 1998).
2. Many ancient Christian sites, such as Ephesus where St Paul lived for three years, are promoted by Turkey's Ministry of Tourism in national advertising campaigns in the United States. (Cruz 1984).
3. During the Jubilee Year in celebration of the 2,000th birthday of Jesus, the Vatican and the Italian government spent millions of dollars refurbishing infrastructure and religious and historical sites (McGurn 1999), and Israel invested US$300 million to accommodate over four million visitors (Beckwith 1999).
4. Between 2002 and 2007 the central government of China spent 330 million RMB Yuan (about US$43.9 million) for a five-year restoration project on the Potala Palace, Norbulingka and the Sagya Monastery and the local government invested heavily in improvement of the urban infrastructure of Lhasa to accommodate the ever-increasing tourists (China Tibet Information Centre 2005).

One can thus conclude that, worldwide, governments utilize religious sites within strategic tourism promotions to increase income derived from visitors. Such income may be used for varying purposes, ranging from maintenance of the religious centres to being simply part of taxation revenues for national exchequers (Harrison 1995).

In many places, religious sites are the main visitor attractions and entire local economies are built on them (Olsen and Timothy 2006). For example, Mack (1999) found that the temples in India serve as economic centres, for in some towns the entire socio-economic structure relies on the temples' visitors and the income they generate. The development of the El Rocio shrine in Spain was specifically identified as a strategy to resolve unemployment in a rural town in peripheral Europe (Crain 1996). The Holy Land has been supported by the pilgrimage economy and local businesses and institutions, including relic-vendors, restaurateurs, royal coffers and churches, have derived considerable profit from the visitors to the Holy Land (Feinberg 1995). Other examples can be found in Santiago de Compostela, Medjudorje, Lourdes and Makkah (Olsen and Timothy 2006).

Despite the sizable economic benefit, excessive tourism development may spoil the sacred spirit of religious sites and prevent them from performing their normal religious functions. Additionally such alternation can lead to a negative impact on the visitor experience at religious sites, as the commercialization associated with tourism development may reduce the authenticity of visitor experience (Cohen 1988). Destination planners and decision-makers at religious sites have to face the management dilemma that demands a balance between accommodating tourists' needs and maintaining the normal religious functions and spirit of a place. On the one hand, taking into account the lucrative tourism revenues, some religious groups are willing to tolerate ignorant profane visitors with indiscreet behaviors (Baedcharoen 2000). At the same time, some care-takers of religious sites view tourists as potential converts and may engage in proselytization (Olsen 2006), which leads to the missionary challenge arising from the integration of a traditional religious focus with secular interests (Nolan and Nolan 1992). Shackley (2001) suggests that in many cases the economic benefits of tourism to religious sites outweigh the negative impacts associated with tourism development, especially for those sites without steady incomes.

For many commentators, however, in spite of the economic benefit derived from religious tourism, there is a perceived incompatibility between religion and economics and between the material and the spiritual (Vukonic 2002; Zaidman 2003). The consumption or indulgence associated with tourism is regarded by some theologians as debasing the spiritual nature of travelling (Vukonic 1998), and because of this some religious groups are very cautious about tourism development, though many site managers encourage visitation to generate funding for maintenance and preservation (Griffin 1994; Stevens 1988; Willis 1994). While religious items and souvenirs have been sold at sacred sites since medieval times (Cohen 1976; Houlihan 2000; Olsen 2003), religious authorities and some academics have shown disapproval of vendors and entrepreneurs who reproduce and sell versions of devotional items and other religious articles, viewing them as tourist trash or religious *kitsch*. As a result, some managers of religious sites try to distance themselves from any trace of commercialization on-site due to their concern about the loss of meaning caused by commercialization of religious rituals and artefacts (Vukonic 1996, 2002).

This paradox is especially acute in the context of Chinese Buddhist sites, for tourism development and the subsequent commercialization phenomenon are contradictory to the Buddhist ascetic principle and need for a secluded environment for meditation and other practices. Concerns have been raised that the widespread commercialization of Buddhist sites are against the precepts of Buddhism as a religion (Tschang 2007). However, supporters for the commercial operation of Buddhist sites argue that Buddhism is to save everyone and not to separate itself from the secular world, and their commitment in secular activities is to achieve this end (Shi 2007). Moreover, it is contended that tourism development is effective in promoting Buddhist culture and boosts its further growth (Yu 2007). Historically, the development of Buddhism in China has seen no absence from material pursuit despite the ascetic monastic ideal. During the Sui and Tang Dynasty (AD 589–AD 907), the heyday of Buddhism in Chinese history, Buddhist monasteries were engaged in farming, trade and moneylending businesses and hoarded precious stones and metals (Ch'en 1973; Jayaram 2000). Chinese monks have always possessed personal belongings, ranging from religious objects like scriptures, to slaves, animals and vast estates (Kieschnick 2003). The variety of possessions has been expanded with regard to the range of businesses in which Buddhist temples engage and the ways in which they obtain donations from lay Buddhists and tourists. Besides operating hostels and vegetarian restaurants, performing religious services and selling incense, some Buddhist sites also derive income by running courses in Buddhism beliefs, have a souvenir trade, hold religious events (Shi 2007) and solicit sponsorship from lay Buddhists for the building of miniature gold Buddhas, halls and roof tiles (Tschang 2007). The tourism development of Buddhist sites certainly facilitates such income-generating activities, as tourists are usually heavy purchasers, especially for hospitality services and souvenirs (Shi 2007).

Shackley (2006) noticed that though it has been an intellectual trend to condemn commercialization at sacred sites (Greenwood 1989; Houlihan 2000; Vassallo 1979), such disapproval usually comes from people unaware of the emotional experience of pilgrims during their visit to the sacred place. While some Buddhist clergy and academics warn against such exploitation of Buddhism, it is necessary to take into account how visitors to Buddhist sites react to the commercial activities present at these sacred places. In particular, understanding how visitors evaluate their experience will aid Buddhist clergy to effectively direct and meet the needs of the *sangha*, and help site managers make more informed decisions on service design, arrangement of visitor facilities, and more importantly, to achieve long-term financial and environmental sustainability of the sites.

WUTAI MOUNTAIN

This study examines the factors considered by visitors during their evaluation of their experience at Chinese Buddhist sites using the case of Wutai Mountain. An interpretive approach was taken, focusing on the views, experience and judgement of the consumer. While most studies on visitor satisfaction use questionnaire surveys as their main data collection method (Jeong and Lee 2006), this research employed qualitative interviews to investigate visitors' subjective experiences and evaluations of their visit. It is believed that people's knowledge, views, understandings, interpretations and experiences are meaningful properties to be studied, and qualitative interviews are more suitable than other methods for obtaining information on these properties from people (Mason 2002). Similarly, Ryan (1995) suggests that the comments of respondents and in-depth interviews can produce a richness of information and feeling about attractions, places and the experience of holidays. Unlike quantitative survey interviews, which require answers to fit into predetermined categories and allow very little room for people to express their own opinions freely, qualitative interviews give the respondents much more freedom and control (Bryman 1988). Considering this research's emphasis on visitors' subjective experience and evaluation, qualitative interviews were considered most appropriate for this study. In total 12 in-depth interviews were conducted with visitors to Wutai Mountain using a convenience sampling strategy. Only Chinese nationals were included in the sample to eliminate the influence caused by cultural differences on visitors' evaluation of their experiences.

As the Interpretivist approach stresses the importance of understanding the overall text of a conversation and seeing meaning in its context (Bryman 1988), observation was used to collect contextual information for this study. Bryman (1988) maintains that a comprehensive understanding can only be obtained by witnessing the context of the event or circumstance to which people refer, which requires the researcher to be there to examine the settings, usually via observation. The following background information about Wutai Mountain was collected by the author either from secondary sources, or based on primary field-work if not stated otherwise.

Wutai Mountain is the most prestigious of the four major sacred Buddhist Mountains in China and is the only Chinese sacred mountain mentioned in Buddhist scripture (China Travel Net 2005). Located in Shanxi Province's Wutai County, the mountain is actually a cluster of five terrace-like peaks, hence the name Wutai (Five Terraces). The peaks are Cuiyan Peak (Peak of Green Rocks) in the middle, Wanghai Peak (Peak Overlooking the Sea) in the east, Guayue Peak (Hanging Moon Peak) in the west, Jinxiu Peak (Splendour Peak) in the south and, the highest, Yedou Peak (Peak of Flourishing Leaves) in the north, standing 3,058 metres above sea-level (Wutaishan Tourism Bureau 2006).

According to the Avatamsaka Sutra, Manjusri, Bodhisattva of Wisdom, once resided and preached Buddhism with ten thousand other Bodhisattvas in Wutai Mountain, and that is why every temple on Wutai Mountain holds at least one statue of Manjusri. Ever since the Southern and Northern Dynasties (AD 420–AD 589), Buddhists from China and all over the world have come to Wutai Mountain on pilgrimage to pay homage to Manjusri (Wutaishan Tourism Bureau 2006). Currently about 3,000 Buddhist monks and nuns live in the temples and monasteries in Wutai Mountain (Wutai Mountain International Cultural Festival of Buddhism Organising Committee 2004).

The first temple in Wutai Mountain was built during the Eastern Han Dynasty (AD 25–AD 220). Since then there have been ups and downs in the number of temples built in the mountain in accordance with the rise and fall of Buddhism in China. There were once about 200 temples during its prime in the period of the Northern Qi Dynasty (AD 550–AD 577). The number reached more than 300 during the Wanli period (AD 1573–AD 1620) of the Ming Dynasty. After three official persecutions of Buddhism and the Cultural Revolution, only 43 temples survive in Wutai Mountain today (Lai 2001). Most of the temples are located in Taihuai Town, which is situated in the middle of the five peaks; others are farther afield in the mountains (Wutaishan Tourism Bureau 2006). The common features shared by the temples in Wutai Mountain are that they are of historic as well as religious significance thanks to the various types of cultural heritage embedded in them. In addition, the beautiful natural landscape of Wutai Mountain provides a pleasant environment for these temples, adding to its attraction as a tourist destination. The Mountain is on the list of National Cultural Heritage and the tentative list of UNESCO's World Heritage Sites (UNESCO World Heritage Centre 2005).

Owing to its rich cultural and religious resources, Wutai Mountain has been vigorously promoted as a cultural tourist attraction as well as a Buddhist sacred site. For example, the Wutai Mountain International Tourism Month lasting from early August to September has been held annually since 1989 (Shanxi Tourism Bureau 2002) and the Wutai Mountain International Cultural Festival of Buddhism since 2004 (Wutai Mountain International Cultural Festival of Buddhism Organising Committee 2004). The effects of such promotion are twofold. On the one hand, they help to attract more visitors and bring more income to the local people and the temples. On the other hand, due to tourism development, the urbanization and commercialization of the central area of Taihuai Town has become more and more marked. Zhang (1999) contends that the original appearance of Wutai Mountain has been changed and the sacred landscape adversely altered by the commercial atmosphere. It is reported that this problem was acknowledged by the local government and an Outline of the General Plan for Wutai Mountain Landscape District has been developed to improve the situation (Gao and Jiao 2004).

The town of Taihuai is the entrance to the Wutai Mountain Landscape District. An admission fee of 90 RMB Yuan is charged for visitors to enter the District. In addition some individual temples also charge entrance fees of three or four RMB Yuan. Located about 120 kilometres north of the capital city of Taiyuan, Wutai Mountain can only be reached by highway transportation. No trains or planes can arrive directly at Wutai Mountain. Tourists can go to Taiyuan or Datong by plane or take a train from the Beijing–Taiyuan Line and alight at Wutai, Shahe, Fanzhi, Daixian, Yuanping, Xinzhou or Taiyuan for bus transfer to Wutai Mountain. Direct buses go to the Mountain every day from Taiyuan Railway Station and the Bus Terminal, running every 30 minutes; transportation in Taihuai Town itself is mainly by taxis and middle bus. Cars are available for hire in Taihuai Town for visitors who want to go to the peaks of Wutai Mountain, the price of which is negotiable depending on the distance.

The accommodation facilities in Wutai Mountain are concentrated in Taihuai Town. There are many family-run low-cost hostels, with prices starting from 10 RMB per person per night. There are also larger scale hotels that provide better facilities and services, including three three-star hotels and nine two-star hotels, charging between 120 and 1000 RMB Yuan per room per night. Because the temperature in Wutai Mountain is never too high, very few hotels are equipped with air conditioners.

Tourist information is adequate in Taihuai Town as far as domestic tourists are concerned, for tourist maps are available from most convenience shops and travel agencies are easily accessible thanks to their location in the central area of Taihuai Town. There are 17 travel agencies in total (Wutais 2005), with only one dealing with international visitors. Most of the signs for visitors in Wutai Mountain are written in Chinese, and few foreign visitors were seen in Wutai Mountain in September 2004 when the author was there for field-work.

There is a strong atmosphere of commerce in the central area of Taihuai Town. The main street, Taihuai Street, is lined with restaurants, shops and family-run hostels. Yanglin Street is a specialized commercial street with rows of stalls selling local products and various souvenirs, such as local mushrooms, ink stones, accessories, prayer beads and small Buddhist articles catering for tourists' needs. In general, the goods are inexpensive and not of the best quality (see Figure 12.1).

The sense of commercialization is not lessened as one approaches the temples. Along the path to the cluster of temples in the central Taihuai area stand quite a few beautifully decorated shops built in the style of the Qing Dynasty (see Figure 12.2). Their names end with 'Ge', literally meaning a place for worshipping the Buddha, but used here probably to indicate that they only sell 'decent' religious articles of Buddhism, which are different from the *kitsch* sold by the street vendors outside. The articles they sell include prayer beads, talismans, incense, Buddhist scriptures, clothing, ritual articles and statues and figurines of Buddhas and Bodhisattvas of different

Figure 12.1 Souvenir stalls in Yanglin Street, Taihuai Town.

Figure 12.2 Foyou Ge, a shop selling Buddhist articles in Taihuai Town.

sizes and materials, the prices of which are usually preset and can seldom be negotiated.

There are countless street hawkers with their mobile stalls scattered along the paths, trying to attract the attention of visitors by showing their goods and offering a low price. The goods they sell are mostly cheap mass-produced souvenirs with little or no relevance to Buddhism. Different from them, the stalls clustered near the entrance to the temples mainly sell incense. The vendors would warn the pass-by visitors that the incense sold in the temple is unreasonably overpriced and ask the visitors to buy it from them at a fair price.

Indeed, it is true that the incense sold in some of the temples is much more expensive than that sold by the outside vendors. The author was charged 199 RMB Yuan for 10 bunches of incense, which was available for no more than 19 RMB Yuan from the vendors outside the temples. The incense was sold by people who are dressed like monks (but were not necessarily real monks). It was said that the purchase of incense at the temple was considered a donation and the purchaser's name would be recorded in the donation book. Adjacent to the temples, there are some shops claiming to be operated by lay Buddhists, which mainly sell incense and Buddhist publications. The price of incense sold there is between that of the street vendors and the monks in the temples.

There seemed to be some hidden deals between some tour guides and the temples, as many tour guides were heard to tell visitors that only the incense sold in the temple is authentic and can be received by the Buddhas and lets them hear the prayers. The incense sold outside the temple was said to be inauthentic and of no help to the prayers. While such words sound ridiculous to knowledgeable Buddhists, it took in a lot of innocent laypeople who were in awe of the religion but knew little about it.

EVALUATION OF VISITOR EXPERIENCE
AT WUTAI MOUNTAIN

An investigation into visitors' reflection on their experience at Wutai Mountain revealed that a variety of factors were taken into account during their visit. These can be categorized into four groups, i.e. people, place, facilities and critical events. The elements in the 'people' group were mainly related to the presence of other tourists and the host community, which included front-line service staff and Buddhist monks and nuns in the temples. Specifically, the presence of a large number of tourists was a cause for complaint because they were found to cause a sense of chaos while competing for space and resources. This is in accordance with Graefe and Vaske's (1987) opinion that usage levels at a visitor attraction can affect perceptions of quality, as increased numbers of contacts lead to increased perceptions of crowding and hence increased dissatisfaction. Moreover, there were also reports about the profit obsession of some service staff, who persistently attempted to charge extra money, and about dubious monks, who coaxed worshippers into buying incense for exorbitant prices by claiming it possessed extraordinary power. It has been suggested that the interaction between salespeople and customers is a vital component of product delivery and thus influences customers' consumption motives and satisfaction (Chang, Yang and Yu 2006), and that the human element of the service interface, especially the attitude and behaviour of the service provider, is one of the essential aspects of quality management in tourism (Witt and Mühlemann 1994). While the respondents had expected honesty and sincerity at the Buddhist site, unethical acts like these led to distrust and distress on the part of visitors. In addition, some Buddhists were disappointed to find out that there were few opportunities to meet and discuss Buddhist teachings with knowledgeable clergy. Since many visitors visit religious sites as an opportunity for educational experiences (Winter and Gasson 1996), the management of religious sites should consider the needs of this group of visitors and provide suitable services accordingly.

The place factors considered by the respondents included natural environment, atmosphere, security, admission and access. The natural landscapes were part of the attraction of Wutai Mountain, which appealed to many nature-loving tourists. Cool all year round, Wutai Mountain is a place where one can escape the heat of a hot summer. Additionally, its height protects it from the polluted air experienced in the lower plains of Shanxi Province. The beautiful landscapes and the pleasant climate were appreciated by tourists as well as pilgrims to Wutai Mountain. Yet unlike the much-favoured natural environment, the artificial setting received criticism for being over-commercialized. Dissatisfaction was reported because of the lack of religious ambience and the apparent commercialization.

The impact of environment and atmosphere on visitor experience has been discussed extensively in the field of environmental psychology related to customer behaviour studies. Ward and Russell (1981) make the point that environmental factors impinge upon the visitor experience and sense of a destination. With reference to Chinese Buddhists sites, the sense of such destinations is traditionally related to serenity, as most Buddhist sites in China were specifically built in quiet locations away from busy commercial areas to separate themselves from the earthly life and for the ease of meditation. Their tranquil environment is valued by visitors who seek refuge from the hustle and bustle of modern life. Aside from the natural environment, Bitner (1992) notes the effect of the built environment, specifically the man-made contextual surroundings, or servicescape, on consumers' satisfaction with the service experience. It is suggested that the servicescape has a strong potential to elicit emotional and subjective reactions (Wakefield and Blodgett 1994) and hence its influence on visitors' evaluation of their experience. Prabhu (1993) indicates that an essential condition to enhance visitor experience at sacred sites is an atmosphere of silence and recollection which facilitates worship.

The issue of security was mainly associated with the mountaineering activity involved in the visit to the sacred mountain. Another problem identified was related to admission and access, as there was concern that the entrance fee was too expensive (90 RMB Yuan) for some poor pilgrims and deprived them of the opportunity to visit the holy mountain, which was perceived as being unfair and against Buddhist teachings. It can be seen that visitors, especially Buddhists, did bear in mind religious principles in their evaluation of the Buddhist site.

Facilities were another group of factors considered by the respondents during their evaluation of the visit experience, including transportation, food, accommodation, parking and shopping. Parking was brought up mainly because half of the respondents came by car. It was found that the respondents attached different levels of importance to different factors depending on their main purpose of visit. The most obvious distinction was found between pilgrims and leisure tourists. Pilgrims were preoccupied with completing the journey and treasured the opportunity to communicate with Buddhist masters. They did not care much about material comfort such as food and accommodation, as to them the journey was essentially about spirituality and they were ready to overcome physical hardships. In contrast, leisure tourists regarded visitor facilities and service quality as vital to their trips.

Another type of factor that had significant impact on the respondents' judgement of their experience were critical events, which could be related to one or more of the factors mentioned previously, but as they left such a strong impression on the respondents' mind they became the stories that had to be told probably every time their trip to Wutai Mountain was mentioned. Critical incidents had a major influence on the respondents' evaluation of their experience, as they appeared to be the primary factor that changed the mood of the respondents during their experience and set the tone of the report. It was noticed that all of the critical incidents reported were direct personal experiences and the relevant effect, either positive or negative, on the respondents was in accordance with their overall evaluation of their entire experience. One example was a young lady's haggling experience, when she was initially asked for 480 RMB for a box of incense and was later offered one for 60 RMB. She felt she was being cheated and was so disappointed that she immediately left Wutai Mountain on that day. A positive example was from a Buddhist nun, who was happily impressed and described vividly a grand Buddhist event she attended—the Feast for a Thousand Monks. She said it was quite a spectacle and no other event elsewhere could compare with it.

Critical incidents usually emanate from the interaction of people, place and facility factors mentioned earlier; they induce strong emotional responses that may redirect thoughts and actions. The impact of critical incidents may be understood when considering the function of the mood owing to the emotional reactions involved. A primary feature of critical incidents is their ability to engender intense emotions, such as anger, annoyance or excitement, depending on the negative or positive nature of these events (Cope 2003). In other words, the occurrence of critical events can result in the change of visitors' psychological conditions, in particular, mood, which may have a pervasive influence on the entire subjective experience (Isen 1984). This could be illustrated by the case of the young lady mentioned earlier.

EXPECTATION AND SATISFACTION

Expectation is regarded as an essential element in experience evaluation by many researchers, where satisfaction outcome is believed to be the result of a comparison between expectation and the actual experience (Bolton and Drew 1991; Chon 1989; Martilla and James 1977; Oliver 1980; Peterson 1974). In this study, it was observed that the respondents did not evaluate their experience precisely against their expectations, but dissatisfaction did occur if the respondents' belief in the essential religious characteristics of Buddhist sites was upset by reality. The latter case could be illustrated by the negative feelings in some respondents caused by the over-commercialized environment and dishonest business behaviours, which were against the religious nature of the site and Buddhist teaching in honesty and denouncement of worldly pursuits. While such a situation lent partial support to the expectancy–disconfirmation theory (Bolton and Drew 1991; Oliver 1980), demonstrating that negative disconfirmation of expectation could lead to dissatisfaction, there was also evidence that challenged the efficacy of this theory.

First, contrary to the claim that positive or zero disconfirmation of expectation indicates satisfaction (Bolton and Drew 1991; Oliver 1980), it may lead to dissatisfaction if the expectation is considered unfavourable in nature. This weakness of the expectancy–disconfirmation theory has been noticed by La Tour and Peat (1979), who raised the point that the theory suggests that if a consumer expects and receives poor performance, he or she will be satisfied. A respondent in this study who had previously visited Wutai Mountain could illustrate such a case. He said that he was aware of the overpricing phenomenon and the profit obsession of some local people, and never failed to be upset by them on each visit. In other words, he had expected the undesirable situation, and his subsequent unpleasant encounters indeed formed positive confirmation of his expectations, but what he had was nevertheless an unhappy experience.

Another problem with the expectancy–disconfirmation theory and all expectation-related satisfaction theories/models, such as importance–performance theory (Martilla and James 1977), SERVQUAL (Parasuraman, Zeithaml and Berry 1985) and HOLSAT (Tribe and Snaith 1998), are with the existence of expectation. It was found in this study that not necessarily every visitor had expectations prior to his/her visit—some regular visitors expressed that they were too familiar with Wutai Mountain to hold any new image prior to this visit, and some first-time visitors simply did not have any specific expectation. Arnould and Price (1993) and Laws (2002), who suggest that travellers' expectations are often vague and non-existent, especially in unfamiliar and unpredictable travel encounters, have noticed this lack of defined expectation. For those who do not have any expectations the expectancy–disconfirmation theory loses its explanatory power, as there are no expectations to compare with to assess the actual experience.

La Tour and Peat (1979) suggest that consumers' evaluations of product attributes themselves may account for more of the variability in satisfaction than would the confirmation or disconfirmation of expectations about those attributes. Oliver (1993) defines attribute satisfaction as consumers' subjective satisfaction evaluation based on observations of attribute performance, and claims that attribute satisfaction has both direct and indirect effects on overall satisfaction. Moreover, Petrick and Backman (2002) remark that attribute satisfaction is an antecedent to overall satisfaction, which will not be high if the visitor is disappointed with a particular attribute performance. In this study it was found that respondents did reflect on the performance of a number of factors, or attributes, when they were asked to evaluate the visit experience, but the weight attached to these factors varied between respondents because primary concerns were usually linked to their main purpose of visit. For instance, pilgrims who concentrated on the spiritual experience could still have an overall satisfying visit even though they were not particularly pleased with the hospitality service; in contrast, leisure tourists regarded visitor facilities and service quality as important to their trips, but did not care so much if there was insufficient space for meditation in the temples. Such differentiation renders Petrick and Backman's (2002) view subject to further scrutiny in that not all attribute satisfaction is an antecedent to overall satisfaction.

VIEWS ON COMMERCIALIZATION

As much as governments and religious clergy differ in their views on the tourism development of sacred sites, the respondents also varied with respect to their opinions on the commercialization present at Wutai Mountain. One group generally accepted it with different levels of tolerance, and another showed strong resentment. Those who accepted the commercialization agreed that it was a common phenomenon at most tourist attractions and understood that it was a source of income for the local people, and that it gratified the needs of some visitors. As to more detailed attitudes, some were open-minded and took pleasure in the lively leisure atmosphere; some thought it was all right as long as it was separated from the temples; some thought it helped to create interest in and introduce Buddhism to laypeople; and some were confused by the mixture of the sacred and the secular.

On the other hand, the remaining respondents disapproved of the commercialization for various reasons. The primary reasons were that it was incompatible with the religious atmosphere of the Buddhist site and the secular trade of religious souvenirs was disrespectful to the Buddha. Comments were given that the commercialization was a sign of Wutai Mountain's assimilation with other tourist attractions and that the sacred mountain was losing its own features. Moreover, it was pointed out that the commercialization was worsened by the haggling process involved in souvenir sales and by lack of proper management. Haggling usually started when the retailer offered an irrationally high price at first, claiming a certain special quality for the item (or its possession of magical power if it was a religious souvenir), and ended when the seller agreed to sell the item at a fraction of the initial offer. Such things even happened in some temples where people dressed like monks made initial offers and were in charge of the sale. Some respondents advised that the retail trade at Wutai Mountain was in urgent need of regulation and proper management regarding its location, scale and operation to let it be more properly integrated with the religious atmosphere of the sacred mountain.

An analysis of respondents' attitudes towards commercialization at Buddhist sites revealed that their opinions were possibly related to their travel experience, age, religious background and level of Buddhist knowledge. It was noticed that experienced travellers were less likely to be shocked by the commercialization than the inexperienced travellers, as the former were used to it and saw it as a common phenomenon. At the same time, older respondents showed more tolerant attitudes towards commercialization at Buddhist sites than younger respondents, probably because they had more life experience and were more tolerant of even unfavourable conditions. In addition, Buddhist respondents generally resented the commercialization, whereas non-religious participants did not show as much concern. However, even among those of the Buddhist faith differences could be found. More senior Buddhists were disposed to distinguish things by their purpose and nature, while lay respondents and Buddhist noviciates were apt to judge by immediate impressions. Ross and Iso-Ahola (1991) suggest that individuals' levels of cognition, or understanding of the true state of a phenomenon (Johnston, Gregory and Smith 1988), play a significant role in their evaluation of visitor experiences. So, it can be argued that the level of knowledge about Buddhism may explain these attitudinal differences between senior Buddhists and their junior counterparts.

CONCLUSION AND RECOMMENDATIONS

The commercialization resulting from tourism development at Wutai Mountain is not unique to Chinese Buddhist sites. Buddhist sites in other countries, such as Burma, Bhutan and India, have also experienced a similar commercialization process (Hall 2006; Joseph and Kavoori 2001; Orland and Belladiore 1990; Philip and Mercer 1999; Simpson 1993). The commercialization of these sacred sites is largely the result of national and regional governments' influence, who, with their economic mandate to increase tourism revenues, make great efforts to increase visitation and transform the religious sites into tourist attractions by active tourism promotion (Olsen 2006) and policy influence (Tschang 2007).

Using the case of Wutai Mountain, this qualitative research explores how visitors to Chinese Buddhist sites evaluate their visit experience. A range of evaluative factors was considered by the respondents and could be placed into four categories, i.e. people, place, facilities and critical incidents. It was found that the relative importance of these factors was related to their main purpose of visit, with the most obvious distinction found between pilgrims and leisure tourists. It was also noticed that many respondents had certain assumptions with respect to the essential characteristics of Buddhist sites, such as there should be a holy ambience and genuine local people, and it was found that followers of Buddhism tended to apply Buddhist principles during their evaluation. With reference to the degree of commercialization experienced at Wutai Mountain, respondents showed contrasting attitudes, where one group accepted it with different levels of tolerance and the other expressed strong disapproval for its inappropriateness and lack of proper management.

The findings have practical implications for marketing, management, service design, planning and development of Chinese Buddhist sites. First, pilgrims and secular tourists valued different things. With this in mind, segmentation based on a visitor's main purpose of visit can be undertaken. As pilgrims do not require high levels of material comfort, economical accommodation and catering services should target this group of visitors. In addition, since most pious Buddhists are vegetarian, it is advisable that catering businesses offer special vegetarian dishes to target this group of religious consumers. On the other hand, in order to target leisure tourists who demand higher levels of material comfort, tourism ancillary businesses can collaborate to raise awareness among the potential customers. For example, restaurants and travel agencies can place leaflets in hotels explaining the nature of the destination.

Second, as pilgrims appreciate the opportunity to communicate with well-informed Buddhist masters, the Buddhist sites may consider appointing certain times of the day, or certain days of the week to offer pilgrims and interested visitors the chance to meet and talk with senior Buddhist clergy. Third, overpricing should be avoided in order not to induce feelings of being cheated or having poor value for money. The regulation of prices can be achieved by offering price guides to retailers and visitors. Fourth, the scale and location of retail businesses should be controlled and relocated in such a manner that they will not create a commercial atmosphere that overwhelms the religious ambience of Buddhist sites. This issue should also

13 Brochures, World Heritage Sites in China and Responsible Tourism

The Tourism Services Perspective

Rong Huang and Danqing Liu

INTRODUCTION

In recent years the tourism industry has seen phenomenal growth in the movement of people to numerous destinations and regions, both in China and also elsewhere. According to the World Tourism Organization's (WTO) preliminary results of year 2005 (WTO 2006), the number of worldwide international tourist arrivals grew by 5.5 per cent and exceeded 800 million for the first time ever. The China National Tourism Administration (CNTA) stated that in 2005 China received around 120 million international tourists who spent US$292 billion, while domestic Chinese tourists spent 5,286 billion RMB (equivalent to US$658 billion) in the same year (CNTA 2006).

While the vast majority of travellers still opt for the conventional 'mass' experience that is best characterized by destinations fulfilling the 'three S'– type experiences of sun, sand and sea, there has also been a tremendous increase in travel to destinations having a 'cultural' and/or 'natural' basis; especially those designated as World Heritage Sites (WHS) (Leask and Fyall 2000). It was in 1987 that China started to submit applications for places to be included on the UNESCO World Heritage Site list. By July 2005 China had had 31 sites accepted on to the WHS list out of a worldwide total of 812 WHS (in 137 countries), giving China the third highest number of WHS in the world (UNESCO 2005).

However because of the increased numbers of visitors, some of these sensitive communities and regions have come under increasing pressure and are therefore vulnerable to change and potential damage to their unique qualities (WTO 2003). Historical evidence seems to suggest that the growth in tourism will continue and therefore it is imperative for these WHS to be proactive in order to maximize tourism opportunities while at the same time maintaining their integrity. Tools to ameliorate the impacts of tourism have included codes of responsible conduct aimed at tourists, largely formulated by host destinations and non-governmental organisations such as the WTO (2003). However, many academic researchers argue that the process of awareness and education should also be a responsibility of the tour operators (Goodwin and Francis 2003; Pennington-Gray *et al.* 2005).

Kotler, Bowen and Makens (2006) argue that tourism brochures are one of the more important tools that tour operators use when promoting their products and services. Indeed, Yamamoto and Gill (1999) claim that the travel brochure is the most important source of information used by tourists when planning their travel. Hence, it can be argued that tourism brochures have an extremely important potential role in educating tourists about behaving responsibly when visiting WHS.

This study investigates the extent to which Chinese travel service companies promote responsible behaviours among tourists who visit Chinese WHS. More specifically, the objective is to examine whether the tourism brochures provided by Chinese travel service companies contain any information that enhances responsible tourism at these sites. China is chosen because it has the biggest domestic tourism market in the world, and its domestic tourism markets have grown so dramatically in a very short time (Xiao 2003). According to CNTA (2006), in 2005 tourism attractions in China received 1.2 billion visits from domestic tourists. At the same time it might be argued that many of these tourists are relatively inexperienced as tourists and perhaps have little concept of the possible results that can arise from their visits to destinations, particularly at heritage sites.

RESPONSIBLE TOURISM

Harrison and Husbands (1996) argue that responsible tourism represents an approach to tourism planning, policy and development that ensures benefits are optionally distributed among impacted populations, governments, tourists and investors. Goodwin and Francis (2003) agree with this and comment that responsible tourism is a management strategy that embraces planning, management, product development and marketing to bring about positive economic, social, cultural and environment impacts. The WTO emphasizes that these responsibilities are held by a combination of stakeholders; such as government agencies, employees of the tourism industry, employees outside of the tourism industry and members of local communities. Recently, different rules of conduct have been created in order to generate awareness and compliance of potential visitor behaviour (Dubois 2001; Sirakaya 1997; International Ecotourism Society 2004). Many countries have recognized the importance of developing guidelines for responsible tourism behaviours and have developed rules concerning how to act appropriately in culturally and environmentally sensitive regions. As a specialized agency for the United Nations and a global forum for tourism policy issues and practical sources for tourism development, the WTO (2003) issued what is perhaps the best-known reference for the responsible and sustainable development of world tourism, its *Global Code of Ethics for Tourism*. This Code covers 10 articles:

1. Tourism as a contribution to mutual understanding and respect between peoples and societies.
2. Tourism as a vehicle for individual and collective fulfilment.
3. Tourism as a factor for sustainable development.
4. Tourism as a user of the cultural heritage of mankind and contributor to its enhancement.
5. Tourism as a beneficial activity for host countries and communities.
6. The obligations of stakeholders (government, companies, local communities) in tourism development.
7. The right to tourism.
8. The liberty of tourism movements.
9. The rights of the workers and entrepreneurs in the tourism industry.
10. Implementation of the principles of the *Global Code of Ethics for Tourism*.

In particular, articles 1 and 3 of the *Global Code of Ethics for Tourism* emphasize the importance of responsible tourism. Article 1 highlights ethical values common to humanity, and an attitude of tolerance and respect. In addition, it focuses on the responsibility of tourists to educate themselves about the countries they plan to visit. Article 3 emphasizes that all stakeholders in the tourism industry should safeguard the natural environment. It requires tourists to preserve endangered species of wildlife and that the stakeholders in tourism development, especially the professionals, should agree to the imposition of limitations or constraints on their activities when these are exercised in particularly sensitive areas. Both authors of this research argue that the Code should guide the behaviour of tourists when they travel to natural and cultural heritage sites, especially WHS, even those tourists who are visiting such sites in their own country.

According to the *Handbook for Tourism Operators* (2002) produced by the International Centre for Responsible Tourism, responsible tourism is about providing more rewarding holiday experiences for guests, whilst at the same time enabling local communities to enjoy a better quality of life and conserving the natural environment. Therefore, according to the Code, the authors of this research argue that educating tourists about responsible behaviour while they are visiting WHS is also the responsibility of tour companies and guides.

TOURISM BROCHURES AS A MEANS OF COMMUNICATION

According to Kotler *et al.* (2006), tourism brochures typically include information on prices, itineraries and company information; and the brochures' objectives are to create attention, interest and action. Thus Wicks and Schuett (1991) argue that tourism brochures are an attractive information source for many decision-makers, while Holloway (2004) notes that a tourism brochure is one of the most widely utilized information sources by tourists when they look for holiday destinations. As far as Voluntary Service Overseas (VSO) (cited in Goodwin and Francis 2003) is concerned, the tourism brochure should provide 'customers with more information about the people and the places they will be visiting, including advice on how they can visit locally owned facilities and resorts'. However, to date, information on tourism brochures which include such details and advice for tourists is not well documented; the one exception being Pennington-Gray *et al.* (2005) with regard to US tour operators' brochure design.

WORLD HERITAGE SITES IN CHINA

It is a cliché that China is one of the oldest civilizations in the world and has diverse and beautiful natural and geographical conditions. These rich and precious cultural and natural heritages of China benefit both the Chinese nation and the rest of the world. China and other nations of the world are trying to find ways to protect internationally significant heritage sites. In 1985, China acceded to the Convention Concerning the Protection of the World Cultural and Natural Heritages, also known as the World Heritage Convention. Between 1987 and 2005, 31 sites in China were entered on to the list of WHS (UNESCO 2006). These 31 sites are detailed in Table 13.1.

In China, CNTA is responsible for developing, promoting and regulating China's tourism industry. CNTA continues to make significant efforts in environmental protection and sustainable development. In early November 2000, CNTA and the UN jointly sponsored the Symposium on Sustainable Development of China Tourism. Representatives from the tourist organizations of China, Thailand, Malaysia, Sri Lanka, Nepal and Hong Kong spoke at the meeting on topics that included 'The Strategy and Principle for Sustainable Development of Tourism', 'Regional Eco-Tour Development and Management' and the 'Construction and Popularization of Non-Obstacle Tourist Facilities' (CNTA 2007). At the same time, CNTA made it clear that during the 10th Five-Year Plan period (2001–2005), China's tourism would continue to persist in its sustainable development strategy by carrying out green development, producing green products, popularizing green management and building green systems.

Unfortunately, although CNTA has a positive management philosophy, its ideas have not yet benefited many WHS in China as is evidenced by examples cited elsewhere in this book. These sites of cultural and natural heritage are threatened with increasing damage, caused by their own antiquity and degeneration and also by changes in social and economic conditions. A number of Chinese newspapers have carried reports about WHS in China and the condition in which they find themselves. A common theme from these reports is that the WHS in China are in danger because of overdevelopment

Table 13.1 World Heritage Sites in China

Type	Name	Year
Cultural	Peking man site at Zhoukoudian	1987
(23)	Mausoleum of the First Qin Emperor	1987
	The Great Wall	1987
	Mogao Caves	1987
	Imperial palaces of the Ming and Qing Dynasties in Beijing and Shenyan	1987
	Ancient building complex in the Wudang Mountains	1994
	Historic ensemble of the Potala Palace, Lhasa	1994
	Temple and cemetery of Confucius and the Kong family mansion in Qufu	1994
	Mountain resort and its Outlying Temples, Chengde	1994
	Lushan National Park	1996
	Old town of Lijiang	1997
	Ancient city of Ping Yao	1997
	Classical gardens of Suzhou	1997
	Dazu Rock Carvings	1997
	Summer Palace, an imperial garden in Beijing	1998
	Temple of heaven, an imperial sacrificial altar in Beijing	1998
	Ancient villages in southern Anhui-xidi and Hongcun	2000
	Imperial tombs of the Ming and Qing Dynasties	2000
	Longmen Grottoes	2000
	Mount Qingcheng and the Dujiangyan Irrigation system	2000
	Yungang Grottoes	2001
	Capital cities and tombs of the ancient koguryo Kingdom	2004
	Historic centre of Macao	2005
Natural	Huanglong scenic and historic interest area	1992
(4)	Jiuzhaigou valley scenic and historic interest area	1992
	Wulingyuan scenic and historic interest area	1992
	Three parallel rives of Yunnan protected areas	2003
Mixed	Mount Taisha	1987
(4)	Mount Huangsha	1990
	Mount Emei scenic area, including Lesha Giant Buddha scenic area	1996
	Mount Wuyi	1999

Source: UNESCO (2006).

by local government and developers, or the irresponsible behaviour of tourists (for example, see Wang, Xu and Chen 2003b; Wang 2003).

THE STUDY

Visits were made to the head offices of China International Travel Services (CITS), China CYTS Tours Holdings and China Travel Services (CTS), and also to some of their branches in Beijing and Tianjin. These three companies are among the largest and more popular travel service companies in China. Due to limited time and manpower, a convenience sampling method was used. The selected travel service companies were personally visited from January to April 2006 and asked for their tourism brochures containing information about WHS in China. The researcher probed the employees of the travel service companies about responsible behaviour at WHS by asking, 'I am very keen to visit the World Heritage Sites in China, do you have any material that I can have a look at?' The researcher also asked supplementary questions such as, 'As they are World Heritage Sites, do you have any material which will tell tourists about how to behave whilst they are at the sites?'

In total, 195 different tourism brochures were collected that contained information about WHS. However, with reference to printed material that specifically dealt with rules relating to the behaviour of tourists whilst at these sites, most staff contacted by the researchers reported that they did not possess such material. They often suggested that their tourism brochures sometimes included this aspect of the information and gave recommendations relating to behaviour at the sites. Six receptionists further recommended that the researchers read some guide-books or visit the web site of the CNTA for information on this matter. Four receptionists from different branches of CITS suggested that the researcher visit only their own company's websites. However, for the purpose of this study that concentrates on printed brochures, no analysis was undertaken of these alternative sources of information. One justification is that while Internet usage is growing fast in China, and in total the Chinese now number among the most numerous users of the Internet, within the country itself it is still only a minority who frequently access the net. In addition, excluding these sources maintained a clear focus on tourism brochures from the tourism service providers themselves.

The collected tourism brochures were then examined using a content analysis technique. McNeil (1990: 112) defines content analysis as 'a method of analysing the contents of documents or other non-statistical material in such a way that it is possible to make statistical comparisons between them'. On the other hand, Krippendorff (1980: 21) is critical of such an approach to content analysis based on the simple counting of a frequency of qualities, be they words, attributes or whatever. He defines content analysis as

Table 13.2 Key Concepts Used to
Examine the Tourism Brochures

Key Concepts	
1	Respect for diversity
2	Religion
3	Moral beliefs
4	Social and cultural traditions
5	Natural environment
6	Local environment
7	Cultural understanding
8	Natural or cultural heritage
9	Rules and guidelines
10	Ethical values

'a research technique for making replicable and valid inferences from data to their context'. In the field of tourism there is a growing body of content analysis research, for example, the use of postcards (Albers and James 1988) and also promotional literature such as destination brochures (for example, see Pritchard and Morgan 1996; Dann 1996).

The contents of the tourism brochures were analysed for responsible tourism practices guided by the principles of sustainable tourism, and numerous 'key' words that are identified in the *Global Code of Ethics for Tourism* (WTO 2003: Article 1 and 3) were used to examine the contents of the brochures (see Table 13.2 for details).

In the 195 tourism brochures, a total of 325 references with regard to the key concepts in Table 13.2 were identified in the printed material. All references to responsible tourism were then coded.

FINDINGS

When examining the collected tourism brochures, it was noted that they all included information about the travel service company, normally including telephone and fax numbers for different branches, and also the web site for that company. However, this part of the information is primarily descriptive, and no mention is made about companies' policies or personnel strengths, such as strength in tour guides' ability and experience. This finding accords with research by Pennington-Gray *et al.* (2005) about travel brochures provided by US tour operators. However, according to Davis (1994), the corporate advertisement should include information about how the company is involved in and committed to the environment. Except for the information about the companies, the other information in the brochures can be divided into two sections: destination information, mostly including a brief introduction to travel itineraries, and tourist information, usually giving detailed explanations about pricing structures and safety measures.

Therefore, the authors divided all the collected information into two sections: one entitled 'destination information' and the other 'tourist information'. Three themes emerge when examining the tourism brochures: two from the destination information section and one from the tourist information section.

Section One: Destination Information

This section of the tourism brochures deals with information about the WHS to be visited. It is very descriptive and its purpose is to present the travel itinerary and provoke interest in the trip. The dominant themes that emerge from the brochures' 'destination information; are labelled 'world heritage' (Theme 1) and 'local traditions' (Theme 2).

Theme 1: World Heritage

Within the brochures the 'world heritage' theme is mentioned 134 times. This total of 134 is comprised of the world heritage sub-theme of 'cultural heritage' which is mentioned 98 times, and the world heritage sub-theme of 'natural heritages' which is mentioned 46 times. Where the sub-theme 'cultural heritages' is mentioned, it covers almost all the cultural WHS in China. The three most frequently mentioned cultural heritage sites are the 'old town of Lijiang' (32 times), 'The Great Wall' (28 times) and the 'Forbidden City' (Imperial Palace of the Ming and the Qing Dynasties in Beijing) (20 times). However, there are no mentions of the 'historic centre of Macao' and the 'ancient city of Ping Yao'. The 'historic centre of Macao' has only recently been listed as a WHS, also it is obviously in Macao and thus often not considered to be part of mainland China, even though Macao is now controlled by the Chinese government. Therefore, the fact that the 'historic centre of Macao' WHS is not mentioned in any of the brochures is understandable. However, the fact that the 'ancient city of Ping Yao' is not mentioned in any of the 195 tourism brochures sends an important message that this site is either unpopular or suffers from a lack of promotion, or there is some other reason for its non-appearance. It is worthwhile for relevant authorities or companies to undertake further research into this fact. Under the sub-theme of 'natural heritages': All of the nature-based

Figure 13.1 Tourist poses for photograph at Stone Forest National Park.

WHS within China are mentioned, particularly the 'Wulingyuan scenic and historic interest area'.

However, although a great deal of information is provided about specific WHS, there are no guidelines provided about how to behave responsibly at these culturally or environmentally sensitive sites. Consequently, behaviour of the type illustrated in Figure 13.1, spotted at Stone Forest National Park in Yunnan Province, might at least be given a second thought.

Theme 2: Local Traditions

This theme appears in 158 tourism brochures. Within this theme two sub-themes emerge, 'food' and 'dance'. Chinese people treat food as one of the most important aspects of their lives; therefore, local 'speciality' foods are emphasized in most of the brochures. Food from ethnic minorities and also the drink, 'tea', are the most frequently mentioned items. This might be because most potential tourists are from Han ethnic backgrounds, who comprise 91.96 per cent of the country's total population (CNTA 2006). The Han people have a long history of drinking and appreciating tea. Furthermore, different areas in China produce different speciality brands of tea, and thus one can understand the frequency with which tea is mentioned as a promotional item in the brochures.

Dances performed by ethnic minorities living in the WHS are also frequently discussed. The peoples of the ethnic minorities generally have different cultural practices than those of the Han majority. Consequently, these differences between the ethnic minorities and the Han are frequently used as selling points by the tourism industry in China. For example, a tourism brochure about Yunnan Province includes a detailed introduction to the tea and dance traditions of the Bai people. However, yet again in all the collected brochures there is no mention about culturally responsible behaviour in the WHS.

Overall, in terms of destination information, although the brochures contain some beautiful descriptions of the different WHS in China, and the travel companies emphasize the uniqueness of these sites, there is not a single mention about the tourists having a corresponding responsibility to behave in a manner that protects these unique WHS.

Section Two: Tourist Information

This section of the analysis dealt with the presentation of general travel information for the prospective tourists. Different travel service companies use different emphases when describing information relevant to this section, but normally include information relating to service standards and price packages. However all of the tourism brochures from the CTS (45 brochures in total), contain some reminders about local weather, clothing and other relevant information such as a need for identity cards, sunglasses etc. Within the information about certain WHS in Yunnan and Guangxi Provinces there are nine references about how to deal with animals and the natural environment.

'Service standards' is the one clear theme that emerged in the analysis and appears in 95 of the tourism brochures. But it is primarily related to detailed pricing matters. It normally states what is and what is not included within a particular holiday package. The brochures discuss shopping activities in some detail. But although Chinese tourists like shopping when they are travelling abroad (Arlt 2006; Huang 2004; Zhang and Chow 2004), Chinese tourists on domestic tours dislike shopping trips arranged by tour guides because they are seen mainly as a way for the tour guides to earn commission from local businesses (Huang 2006). Clearly identifying how many times a tour will contain shopping activities is just a strategy on the part of the travel service companies to promote their tours, with an emphasis on there being an appropriate number to meet needs for family gift buying and giving, but not too many as to raise suspicion about the motives for these shopping trips. Surprisingly, although some brochures do mention endangered species, rare or ancient materials etc. at some WHS, instead of warning the tourists not to buy goods made from these things, the brochures seem designed to increase the tourists' interest in purchasing artefacts made from such materials. Yet again, in this section, there is no mention of service standards in terms of responsible behaviour for either tourists or tour guides.

IMPLICATIONS AND CONCLUSIONS

This research reveals that the Chinese travel service companies examined in this project do not use brochures to communicate a concept of responsible tourism to their customers when their customers are planning trips to WHS within China. These research results also suggest, through the absence of such information, that the tourism brochures distributed by the major Chinese travel companies need to contain more information about responsible tourism to help ensure the long-term sustainability of the sites. There are three areas for development and improvement.

1. In terms of information about the travel service companies themselves the companies should not only include their contact information, but also relevant information about the companies' policies of supporting and protecting the WHS in China. If they do not have this kind of policy, then they should develop them as soon as possible. The CNTA should play a key role in making sure that every company in the tourism industry has relevant rules, regulations and policies with regards to dealing with WHS. Not only can this policy be regarded as one appropriate for the longer term generation of awareness in the domestic market, it may aid the companies to better prepare for an international market, many of whose clientele will expect or demand such information. Thus, good practice in the domestic market may be regarded as a necessity for developing markets and products among an international marketing context.

   The brochures could also promote experienced tour guides who have an understanding of the natural and cultural environment of the WHS, thus showing that the company has ambitions and abilities to guide tourists around the unique WHS in a culturally and environmentally sensitive manner. Different formats of educational tourism brochures, and also information about responsible tourism behaviours, can be produced and given to tourists when they collect promotional materials about tours of the WHS. These guides and the information can also be placed on company web sites.

2. In terms of information about the actual WHS, then the companies should not only describe the unique natural beauty and cultural practices that can be found there, but also emphasize the corresponding responsibilities that the companies and the tourists have towards protecting the uniqueness of those sites. For example, when tourists are visiting the Huanglong scenic and historic interest area, travel service companies should emphasize how vulnerable the natural environment is at that site; and encourage an appreciation of the necessity for protection of the site by tourists. Again this can be justified by the need not only to create longer term sustainability of an asset for future generations, but also by reference to need for conservation for future

financial viability and economic benefits. In the longer term, destruction of a site through overuse, poor visitor behaviour and poor management simply means destruction of a reason for tourists to visit a site.

3. In terms of information, especially for tourists, the companies should not just emphasize their product and price package, they need to give suggestions to tourists on how to behave at the WHS. For example, when tourists are visiting the WHS, they should not throw rubbish into the rivers and woodlands, and they should desist from adding graffiti to the walls etc. of old buildings.

If the proposed actions are adopted by travel service companies in China, it will represent an important step towards the development of responsible tourism at the Chinese WHS. Travel service companies should set their own code of practices with regards to responsible tourism, using the WTO's *Global Code of Ethics for Tourism* as a good reference point upon which to build high standards. As travel service companies play an important role in shaping the tourist experience, promoting responsible tourism is not just a benefit for a company's reputation but will also help improve the quality of the tourist experience at the WHS by encouraging responsible visitor behaviour. This will benefit both the tourist and the WHS itself by protecting the cultural and natural environment. It can thus be preserved and enhanced for future generations of both tourists and minority peoples.

There is significant scope to develop research in this area further than has been possible in the context of the small-scale investigation documented in this chapter. Certainly research is needed on the role brochures, guides and other tourism services play in educating tourists' responsible behaviour at WHS in China at the different stages of actual and pre-travel experiences, and in the post-trip recollection. (For example, do tourists come to feel ashamed about the type of behaviour illustrated in Figure 13.1?) In addition, a comparison of responsible tourism education undertaken in different tourism subsectors (such as in accommodation, transport etc.) would be valuable to assess how responsible tourism initiative varies among tourism sectors. Among other factors that also must be considered far more carefully are the following, for it is not possible to simply transfer western practices straight into a Chinese context. First, as pointed out by Ma, Ryan, Bao, Li and Sofield in this book, the Chinese concepts of harmony between man and nature permit greater human intrusion on natural formations than would be permitted in National Parks in western societies. Second, as again noted by Zhi and Bihan in their chapter, Chinese notions of crowding and psychological carrying capacities may well differ from their western counterparts due to the high densities of population found in many parts of China. Such further research will provide a much needed fuller understanding of responsible tourism practices in general and possible better implementation of responsible tourism education functions by the tourism industry in China.

14 Analysis of Tour Guide Interpretation in China

Yang Hongying and Chen Hui

INTRODUCTION

The purpose of this chapter is to examine the role of the tour guide in the interpretation of place and its associated history, heritage and peoples to visitors. The stories they tell, the manner in which perceived facts are filtered, communicated or silenced—all become important in creating image, knowledge and affective response on the part of the visitor. Their skills are key to many of the understandings held by visitors about a place during and after their visit. Consequently the chapter is divided into the following sections. First, it notes the growth in the numbers of guides working in China and briefly notes their characteristics. Second, it identifies issues that might be found in interpretations offered by guides. Next it discusses the cultural context within which guides work and which may influence the stories told and the manner of the telling. From this the authors then propose a series of principles that can inform the work of guides before summarizing and ending with concluding comments.

THE NUMBERS OF GUIDES
AND THEIR CHARACTERISTICS

China's state tourism authority, the China National Tourism Administration (CNTA) defines tour guides as 'the licensed personnel sent by the travel agencies to provide guiding, interpretation and tour services' (CNTA 1999). The job of the tour guide is described as 'introducing cultural and natural attractions, folk customs and conventions of the places visited and providing tour service'. It is stressed that interpretation is one of the required responsibilities of the Chinese tour guides and that it constitutes a vital part in creating satisfied tourists. Tour guide interpretation is a communication process. It is designed to reveal meanings and relationships of the cultural and natural attractions through involvement with objects, artefacts, landscapes and sites. It aims to translate, explain and reveal the meanings and relationships of a natural or historic site to visitors. So tour guide interpretation is also a process of communication which enables visitors to turn tourist products into tourist consumption, or a meaningful and learning experience for the visitor.

Together with the further opening up of the economy, associated reforms and the establishment of the socialist market in the 1990s culminating in China's accession to the World Trade Organization and its aftermath, tourism boomed as the focus of the tertiary industry and the new growth point of the national economy. Inbound tourism, outbound tourism and domestic tourism all thrived creating three separate but associated markets with an increasing demand for tour guides. Chinese tour guides make up the backbone of China's tourism industry, contributing to its fast-paced tourism development and promoting cultural exchanges and developing friendly ties with other peoples in the world. To meet the growing demand, the number of tour guides in China has been expanding at an exceptional rate, reaching 320,000 by the end of September 2006, the largest recorded number as at that time.

As the fourth largest global tourism destination, China has experienced the tourist boom at the beginning of the new twenty-first century. In 2006, inbound tourist arrivals amounted to about 125 million, which earned US$33.95 billion, and the number of domestic visits rose to 1.39 billion, which generated 623 billion RMB (Shao 2007). The 11th Five-Year Plan for Tourism Industry explicitly affirmed the goal to develop tourism as the main industry of the national economy and the primary tasks include accelerating the optimization of the tourism industrial structure and completely lifting the quality of resource, service and assets so that the industry can fully exercise its functions in an optimal manner. China's tourism was expected to reach a new stage with the advent of the Beijing Olympics in 2008 and Shanghai Expo in 2010.

The World Tourism Organization (WTO) (now United Nations World Tourism Organization) estimates China will become the largest destination and the fourth largest tourist generating country by 2020 (WTO 2003). The CNTA also aims at building a world tourism 'giant'. Consequently, China's tourism industry has placed much emphasis on optimizing its structure and improving service quality. As workers at the front line, Chinese tour guides apparently occupy a prominent place in the course of building the industry's core competitiveness. And market competition largely consists in quality guide service. From this it follows that improving tour guide interpretation becomes the key to the success of China's tourism development and its longer term sustainability.

Yet the current fast expansion in the number of tour guides is inevitably accompanied with some problems that inhibit the desired improvement of China's tourism service. According to an investigation conducted by CNTA in 2002 (CNTA 2003), the composition of Chinese tour guides shows some weaknesses in the following respects:

1. *Age*. Chinese tour guides are relatively younger, with 70 per cent below 30 years old, implying a lack of not only guiding but also perhaps life experience.
2. *Educational background*. Tour guides, especially the Chinese-speaking guides, have a relatively weaker educational background. Of the Chinese tour guides, 41.7 per cent only have a senior high school and secondary vocational school diploma. As for tour guides speaking foreign languages, only 18.9 per cent have received university education while 52 per cent are graduates from senior high schools, secondary vocational schools and colleges of three-year programmes.
3. *Specialty*. Of Chinese-speaking guides, 34.9 per cent are tourism majors and 12.1 per cent and 17 per cent are respectively foreign language majors and management majors while 36 per cent are graduates from other subject streams. There is a lack of tour guides who speak excellent foreign languages and acquire top guiding skills.

It is evident that such young tour guides with a far from good educational background and less specialized knowledge find it hard to obtain the required cross-cultural communication information or foster awareness of different cultures in a knowledgeable and empathetic manner. Additionally, few have had experiences of living overseas. Then they may encounter difficulties in meeting the demand and expectations of the rising number of inbound tourists who are often amazed at the sights of Chinese history and culture. What they currently need is professional tour guide interpretation training provided by the institutes of higher learning and training organizations concerned. Hence, the authors would suggest that research on tour guide interpretation is of great significance and demands more attention and continuous exploration for many reasons that include a need to ensure a more knowledgeable profession and to better enable visitors, especially international ones, to understand China's past, present and future needs and culture.

The past two decades have witnessed an evolution of tour guide interpretation from rather simple information provision at an earlier stage to the present ideal of effective interpretative communication with tourists through utilizing a diversified range of knowledge related to different subjects. Today, in many higher education institutions and elsewhere, tour guide interpretation has been established as a special field of study in the art of communication of place, history, heritage and culture.

For years the authors have been engaged in tour guide interpretation research and have offered consecutive lectures to international tourists to Xi'an. Based on hands-on experience, observations and interviews, the authors have attempted to bring attention to current issues in tour guide interpretation in China and to propose some principles for effective tour guide interpretation by borrowing and integrating recent findings of studies in the field of interpretation and cross-culture studies. This chapter represents a continuation of that work.

ISSUES IN TOUR GUIDE INTERPRETATION IN CHINA

A systematic study of China's tourism interpretation has been initiated (Wu 1999), yet with regard to the application of interpretive theories to tour guiding, the research is still in its early stages (Wang 2000; Fu 2002; Huang and Lin 2003; Yang 2005; Yang 2006), and the number of completed relevant treatises still remains comparatively small, albeit now beginning to grow in number (Wang 2002). Indeed, a large number of tour guides from China's travel services, in spite of a lack of effective tour guide interpretation training, do attempt factual interpretations while conducting tours. However the information offered through these guide interpretations is rather incomplete and unsystematic, leading to cultural misunderstandings and communication barriers with and among tourists. In many instances international tourists fail to obtain a quality and satisfactory visitor experience (Yang 2006).

There are a variety of ways to define 'interpretation' since people approach this concept based on their own backgrounds, training or experience in their interpretive professions. One of the earliest and most authoritative is attributed to Freeman Tilden. According to Tilden (1977: 8), 'Interpretation is an educational activity which aims to reveal meanings and relationships through the use of original objects, by first-hand experience, and by illustrative media, rather than simply to communicate factual information'. Since the 1980s, scholars and experts in tourism circles have introduced this idea into tourism studies (Alderson and Low 1985; Moscardo 1999; Pastorilli 2003) enabling tour guides to help visitors acquire new knowledge, learn exotic cultures, experience tours, foster environmental protection awareness by stimulating all the senses of visitors and offering interpretation in an entertaining and enlivening manner.

After observing over 20 guided tours in major tourist cities in China such as Guilin, Xi'an, Kunming and Guangzhou and interviewing a number of full-time tour guides from Xi'an, Guizhou, Kunming and Shijiazhuang, the authors have found some urgent issues in tour guide interpretation in China. They can be summarized as 'five stresses and five absences' as follows:

1. Stress on history and an absence about present everyday life.
   Tour commentaries only focus on historical events and past experiences and seldom cover the everyday life of 'average' people. Interpretation tends to overlook the daily life of contemporary people.
2. Stress on cultural resources and an absence about natural resources.
   China abounds in both natural and cultural resources, but interpretation exclusively attributes more weight to cultural ones and ignores

information concerning nature, such as topography, terrain, flora and fauna, etc.
3. Stress on the stories of eminent people and an absence of common folk.

   Stories of renowned people are widely spread and rather frequently referred to in tour guide interpretation but there is less mention of the life and stories of 'ordinary people'. Therefore tourists have the feeling of being separated from the real life of local residents and thus have less opportunity to develop 'a feel of place'.
4. Stress on legends and an absence of scientific principles of phenomena in tourist attractions.

   The tourist attractions are always places of legends and fairy-tales in guide interpretation. China, one of the ancient civilizations, seems to have developed without any scientific accomplishments. And many unusual phenomena are explained according to tales based in fantasy rather than from a scientific perspective.
5. Stress on superficial phenomena and an absence of examination of the causes that lie behind them.

   Guides tend to speak of some phenomena in Chinese society and life but fail to uncover the causes behind them, and are thus unable to help tourists develop a cognitive understanding about the host community and its local people.

CULTURAL REASONS FOR SUCH ISSUES IN GUIDE INTERPRETATION

These five stresses and five absences are quite common in China's tour guide interpretation and are closely associated with the unique culture and conventions of China. Continuing the theme of 'five' it is possible to identify five reasons for these articulations and silences in guides interpretations.

1. China is a country with a time-honoured history where the importance of gaining lessons from history has always been stressed throughout much of its cultural tradition. 'Review the past and open up the future', 'History mirrors the present' and 'Make the past serve the present' are all common household sayings. This deep-rooted idea is reflected in tour guide interpretation by providing too much historical information about tourist attractions and relatively inadequate introductions to people's everyday life. This causes tourists to complain that listening to tour guides is like attending boring history lectures.

   On the other hand, it is recognized that ancient heritage sites remain largely featured in lists of Chinese tourism resources given the 5,000 years or more of history associated with Chinese civilization. That civilization and history has attracted significant archæological and historical research and thus it is now quite well known. It is also perhaps less controversial than more modern times and the guides will often feel at ease talking about the ancient past. Consequently, be it a cultural or natural attraction, tourists are told many things that took place in the past and in Chinese classical period. Consequently, few tour guides will take the initiative in touching on occurrences of everyday life that individuals from every culture experience, such as daily routine, family life, marriage and burial, schooling and employment, medical care, dietary customs and the like, which is the very information about which many tourists are curious but can rarely obtain from books. Their only possible access is the guides who are born and raised in China who have plenty of examples and life stories on hand.

Such phenomena are similarly found in guide-books. Guide-books to major Chinese cities are readily available in many bookstores. Whether it be a guide-book to Beijing or a guide-book to Hangzhou, the main introduction tends to exclusively concern places of historic interest but rarely spares even a chapter on everyday contemporary life. One example is *Travel China—Best Selections of Chinese Tour Commentaries* (comprehensive edition) (CNTA 1997), the first book of compiled tour commentaries published by CNTA in 1997. The book includes 31 tour commentaries by the best tour guides from different parts of China. Only one commentary, of a day tour to the Stone Forest in Yunnan Province, introduces the life and customs of the modern native *Sani* minority people, while the remaining commentaries unexceptionally deal with events from history. Under such circumstances, it is not surprising that Chinese tour guides are more likely to stress history and overlook present everyday life.

2. China is rich both in cultural and natural resources. But traditionally more studies are conducted on the cultural heritage and the same is true when looking at the content of tourism studies. Books on these topics are easily found in bookstores which provide tour guides with good information on the interpretation of these attractions. Consequently tour guides tend to interpret cultural sites in much greater detail and to a large extent overlook the natural environment of a given site. This means that visitors tend to know less about the natural environment, and hence perhaps the stresses that large numbers of people can cause on those environments.

This is illustrated by observations at the Three Gorges area, which is a fairly popular nature-based tourist attraction. Even here the majority of tour guides place the natural qualities of the area in last place in their presentations. Instead of presenting information about China's major watercourses, the impacts of the Yangtze River on the climate of the area, flora and fauna and other essential knowledge concerning natural phenomena, tour guides deliver quite lengthy commentaries on the history and folk-tales of the area. It results in a situation where experienced tourists can hardly have their expectations met and inexperienced visitors tend to develop a rather superficial understanding of the attractions. Their impression of the area is none other than fabricated tales and historical events presented in haphazard order.

3. China has a culture which favours collectivism and does not advocate individualism. Collectivism means greater emphasis on (a) the views, needs and goals of the in-groups rather than oneself; (b) social norms and duty defined by the in-group rather than individual behaviour seeking pleasure for self; (c) beliefs shared with the in-group rather than beliefs that distinguish self from that in-group; and (d) a greater readiness to co-operate with in-group members (Trandis 1990). When a 'we' consciousness prevails, individuals will seek to belong to organizations and shun individual publicity. Even though a person may have changed the history or the convention of a region by his own wit and strength, it is unlikely that he will be introduced individually unless listed among historical celebrities.

   Tour guides in a collective culture have a tendency to give more attention to the histories of prominent people associated with a place and are reserved about telling their own personal stories which, in fact, are equally if not more attractive to tourists, especially for visitors from overseas. This phenomenon can be misinterpreted by tourists from a different culture as tour guides trying to maintain a distance from them and viewing their commentaries as not being sufficiently personal and indeed often not engaging. And yet many visitors can learn more about a place from hearing guides talking of daily lives of less exalted people and their daily trivial experiences and the changes in their lives that they have experienced in contemporary China.

4. It has long been a tradition in China that the liberal arts are superior to science and technology. This idea has only just recently begun to change. But this traditional view has greatly influenced the introduction and tradition of tour guide interpretation in China. Tour guides are all well armed with various legends as to how the shape of a lake has changed and how the stalactites and stalagmites are formed, but find it hard to explain these in a scientific manner. And there is a mistaken idea among the guides that tourists join a tour only for fun and that any scientific interpretation that might be offered would not be appreciated.

   Almost regardless of different audiences and occasions, tour guides tend to always associate natural formations and phenomena with fairies and gods and monsters, thus giving international tourists an impression of publicizing superstitious ideas. A possible resultant misunderstanding on the part of tourists is that tour guides in China are lazy and they all try to bamboozle tourists by telling children's stories. Interpretation should be both entertaining and educational. Some visitors and scholars have noted this and expressed concerns and suggested that tour guides consider 'changing their way of interpretation' (Jiao 2000).

5. Tour guides take for granted many Chinese everyday practices without trying to find the causes for those practice. When asked about some unusual practices such as why the children wear open-seat pants and why Chinese people prefer boys, most would answer that it is a tradition or people are still old-fashioned. But these answers are not convincing to many tourists. Tour guides may interpret a site or a social phenomenon at three levels, namely, site description, background introduction and cognitive content. But most tour guides only get to the second level, unaware of how to lead tourists to analyse an object and acquire their own knowledge that enables them to experience pleasure from understanding the cultural differences. Fu (2002) argues that it is this that represents the best guiding practice.

In addition, most tour guides in China are not well travelled; many have not been overseas and thus are not cross-culturally conscious. In cross-cultural communication, one should know one's own culture well before learning a different culture. This is possibly one reason why many guides do not perceive the importance of cultural differences and therefore are often not prepared by having more information behind seemingly simple questions. For example, many guides tell international tourists that, in rural China, men are traditionally considered superior to women, but they make no reference to the reason that in an agrarian culture with little use of modern farming machines, men still constitute the principal labour force in the farm work. Often, instead, international tourists may picture Chinese farmers as stereotypically stupid, conservative and as a less caring people.

PRINCIPLES OF EFFECTIVE
TOUR GUIDE INTERPRETATION

Interpretation is not a simple matter of information providing but rather is a matter of translating information in specific ways relevant to the audience. And effective communicative interpretation helps tourists to appreciate a natural or cultural heritage and learn its meaning. It is worth mentioning that *how* information is delivered is more important than *what* the information is. In *Interpreting Our Heritage* (1977), Tilden set out 'six interpretive principles':

1. The interpretation must relate to what is being displayed or described to something within the personality or experience of the visitor.
2. Information, as such, is not interpretation. Interpretation is revelation based on information. However, all interpretation includes information.

3. Interpretation is an art, which combines many arts, whether the materials presented are scientific, historical or architectural. Any art is in some degree teachable.
4. The chief aim of interpretation is not instruction, but provocation.
5. Interpretation should aim to present a whole rather than a part, and must address itself to the whole man rather than any phase.
6. Interpretation addressed to children (say, up to the age of 12) should not be a dilution of the presentation to adults, but should follow a fundamentally different approach. To be at its best it will require a separate programme.

In all, the basic principles of interpretation are summarized as Provoke, Relate, Reveal, Stress message unity and Address the whole (Veverka 2007). The interpretation must provoke curiosity, attention and interest on the part of the audience. The interpretive communication must find a way to relate the message to the everyday life of the visitors. A good interpretation must reveal a new insight into what makes a place special and give people a new understanding. The presentation of the message should be supported with the right designs, colours, music and the like to guarantee the unity of the message. The interpretation should address the main point or theme in which visitors are interested. And for interpretation to be successful, it should arouse the audience's imagination, be relevant to its needs, propose a new insight and be suitable to the site or object (Ham 1992; Tilden 1977). Such development and findings in the theories and principles of interpretation may provide very helpful insights into tour guide interpretation in China.

Tilden's theories were first proposed half a century ago and were suited for the National Parks service system in western countries. His interpretation is visitor-oriented and aims to provoke visitor curiosity and help them find meanings and come to a new perception themselves. Visitor involvement and interaction is highlighted throughout the tour. Those interpretation theories work well because tourists there are nurtured in a similar cultural background as the one that gave rise to Tilden's principles and are well acquainted with related history, free of any communication barriers. Such principles are especially effective when tours are conducted at natural sites and the communication channels between the two sides are always open and function well in transparent fashions. But the same principles may not necessarily apply to the Chinese situation.

Most tourism resources in China are cultural in nature with historical origins and distinctive national features. And cultural shock and uncertain perceptual differences can inhibit successful communication and effective interpretation. Actually both tour guides and tourists need to develop some cross-cultural consciousness for effortless communication and personal interpretation. Even though the world is becoming a 'global village', quite a number of inbound tourists live in countries far away from China and have inadequate preliminary knowledge and understanding about the destinations they visit. It is also common to find some international tourists with stereotyped perceptions about China who may judge it through pre-existing perceptual filters that also impede more open communication. Considering the tenets of communicative interpretation, the authors maintain that tour guides in China should be cognizant of the previously mentioned six principles but also need to keep the following principles in mind while offering interpretation of a site.

1. Relate to the audiences' expectations and needs.
   Good tour guide interpretation is relevant to tourists' expectations and needs and arouses their interest and imagination. No audience will be enthusiastic about information that fails to cater to their desires and preferences. Then, what is the point of doing interpretation if the target audience is not interested in it? Tourists vary in age, gender, education, religions, occupation, etc., so an experienced tour guide should spend some time finding why tourists join a tour and what they expect to learn, and so select the information related to them to get their attention. For example, interpretation for adults and children should be approached in different ways and the information presented needs to be provided in an acceptable manner for their age. Then perhaps the interpretation may provoke their curiosity, interest and exercise their imagination and perhaps jolt them into a completely new understanding of what they have been shown. Only when the tour guide builds up connections between tourists and the cultural or natural heritage or object can the tourists come to appreciate and understand what makes that attraction special.

2. Refer to both historical facts and the modern everyday life.
   Every place has a history and every culture has its historical events and eminent people from different periods of its development. History is linked to the past but development leads to the present. Tour guide interpretation should pay equal attention to both historical facts and the occurrences of modern everyday life. Ham (1992) suggests that if something is to be relevant to the audience, it must be both meaningful and personal. Areas with ancient remains not only captivate visitors with that glorious past history but they can also celebrate present lifestyles and consider the future. Many international tourists are particularly interested in the social changes taking place in China with the advent of reforms and the opening up of society and these stories also need to be told.

   In this vein popular topics include average salary differences between the coastal and interior areas, the living conditions of the huge numbers of migrant farmer workers in urban China, issues about pensions, retirement and employment, housing and schooling, etc. Introductions to those changes in everyday life present the on-going prosperity of the host culture and its recent development. Referring to real people, quotations and stories also give life to and reinforce a genuine sense of a place and its culture. Such an interpretive approach promotes a fuller understanding of the culture and customs of the locals and the destination.

3. Reveal the meanings behind in a scientific way.

Interpretation is about sharing meanings about a place or object. While good interpretation is inevitably entertaining, interpretation without information is little less than entertainment. Yet information that does not reveal meaning is not interpretation at all. Revealing meaning is to connect the physical aspects of a site or object with associated concepts (Beck and Cable 2002), and the concepts must be revealed in a unique or unusual way so tourists can better benefit from the experience.

Furthermore, the concepts should be explained in a scientific way. Legends and tales are an integral part of every culture and usually bring some fun to tour guide interpretation, but the science and technology behind the physical site, object or natural phenomena makes a tour a fantastic learning experience. International tourists generally respect science and technology, so an introduction to natural phenomena, watercourses and topographical features along with the legends and tales can be both recreational and more convincing to tourists. It is essential for Chinese tour guides to acquire a much wider scope of knowledge ranging from geology, meteorology and biology to archæology.

4. Address the whole part with a unifying theme.

Interpretation is not a range of collected information and unconnected facts about a place or object. Instead, there must be a theme organizing interpretation and expressing an overall idea that tour guides want tourists to take away with them. A well-armed tour guide has good access to facts and stories about a place, but it is the way he or she addresses them that matters more in communicative interpretation. Rather than randomly getting across messages about a place to an audience, the tour guide should address it as a whole with a main point and in a unifying way. Given a mass of facts and information, the interpretive guide only needs to focus upon those which support the theme and put them across to bring forth the 'big idea'. What is the big idea? It is this that makes the place more distinctive and important. Sensitive choices of colour, style and materials can add to a sense of a place and help reveal its main character.

SUMMARY

Tour guide interpretation promotes understanding and enthusiasm, and helps tourists better learn a natural or cultural heritage. Communicative interpretation theories developed by Tilden and other scholars place an emphasis on the function of communication with the audience and effective design and delivery and therefore offer helpful insights into improving tour guide interpretation in China. China is a multi-ethnic country with diversified cultures and abundant tourism resources featuring both natural and cultural heritage. Effective tour guide interpretation is of significance. The phenomenon of 'five stresses and five absences' reflects the current weaknesses in interpretation that require more attention from tourism educational organizations and demand more research work.

Compared to earlier empirical interpretations in tour guiding, the four principles proposed are thought to be practical and helpful for tour guide training organizations and to contribute to creating more competent tour guides in China. Yet the authors also admit that the paper might be biased since 60 per cent of tour guides in China are freelancers and their observations are based upon a relatively small number of tours within the time available to them. The samples and interviews used in the research cannot be said to be representative and thus readers need to bear this caveat in mind when coming to a judgement about the findings and proposals that the text reports. Nonetheless, it is felt that the ideas have value, and they also represent a step in better understanding how visitors may perceive the sites visited—which in itself represents a means of site and destination management.

Part III

Community Participation and Perspectives

Hall (2000) comment, the nature of communities become in themselves a means of motivating travel as people seek difference, nostalgia or temporary immersion into a different pattern of life. Communities also shape the landscapes consumed by the tourists through not only their architecture and land usage patterns, but through what they select and frame for the tourist gaze. In undertaking this process, communities both explicitly or implicitly react to changes and thus either seek to establish their own priorities and values, or have values thrust upon them by external forces, or, thirdly, negotiate understandings through interaction between internal and external social dynamics. Communities possess identity from the places they occupy, the local social structures they evolve, the feelings of *communitas* and place attachment they engender, and the ideologies, networks and power structures that bind them together. Consequently analyses of community tourism are often associated with stakeholder theory, network analysis and issues of equity, and thus with concepts of sustainability in tourism (e.g. Weaver 2008). Equally, however, communities are not isolated from their wider societies, and this is particularly true in China where centrally determined legislative frameworks determine the discourse of local application of laws and regulations associated with tourism and local community planning. For a time it became fashionable to assess change through the lens of globalization wherein local communities reacted to the emergent global nature of markets and modes of production by absorbing the global brands into the local economy, and then slowly losing sense of individual identity as those global names and modes of production took over from the local. A linear process was envisaged where the local position was solely reactive, and possibly in the tourism literature this position might be said to be represented by Go and Ritchie's (1990) and Hjalager's (2006) model of tourism development. Subsequent to this approach the transformative aspects of globalization upon local communities became increasingly analysed, and commentators utilized concepts of demonstration affects and acculturation to assess socio-economic changes on such processes as gender role change (McGibbon 2000), intergenerational change (Lever 1987) and spatial patterns of land use (Smith 1992), among other things. More recently counter-arguments to the McDonaldization theory of Ritzer (2004) have emerged, and local communities are perceived as being potentially empowered to make choices and to be truly pro-and interactive with external forces. This has possibly been recognized for some time in the case of indigenous tourism, and many of the same concepts are now being utilized within the wider and associated literature of sustainable community tourism, while links are also now being made with the literature on destination branding (for example, see the Edwards *et al.* 2000 analysis of Alto Minho in Portugal). With specific reference to China, Ryan and Gu (2007) argue that 'glocalization' is an inherent potential process of the destination life-cycle as local communities seek to gain competitive advantage in global markets by taking, selecting and using local identities and customs as unique sales propositions.

While Ryan and Gu's concept is derived in part from management perspectives, it sits comfortably with an emergent literature that emphasizes local social perspectives of landscape, both natural and man-made. Landscapes have long been an important component of destination image (Gallarza, Saura and García 2002), but it is increasingly being appreciated that landscapes are not solely issues of topography and geomorphology, but are interpreted environments upon which cultures impose their various meanings. This is internationally recognized as evidenced by the International Council of Monuments and Sites (ICOMOS) International Cultural Tourism Charter (Managing Tourism at Places of Heritage Significance) of 1999, which states:

> Heritage is a broad concept and includes the natural as well as the cultural environment. It encompasses landscapes, historic places, sites and built environments, as well as bio-diversity, collections, past and continuing cultural practices, knowledge and living experiences. It records and expresses the long processes of historic development, forming the essence of diverse national, regional, indigenous and local identities and is an integral part of modern life. It is a dynamic reference point and positive instrument for growth and change. The particular heritage and collective memory of each locality or community is irreplaceable and an important foundation for development, both now and into the future.

The charter then proceeds to state that 'At a time of increasing globalisation, the protection, conservation, interpretation and presentation of the heritage and cultural diversity of any particular place or region is an important challenge for people everywhere. However, management of that heritage, within a framework of internationally recognized and appropriately applied standards, is usually the responsibility of the particular community or custodian group' (ICOMOS 1999).

Stephenson (2008) points out that landscape and its management is often subjected to 'expert' management, but the different disciplines have different approaches to landscapes based upon sociological, psychological, geographical and physical sciences (see Figure 15.1). This is possibly even more true of China than elsewhere with, as noted, a planning predisposition for top-down approaches based in part on the recruitment of academic 'experts'. However, an increasing awareness has emerged that if the cultural values associated with a place are to be respected, and indeed understood, then those values emerge from a 'sense of the past' (Daniels 1989) on the part of those who live and use the landscape. From a historic perspective many of those values emerge from communities local to the landscape, but in the postmodern era those values become utilized for purposes of place promotion to tourists, who visit places complete with their expectations and value systems. Stephenson (2008) poses what she terms a Cultural Values Model of Landscapes, which is modified as illustrated in Figure 15.2.

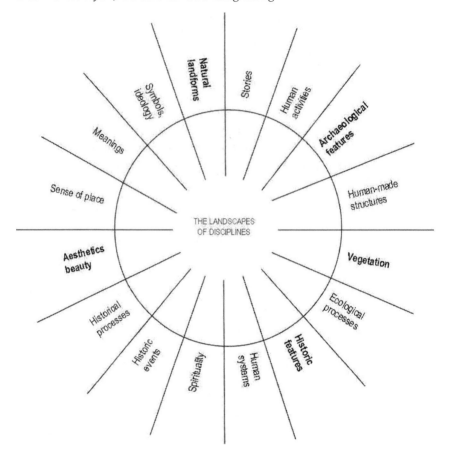

Figure 15.1 Landscape, community and interpretations.
Source: Stephenson (2008).

Thus, in the present day, there exist what she terms 'surface values' of natural form, relationships and practices, but these are embedded through histories and management of historic sites bequeathed by the past. Relationships comprise meanings, symbols, spirituality, senses of place and aesthetics while practice can incorporate ritual and human practices as well as the ecological processes associated with geomorphology. Equally, forms are both natural and man-made. However, the use of culture, tradition and heritage for purposes of tourism imposes new meanings on place that are derived from outside of the local community, and which are selective in nature. Tourism and destination promotional bodies will seek to emphasize the unusual and the extraordinary in order to promote one place as different from other places, and thus communities become themed in representations directed at tourist-generating zones. Tensions thus result between communities seeking to derive economic value from their patterns of life and those seeking to be

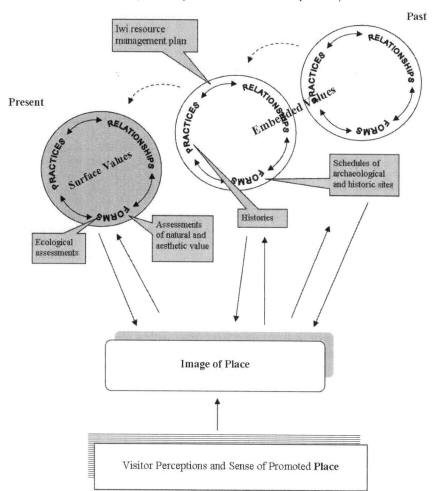

Figure 15.2 Tourism and the Cultural Values Model of landscape.
Source: Adapted from Stephenson (2008).

selective in what they take from the community as consumers of place and its culture. The community itself has its own tensions, seeking to retain traditional values while socially, politically and economically advancing their well-being. There is a danger that the traditional values become walled-off, become a performance to be played for the visitor, but also increasingly become an anachronism within the daily routines of life in a community that becomes more open to the modern world and its imperatives of business, time zones and functional relationships. The visitor too is potentially experiencing tensions, between that which is being placed on display and performed, and a wish to better understand realities of different cultures. Questions arise as to what is 'authentic' and 'authorized' (Ryan 2003), and a tension exists between the desire for a difference and possibly a growing realization that the communities' daily realities may not be far different from those of the visitors. As Fan, Wall and Mitchell (2008) indicate, community planning and tourist place becomes a site of creative destruction.

THE CHINESE EXPERIENCE

Compared to other developing countries, China is distinctive in its size, diversity of land and people, and its centralized bureaucracy which excises a high degree of economic and social control through its essential role in policy-making at all government administrative levels. In developed western countries, a sustainable approach to community tourism promotes a dynamic tourism–community relationship which emphasizes the 'community' and 'stakeholder' involvement in the tourism development of destination communities (Stone 1989; Murphy 1985; Ryan and Montgomery 1994; Li 2006). It is argued that ideally, the community and each stakeholder have a legitimate voice and equitable participation in the tourism development process, and can share economic, psychological, social and political benefits to sustain both the community and the industry (Til 1984; Timothy 1999; Li 2006). However, in the Chinese context, community involvement in the tourism development is considered 'a political as well as a social, economic and moral issue associated with the modernisation process' (Li 2006: 47). China has been experiencing a tremendous modernization and development process in its economy, social norms and values over the past 30 years under its current political form and system. Under this circumstance, a stable tourism–community relationship is emphasized in line with the Chinese government's main priority of maintaining social harmony and promoting a socialist market with a Chinese perspective (see the final chapter of this book). China's tourism industry has long been seen as an effective source of foreign exchange that contributes to local, regional and national economic development and represents an administrative focal point for governments at all levels. As Li (2006) proposed, western scholars consider community tourism a democratic process with the ultimate goal of sustaining the community rather than tourism or some specific aspects of tourism development, whereas China's community tourism remains predominately a planning strategy with an ultimate goal of sustaining the tourism economy in order to enhance communities through economic development. This is perhaps best illustrated by Li's (2006) research in Wan'an County in Huangshan City (Anhui Province). He argued that local government officials viewed their primary task as the attracting of capital investment thought vital to the survival of the county's economy. Specifically, the officials' working agenda was to ask whoever arrived in the county, from abroad or other wealthier regions in China, to bring investment capital to help transform the county into a popular tourist site. Wan'an had not wholly benefitted from the tourism associated with Huangshan National Park because of its geographical position peripheral to the tourist flows generated by the Yellow Mountains, yet the local residents were keen to join the tourism ventures of Huangshan to share the economic benefits. Li notes that many had left their hometowns for urban areas, for 'working on agriculture only produces poverty' (direct quotation of interviewee respondents; Li 2006: 53). As evidenced elsewhere in this book, and by this example, the main contribution required of tourism was primarily wealth generation. However, given a changing and renewed emphasis on matters of cultural expression and natural environments voiced at the 2008 NPC and CPPCC sessions, modifications to these policies are beginning to emerge. These are potentially important for rural China as elsewhere.

The Importance of the Village

China has approximately 800,000 villages, and it is estimated that about 8 per cent of the world's total population lives in a Chinese village (Dumreicher 2008). Given the poverty that exists, meaningful questions include what to develop, what to retain and how to develop within the resources accessible to such villages if targets of economic progress are to be achieved. At the same time, the issue of villages retaining people and being able to offer opportunities is made all the more urgent as people, both legally and illegally, are drawn to the cities in order to better themselves. Issues of sustainability are thus writ large when considering tourism and its planning in the rural zones of China. One definition of sustainability adopted by the European-funded SUCCESS project in rural China was that of Levine *et al.* (2005), namely, 'Sustainability is a *local, informed, participatory, balance-seeking process*, operating within a Sustainable Area Budget, exporting no harmful imbalances beyond its territory or into the future, thus opening the spaces of opportunity and possibility' (emphasis added by current authors).

In 2006 the China National Tourism Administration (CNTA) announced that rural tourism was to be the theme of that year, with promotions and initiatives additional to those already existing to be undertaken. Later, in 2007, the Beijing Tourism Administration (BTA) commissioned additional research from Chinese academics to better understand rural tourism within its region. However, even prior to these years rural tourism had, as already noted in this chapter and elsewhere in this book, been utilized to generate economic growth, and had matured quickest in the peripheries of many large cities, notably Beijing, Shanghai and Chengdu. In the wider Beijing municipality that includes the rural suburban areas, Fan and Huang (2005) estimated that by 2005 income from tourism accounted for over 30 per cent of income for the agricultural sector. Zhang, An and Sun (2007) cite data derived from BTA that indicated that arrivals in these suburban areas numbered 4.62 million in 2002, climbed by a third in 2003, had almost doubled by 2004 and all but trebled by 2005 to reach a total of 12.28

million. Spending by these tourists was, in 2002, 371 million RMB, 610 million RMB in 2003 and by 2005 had exceeded the billion RMB figure, being 1,098 million in that year.

In an analysis of changing patterns of tourist industry activity in this region, Zhang, An and Sun (2007) note the development of tourism originally based primarily on accommodation and catering has become a more diverse product based on a range of activities that include fruit picking, fishing, spa-based tourism, meetings and conferences and adventure products. The contribution made by tourism to family incomes is clearly shown by the experience of Mayufangzi Village in Changping District, Beijing. In 1998, 40 households were licensed as providers of accommodation to tourists, and they had a combined income of 412,000 RMB. By 2004 there were 56 such providers, and gross income was 3,302,000 RMB. BTA data indicate that in Haairou District, by 2005, tourism was directly employing 10,000 local residents and 50,000 indirectly. The success of these areas is based primarily on two main factors according to Zhang, An and Sun (2007). These are, first, the natural resources that exist, which include not only scenery but local cultural customs, and the second is organization and entrepreneurial initiatives. Five forms of organization are noted, and these are:

1. Individual households—here products are primarily those of the provision of accommodation, picking fruit in orchards or fields, and the selling of local produce.
2. Collective operations—these are of three forms: Initiatives by strong Village Committees where revenues derived from tourism are invested in improved infrastructure, marketing and the development of new resource under various modes of ownership as, for example, described by Zhou and Ma in this section of the book. Second, village-based consortia or committees are specifically established to run rural tourism, but under this scenario not all members of the local community may be involved in participating in, or deriving benefit from, tourism. Third, co-operatives may be established whereby the community as a whole gains benefits from revenue derived from the sale of licences to home-stay households, and where tourists are assigned to those households by the co-operative, a system motivated in part to ensure some degree of equity in the distribution of income that results from direct tourist expenditure.
3. Outside capital. Non-local residents or companies purchase assets from residents and establish businesses in the locale. Having more capital to invest, and often having more sophisticated management skills and coming from the tourist-generating zones, and thus better understanding visitor needs and preferences, these businesses are often able to provide better service and physical resource than can the residents.
4. Joint-venture operations. These are initiatives developed by individual villagers working together, or by the village working as a whole.

5. External investment, whereby local villages lease assets to enterprises external to the village. For example, Changdian Village in Jinzhan Township leased land to the Beijing Xiedao Green Ecological Resort Company at 1,500 RMB per mu (approximately 0.16 of an acre) and the company seeks to encourage village shareholders in the company (Zhang, An and Sun 2007).

Inhibiting and Facilitating Factors

Past research of rural communities (and minority peoples) engaging in tourism have illustrated a series of factors that inhibit or facilitate tourism development and its impacts on local communities. These include:

Lack of Management Skills

Initial tourism products in rural regions are often based upon accommodation and catering, with possibly some visits being organized to local heritage spots. Such activities are generally within the skill base of farmers, but issues arise when such a business seeks to grow. There may be issues of employment of non-family members, how can an enterprise progress its marketing beyond word-of-mouth recommendation, how does it access travel companies located in tourism-generating zones and how can it begin to develop new product? There is also the danger of successful ventures being copied by other local people, thereby possibly generating oversupply of a given form of product (e.g. farm-based accommodation) and a threat to financial viability of existing businesses. One potential consequence is that an over-dependence on external resources occurs, or business people from outside the location are attracted. For example, in 2006, in Haizi Village in Pinggu District, it was estimated that 60 to 70 per cent of tourism businesses are run by non-local businessmen who opt for more commercialized approaches (Zhang, An and Sun 2007).

In other locations, not only do non-local businesses begin to commercially marginalize the local, but local communities also find that their traditions and cultural symbols are appropriated by 'outsiders'. For example, Yamamura (2005) describes how outside Han entrepreneurs moved to Lijiang to sell replica Dongba art forms. On the other hand, in some instances a village community is able to access managerial skills through extended networks. Dumreicher (2008) provides the example of Xiao Qi in Jiangxi where residents were able to call upon extended family networks and utilize the skills of those who had previously migrated from the village into the cities.

Mixed-Management Hierarchies

As a multiple activity, tourism comes under the headings of different ministerial departments as discussed by Ma, Si and Zhang in this book with reference to the situation in Qufu. The same sort of situation arises in the case of rural tourism. In the case of Beijing's rural areas, orchard visits and fruit picking in orchards comes under the auspices of the BTA, but orchards and fishing ponds are the responsibility of the Forestry and Fishing Departments respectively. Potential entrepreneurs are thus faced with a maze of regulatory frameworks to acquire the necessary licences, and overall standards are not necessarily improved by some farmers offering products illegally. The issue can also be complex with reference to heritage and cultural assets. Private property rights may exist for an individual family that occupies a 'heritage house' but the whole community may exercise ownership over other historic and cultural assets. The situation is not aided by a lack of clarity in property regulations and law.

Encroaching Urbanization

The attraction of rural villages lies in their rustic patterns of buildings and the rural life-style, but the desire for additional income is accompanied by a wish for better housing. One consequence is that additional income derived for tourism may be used for the tearing down of old houses and the building of new—and indeed this can be cheaper than an alternative policy of refurbishing and restoring older houses not designed for modern conveniences such as running hot and cold water and proper sewage disposal, much less the use of air conditioning and other features that may make the lives of hosts and guests more comfortable during hot summer months. Additionally, outsiders attracted to rural living may well build in non-traditional ways—and the overall effect is that the village begins to lose the architectural homogeneity that attracted tourists in the first place. A second danger is that in attempts to retain that traditional style, properties change their use and become, for example, small hostels, restaurants or retail units while the family moves out into more modern buildings. This has a private benefit in increasing any rate of return on the refurbishment, but if a cluster of such buildings emerges the locale begins to take on the appearance of a commercialized tourism zone, or even a theme park appearance. This is not only a problem for rural areas, but also for the more historic zones of older towns. For example, the main thoroughfares of Pingyao in Shanxi Province as illustrated in Figure 15.3. In this instance many of the main routes from the gates of the old walled city to the centre have become a row of retail, catering and accommodation units that began to take on the appearance of a 'themed classical Chinese tourist old-town'. Indeed Pingyao's 'old city' represents an interesting case of touristification, given plans to resettle half of its 40,000 residents into the surrounding 'new city' to meet plans of tourism development, heritage protection and life enhancement for residents through the provision of better living accommodation.

A third possible problem is that the local takes on a hybrid appearance, mixing old and new, thereby again losing the presumed authenticity of its

Figure 15.3 Café and retail areas in main street in Pingyao.

built heritage. The overall effect is one that endangers that which originally attracted the tourists.

However, there are examples where tourism now allocates value to premises that previously were regarded as having little value and which were falling into disrepair and being allowed to fall into ruin, but which were not demolished because incomes were insufficient to finance an alternative building. The paradox thus arises that it is a lack of income that has helped retain the older buildings. One example where this has occurred is in Chi Qiao in Shanxi Province. Here an old village temple that had become almost totally destroyed was rescued and rebuilt under a village regeneration scheme that permitted a group of Buddhist nuns to restore and use the temple. This has also been supported by a co-operative of female workers who now make clothes that are sold in the local capital, Taiyuan, while that town is also a market for a growing rural tourism (Dumreicher 2008).

Encroachments on the Environment

The very success of rural tourism in some areas has created problems in handling the numbers that visit the zones. Zhang, An and Sun (2007) report that in the Golden Week following National Day in 2005, 250,000 different tour buses were present in the Hauirou District. A major rural attraction in the

Figure 15.4 Backpacker's hostel in Pingyao.

Yanxi Lake was reported as receiving 82,000 visitors with 3,000 to 4,000 cars parked there each day. The extra demand from visitors is having an effect on waste disposal. Landfill sites are in danger of being overwhelmed with potential problems of toxic water run-off into groundwater and natural drainage systems. In many rural areas water is taken from wells, and apart from perhaps boiling the water no other system exists for the removal of impurities. As a result, rural tourism, especially that close to conurbations, may gain little economic benefit from tourism, but may well incur cost externalities due to environmental pollution. Gu and Ryan (2007) reported on the situation in Hongcun where a small moated, walled village of some 700 households from the Shaoxing period of the Southern Song Dynasty now host over 500,000 visitors a year with associated social disruption due to noise and crowding. However, as in the case of Zhang, An and Sun's study, in spite of evident problems of overcrowding and potential environmental damage, the local communities want even more tourism because of the economic improvements they see in their incomes. This represents a continuing *leitmotiv* when considering rural tourism in China—the wish to escape grinding poverty for many millions means a willingness to grasp at any means of escape. The harvesting of tourists represents a far easier means of earning an income than the tenuous harvesting of crops, and the outcome is that the current generation will accept the development of tourism while leaving the consequences of overcrowding and environmental degradation to be dealt with by the next. While this might be seen as short-sighted from a western perspective, the issues can be stark in the rural communities, where it is only by developing tourism that individual families can provide a future for their child that can empower them to deal with future problems. Failure to act in such a way may mean a condemnation to even worse ways of earning a living as is demonstrated by the well-documented ills of the Chinese coal mining industry.

In many instances of studies of resident attitudes towards tourism in China one finds evidence of enthusiasm for yet further developments even where respondents can identify negative effects (Ryan and Gu 2007) or even where sometimes initial expectations have not been fulfilled (Wall and Stone 2005). Wall and Stone (2005) note, in the case of Jianfengling National Park, that optimism as to future benefits were high in spite of evident weaknesses in tourism planning and actual visitor flows, such as a lack of linkages between the park and the town of Jianfeng, lack of accommodation and profitability due to numbers of visitors failing to meet expectations, lack of involvement of the local community in planning and a lack of experience on the part of park managers. That optimism remained high they explain by 'the early stage of development' (Wall and Stone 2005: 83) although it might be added that the very lack of experience might also have been a factor in that managers failed to recognize they had a problem! Indeed, further study by these authors found confusion over what actually constituted ecotourism, which was supposedly the centre-piece of the Provincial Government's Department of Lands, Environment and Resources EcoProvince proposal for tourism development in Hainan. Wall and Stone report that the details of the plan not only incorporated remote wilderness areas, but also botanical gardens, beaches, a zoo, wildlife, ocean and cultural theme parks. They note that some of these do not meet the usual definitions of ecotourism that was being touted, but it can be argued that the concept of 'EcoProvince' goes beyond restrictive definitions of 'ecotourism' as better representing a branding of a portfolio of products under a heading. However, significant deficiencies were identified that included a failure to recognize that ecotourism resources required investment of earned monies—that, in short, they could not be used simply to subsidize other activities, while few cited the cultural aspects of ecotourism—implying that local community involvement was either understated or undervalued. Wall and Stone conclude that Jianfengling National Park is not a success, and nor is the EcoProvince policy, but that the fault lies not so much in the concept, but in its management. Under these conditions improvements in management structures, skills and decision-making become means of tackling the issues revealed by this study.

Lack of Community Involvement

Currently among Chinese researchers, the concept of community involvement in tourism development is mainly from the government and developers' perspectives, that is, tourism needs community support and therefore developers should seek the co-operation of the community to accomplish projects (Li 2004). However, in practice it often appears community involvement is generally missing in the initiation of tourism projects. Li's (2004) research on community tourism in Nanshan Cultural Tourism Zone (NCTZ) in Sanya, Hainan Province, revealed various problems and barriers. For example, tourism development in China is, to a large extent, planned and manipulated by local government that, on many occasions, supports the developer because of the investment being attracted. Often the government owns the land and has the authority to designate anywhere for tourism development, and thus communities can only passively allow developers to use their land for any state-supported projects, receiving only a one-off payment and land compensation fee. The developer's understanding of tourism–community relationship is limited to simply maintaining a good relationship with local residents and has little to do with tourism planning and product development. This approach is often sustained by a belief that local residents are unqualified to be involved in tourism projects. Commonly, there is a perceived need to obtain quick returns from the investment without the 'trouble' required to organize and sustain community participation (Tosun 2000). On the other hand, due to limited experience and lack of education, local residents (villagers) are often unaware of tourism impacts and this too causes some apathy towards participation in tourism. Li (2006) also suggests that the absolute authority of government and their manipulation of tourism developments in NCTZ also caused corruption in government and academic institutions.

Ying and Zhou (2007) support this perspective, pointing out that the current socio-economic and political settings of China make it almost impossible to completely copy western society's community participation principles into China's rural areas. However, the case of Xidi (Anhui Province) discussed in their study (and the next chapter of this volume) indicates that an elementary involvement of community on the benefit-sharing can be achieved. Currently in rural China, local residents generally lack democratic awareness and monitoring systems, and do not possess sufficient capabilities to participate in the tourism planning process. Therefore, the decision-making phase of community involvement commonly tends to address economic interests rather than democratic or political rights (Ying and Zhou 2007).

The Urban Environment

It can be noted that most studies on tourism development and community participation in China are related to rural environments. Few seem to exist with reference to urban locations, unless specific to heritage locations. Generally speaking, the very numbers of China's urban populations create tolerances of crowding—indeed, it is almost to be expected. Complaints exist about traffic congestion, noise and polluted air, but these are perceived as being more a problem associated with fast economic growth, a building boom that has been sustained for over a decade, and continuing industrialization. A few studies do exist. For example Wang, Li and Bai (2005) undertook a small-scale study of residents' perceptions of tourism in Harbin, which has a population of 4.64 million in the city area itself. The city is also famous for its Ice and Snow Festival, one of the largest of its type in the world. Generally it was found that tourism made Harbin a more exciting place, and that it enriched the local culture, but generally the potentially negative impacts of tourism were not regarded as a problem and there was little evidence that tourism added significantly to the urban problems. Similarly, there was a perception that while tourism generally did create employment opportunities, within the totality of economic life tourism was associated with problems of seasonality and other industries made far greater contributions to income and employment.

In two studies of Beijing's Shi Cha Hai hutong, Ryan and Gu (2007) and Gu and Ryan (2008) found evidence that tourism services were primarily provided by non-Beijing residents. Again complaints existed by the residents of the hutong about noise and congestion, but opinions differed as to the degree to which residents felt able to influence Beijing's planning authorities. However, visits over a period of three years prior to August 2008 provided plenty of evidence of the BTA's determination to ensure that the hutong would be 'spruced up' for the Olympics held in that month, as shown in Figure 15.5, which shows workmen plastering external walls while many of the alley-ways had been paved, and in some instances widened.

In many respects, tourism and its impacts on China's major cities pale into insignificance when considering the major boom in urban redevelopment. It is estimated that in Beijing, from 1990 to 1998 some 4.2 million square metres of housing in the old city was demolished. By the end of 2000 it was estimated that about 32,000 families, comprising about 100,000 people, had not yet been accommodated, although some had been waiting up to five years to move into a new house (Fang 2000). Similarly in Shanghai, from 1995 to 2004, more than 745,000 households were relocated and over 33 millions of square metres of housing have been demolished (Shanghai Statistical Bureau 2005). This has not been without significant social disruption as residential neighbourhoods have been displaced with the loss of social networks and, in some instances, lost life chances (Leaf 1995; Gaubatz 1999; Fang 2000; Zhang 2002; Zhang and Fang 2004).

Ways Forward

It has been noted that today there may be even greater differences in income and wealth between urban and rural areas than in the past. From an economic theory perspective, Kuznets (1955) argued that the relationship between economic development and income inequality followed an inverted

Figure 15.5 Preparing for the Olympics in Shi Cha Hai hutong, 2007.

U-shaped relationship whereby, in the early stages of economic development, income inequalities widen, only to close as the economy matures. The basis for this is that, as is arguably occurring in China, as the non-agricultural sector achieves higher efficiencies and incomes, given its initial minority position, income inequalities are at their greatest in the earlier stages of economic development. However, as over time the agricultural sector continues to lose labour to the non-agricultural sector, the latter becomes an increasingly larger part of the economy, and thus the wage and income differentials become less marked, thereby reducing the inequities in income. A similar approach is adopted by the Solow-Swan model which argues that higher yield capital attracts additional investment and labour, and over time there is a shift from low capital–labour ratios to higher capital–labour ratios, thereby creating a convergence of incomes. From this perspective the issue of income inequality becomes one that is soluble over time and depends upon rates of change in economic development. Peng (1999: 239) notes that one of the 'most tangible achievement of Chinese economic reform is the burgeoning growth of the rural nonagricultural sector, primarily township and village enterprises (TVE)' and that TVEs outpaced state-owned enterprises for much of the period since 1980. Given this, one would expect a convergence of incomes over time. Peng (1999) also notes that prior to the reforms of 1978 less than 10 per cent of China's employees were in the non-agricultural sector, and thus it is not surprising that the divergence of income noted earlier has come into being as the economy developed. To this observation can be added a second, namely, that while there existed a surplus of a rural labour force, from many perspectives there existed an under-supply of productive farm land, thereby again depressing rural incomes further. Peng's (1999) own analysis supports much of these contentions, and being based on sub-provincial administrative regions he found that while regional differences existed, the income inequalities were not only of an urban–rural divide, but also an inter-rural and intra-urban nature too. The evidence seems to point to rural areas proximate to urban centres appearing to be the initial ben-eficiaries of this process of a shift towards higher capital–labour ratios and the growth of non-agricultural employment, and it might be argued that tourism is both determined by, and determines in an interactive nature, this process. First, levels of tourism are determined by the growing incomes of an urban class that then seek periods of relaxation, often in rural locations within easy access of the centres of population. Second, the beneficiaries of the resultant income obtain not only more income, but tend, at least initially, to have high leakages into their local economies by buying locally, thereby adding to local income and employment levels.

In some ways the issue is akin to the classic Rostow (1956) take off point. Rostow argued that the process of economic growth effectively occurs within a relatively short period of two to three decades within which a soci-ety and economy transforms itself. During this period the rate of investment 'increases in such a way that real output *per capita* rises and this initial increase carries with it radical changes in production techniques and the disposition of income flows which perpetuate the new scale of investment and perpetuate thereby the rising trend in *per capita* output' (Rostow 1956: 25). In Keynesian terms it might be argued that a series of positive multi-pliers are set up and in due course a self-sustaining point is reached where economic growth becomes the norm. However, as Rostow himself is careful to point out, this requires significant social and political change, and thus, at the local level of the Chinese Village Community an imperative is that the inhibiting factors identified earlier are overcome. Cottrell, Vaske and Shen (2007) implicitly support this proposition in their study of residents' perceptions of tourism and its impacts in Chongdugou Village in Henan Province, which is subdivided into four sub-villages with a total popula-tion of 1,263. Initially dependent on mining, bamboo and timber harvest a local government-sponsored tourism company manages a project whereby local residents rent rooms for bed and breakfast guests. Shen (2004) reports that after four years substantial increases in income were being reported. Cottrell, Vaske and Shen (2007: 221) found the most important factor in shaping respondents attitudes towards tourism and their support for tour-ism initiatives were what they termed the 'institutional dimension' which 'stresses participatory decision-making processes and public participation'. It can be argued that it is the act of debate, being valued and listened to, and through participation that the social and political changes important to securing an economic take off point are secured.

The issue thus arises as to how these changes can be secured. Dumreicher (2008) refers to creating a frame of mind that considers seven different stages of mindset wherein the individual commences with an appraisal of self and through stages of house/family, street, village, region, China and a national identity and the globe progresses to an appreciation of stages towards future sustainability. Referring to the same sets of village projects, Marschalek (2008) might be said to offer more concrete models. Based on a six-year European Union funded project in seven villages in six Chinese provinces, the process was envisaged as a series of loops that began with the collection of ideas from villagers as to what they wished to achieve, how they thought their goals might be obtained, and almost just as importantly, what it was that they did not wish to happen. With a team leader the ideas were evaluated and then subjected to assessment by a team of supporting scientists. Potential outcomes were assessed, projects commenced and a process of review instituted with an important criterion being the capability of a project continuing once European funding and external support were removed. The process is illustrated in Figure 15.6.

A number of strengths of the project can be identified. First, there was the sustained duration of the project and the continued presence of a team leader. Second, the presence of external expert and scientific support that

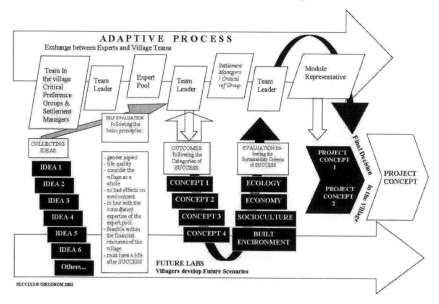

Figure 15.6 Developing village-based tourism, sustainability and participation. Source: Marschalek (2008).

overcame the issues of lack of experience identified by Zhang, An and Sun (2007). Third, Marschalek (2008) also refers to the fact that the 'outsiders' were able to act as intermediaries between villagers and governmental representatives in such a way that both parties were enabled through the defusing of potentially difficult issues. As outsiders the European team could act as mentors, but also as blame takers, thereby permitting villagers and government to retain legitimacy and 'face'. This raises an interesting point as to what degree does such a party have to be 'non-Chinese' to fulfil this role. Another issue is whether such processes need to take six years. In this instance a lack of familiarity with place and people on the part of the European project personnel inevitably meant that time was required for knowledge about each other to be gained by all parties, and it can be argued that similar Chinese-sponsored projects would not require so much time.

THE FOLLOWING CHAPTERS

Many of the themes of this chapter are illustrated in the following chapters. In the first of these, Gu and Ryan draw upon research undertaken in Anhui Province at the UNESCO designated heritage site at Hongcun to show ambiguous attitudes existing whereby local residents are supportive of tourism development, but also wish to have better control of it because of the way in which it has been introduced by control being passed to an external Beijing-based company. Comparisons are made with Xidi in the same province where a Village Committee has retained control and where in consequence, higher incomes are being obtained from tourism. In Chapter 17, Zhang, An and Liu present evidence as to how communities have become involved in tourism in the Pinggu, Yanqing and Miyun districts, Beijing Province. They find evidence of a new appreciation that tourism needs to not simply be a force for the creation of income and employment, but also must seek long-term sustainability through a respect for the community and its interest in local cultural, heritage and natural environments. As indicated previously, one of the key features that has emerged in China within the past few years has been a growing appreciation that tourism growth at all costs is not sustainable, and this work demonstrates how these concerns are being faced at the village level. In many ways their work confirms that of the European SUCCESS (Dumreicher 2008; Marschalek 2008) projects in requiring a reinforcement of local democracies and the provision of educational support.

In Chapter 18, Zhi and Wang return to the vexing issue of carrying capacities, but in this instance look at the psychological tolerances of both visitors and guests. As Gu and Ryan comment in their analysis of Hongcun, domestic tourism can impose considerable potential stresses through the sheer volume of visitor numbers. Zhi and Wang find that an important determinant of visitor tolerances is the attitude adopted towards them by residents—in short, being friendly evokes friendly responses. In Chapter 19, Zhou and Ma examine how two communities in Baiyang Lake have adopted two different approaches to providing tourism products—the one by providing a replica of traditional life-styles that minimizes intrusion effects, the second embracing tourists within the community's own homes and streets in order to obtain a more equitable distribution of income. In Chapter 20, Gu and Ryan examine the concept of place attachment with reference to Shi Cha Hai.

ACKNOWLEDGEMENTS

Photographs are reproduced by the courtesy of Chris Ryan. Figures 15.1 and 15.2 are reproduced by the kind permission of Janet Stephenson and Figure 15.5 by the kind permission of Ilse Marschalek.

16 Hongcun and Xidi

Rural Townships' Experiences of Tourism

Gu Huimin and Chris Ryan

INTRODUCTION

It has been stated elsewhere within this book, and on several occasions, that the Chinese government is driven by a socio-economic imperative to generate income to address the problems of poverty in China. Much of that poverty can be found in both rural and urban areas, but it can be argued that urban poverty is, to some degree, a reflection of rural poverty in that urban areas attract migration, both legal and illegal, because of the low rural incomes. Solving the issue of rural poverty thus helps address, to a degree, the problems of urban poverty by minimizing the need to migrate and accept low income jobs and poor living conditions in urban sites. For example, many construction workers in cities live on the building sites or in poor housing of a temporary nature as in other similar areas of the world such as Dubai. This permits the remission of monies to families back in the rural areas. The Golden Weeks possess enormous emotional importance as being one of the few times in the year when, for several families, parents can be reunited with their young children who will be looked after by grandparents. In a family-oriented society the pains of economic growth are heavy burdens for the current generation of Chinese of working age, and this cannot be stressed too much. Its very commonality often goes unremarked, but the silence does not mean there is no sense of loss for many Chinese.

Under these circumstances the importance of tourism to rural areas has been restated many times in this book. The purpose of this chapter is to draw upon past research by the authors to provide a case study of how tourism has both advantages and disadvantages in the Chinese context, and it does this with reference to the UNESCO nominated villages of Hongcun and Xidi in Anhui Province. However, to establish a context for the research findings two initial sections are provided. The first relates to the role of Township and Village Enterprises (TVEs) and the second describes the nature of the villages. The third section will briefly pick up some themes from Chapter 20 in terms of place attachment and then proceed to describe some of the results derived from the study before discussing the implications of the findings in this specific case.

TOWNSHIP AND VILLAGE ENTERPRISES (TVES).

Zhang Lei (2007: 92) notes that any policies that seek 'to address issues of Chinese farmers, agriculture and rural society are doomed to be facing dilemmas'. In terms of policy development a key role is allotted to the TVEs and small towns. They are often perceived as an answer to alleviating rural policy, generating income growth to reduce the gap between rural and urban incomes and patterns of well-being, creating non-agricultural business enterprise and avoiding social unrest (Jiang 1998; Zhang Lei 2007). In 2000, the importance of TVEs was reinforced in *Opinions on Speeding up the Sound Development of Towns*, issued by the Central Committee of the Communist Party Congress and State Council, and further reinforced by the Five-Year Plan of TVEs issued later that laid emphasis on zoning and restructuring, partly to help control the emergent environmental degradation that had accompanied some past initiatives within the TVE programme. TVEs evolved from the commune and brigade enterprises that existed in the 1950s, and fell under the aegis of the Ministry of Agriculture. They covered most enterprises in rural areas and a range of ownership forms that include privately owned businesses, collectives, co-operatives, village committees, collectives and remnants of the old communes. The year 1994 was to see the apogee of TVEs in terms of numbers employed, 125 million in that year (Ministry of Agriculture 1997), although Ji Li and Fang Yongqing (forthcoming) put the number as being between 92.6 to 120 million.

By the end of the 1990s the economic contribution made by TVEs to the Chinese economy was substantial. Ji Li and Fang Yongqing (forthcoming) state that total revenue was 4,358.8 billion RMB in 1994, and that they accounted for 47 per cent of the country's output. The same source indicated that by the end of the 1980s, 84 per cent of TVEs were run by individuals. Park and Minggao Shen (2001) argue that one factor behind their expansion was the readiness of the banks to provide funding, and by the late 1990s, TVEs accounted for just over a third of bank lending in spite of occasions when collateral was not forthcoming. That this happened was due to the close relationship of business owners to their local communities and the role of local governmental bodies in the businesses. Often local officials provided a monitoring function that, when combined with *tongshou huandai* (the use of pooled funds to repay loans—the pooling often being between other local organizations and concerns) and the Chinese concern of face meant that funding was often forthcoming even when initial available capital was low. However, by the late 1990s circumstances began to change. Interest rates increased, competition between the Rural Credit Co-operatives and Agricultural Bank of China began to favour more conventional private firms and their patterns of ownership and rapid privatization of TVEs began as a response. A conventional explanation for the decline of importance in TVEs that commenced at about 1997 was a growing reluctance by the Chinese banking sector to accede to politically driven requests for TVE enterprise funding of doubtful profitability. Park and Minggao Shen (2001), however, argue that other factors were in play as shown by a rising amount of failure to repay loans by those able to make such payments, (strategic defaulting) and they argue that performance-oriented lending practices with different emphases by the banking sector account for the changed, and more expensive, funding of TVEs.

HONGCUN AND XIDI

As noted in the introduction, both Xidi and Hongcun are UNESCO designated sites. One translation of 'Hongcun' is 'Chinese Village in Picture,' which is appropriate for its setting and layout as illustrated in Figures 16.1 and 16.2. It is noted for a water-supply system designed in the Shaoxing period of the Southern Song Dynasty based on the digestive system of an ox. The winding dam is the 'intestine' of the ox, the Moon Pond is its stomach, the southern lake its belly, while the villagers who live on both sides of the 'intestine' represent the body. The alley-ways are linked by channels of running water, and traditionally given patterns of water usage were undertaken at different times of the day. Within the walls of the city are more than 140 heritage rural houses dating from the Ming and Qing Dynasties. The village

Figure 16.1 Approach to Hongcun across moat and between summer water lilies.

Figure 16.2 Water cistern—this was the backdrop for a sequence in *Crouching Tiger, Hidden Dragon*.

was assessed in 1999 by the Experts Appraisal Committee of the National Construction Ministry, the Antique Management Bureau and other government agencies, who produced the 'Hongcun Protection and Development Plan' in 1999. In November 2000 it was recognized as a World Heritage Site by UNESCO, and since then has won further recognition from the Chinese governmental authorities including a 4A scenic spot classification.

The community comprises a population of 400 households, about 1,200 people, and first became involved in tourism in 1986, but at that time the levels of visitation were quite small. The authors were informed by a community member and past elected government official that, in 1997 the local government and Huangshan Tourism Development Co. Jingyi Ltd, came to an arrangement whereby the company would be solely responsible for the promotion of the location, and from 2002 the revenues resulting from the sale of entrance tickets to the community would be divided on the basis of two-thirds being retained by the company for the costs of promotion and staffing of the ticket sales outlet, and one-third of revenues would, in turn, be passed to the local Huangshan government. Of this revenue, the local government would retain approximately two-thirds for the maintenance of general infrastructure and development of the municipality and the remaining one-third (approximately 11 per cent of the total revenue) would be spent within the township. Of this one-third, approximately two-thirds again would be spent on development of infrastructure within Hongcun itself, and the remaining one-third (i.e. just under 4 per cent of the original ticket sales) would be directly allocated to the population of Hongcun. This arrangement was signed to last for a period of 30 years. It is possibly worth noting that the contract predates the formal recognition of Hongcun as an UNESCO World Heritage Site by three years. It should also be noted that plans exist for further development. In 2002, the Beijing Zhongkun Investment Group and the People's government of Yi County signed a contract for the further development of Hongcun Village that envisaged the construction of a 0.2 billion RMB resort complex—the Huangshan International Hongcun Qishi Lake Holiday Resort. At the time of writing, the hotel resort complex has not been commenced, but tourism numbers have increased, partly due to both Hongcun and Xidi (some 11 kilometres away) being the location for the film, *Crouching Tiger, Hidden Dragon*, directed by Ang Lee in 2000. Tourist visitor records show that in 1992, 6,700 domestic tourists visited Hongcun. A decade later the figure had climbed to 253,900. In 2006 the domestic visitors numbered over 500,000 while international visitors numbered over 20,000 to yield revenue from ticket sales in excess of 16 million RMB for the Huangshan Tourism Development Co. Jingyi Ltd (unpublished data made available from the Yixian Tourism Bureau 2006).

The contract has been the subject of dispute. In November 2000 some 300 households submitted a report critical of an original contract with Huangshan Tourism Development Co. Jingyi Ltd, by which the company retained 95 per cent of ticket sales after a fixed payment of about 160,000 RMB. The county government rejected this and more than 730 villagers then appealed to the Anhui Provincial Court in September 2001: Again the appeal to take back control of local tourism proved abortive. Finally, the agreement described previously was reached in 2002, but it still leaves many dissatisfied (Ying and Zhou 2007). For example, as seen by the authors in this current research, non-local guides are employed, and there is a feeling that the guides are insensitive to local feelings.

This situation compares unfavourably with the situation in Xidi. There the Village Committee as a TVE have taken control of tourism. Xidi Tourism Services was established by the Village Committee and it resisted attempts by the Tourism Bureau of Yixian County to take control of Xidi's tourism. Initially visitor numbers were low (in 1986 there were but 147 visitors; Ying and Zhou 2007), but in 2005 the numbers exceeded 600,000. As Chen and Wen (2002) describe, in Xidi tourism provides 80 per cent of the jobs available in the village and a similar proportion of the village's total income when taking into account both direct and indirect multiplier effects. Tourism has paid for the installation of tapped water supplies and gas for the entire village, and there are 146 souvenir shops and 20 restaurants (Ying and Zhou 2007). In recent years more than 7 million RMB have been spent on road construction, new parking areas, touring routes, viewing spots and additional funds are set aside for restoration and maintenance projects. In total it has been estimated that approximately 62 per cent of Xidi's tourism income remains within the village. Indeed, the village has to cope with very different problems to those experienced by Hongcun. Criticisms are now being made that the village has failed to properly control the number of tourists and that the original nature of the village has been undermined by the very evident tourism-oriented nature of the main streets lined with competing souvenir shops, restaurants and hotels. That this has occurred is due to a very different spatial arrangement of buildings. Xidi comprises a main street with roads in parallel and connecting smaller streets and alleys, along some of which lie other houses of historical and cultural significance other than simply their age. The village lies open to the fields that lie behind it and which until two decades or so ago were the major source of revenue and livelihoods for the community. Unlike Hongcun it is not surrounded by either a moat or a wall, but lies open to pedestrians, although access was and remains primarily by a road that comes from Huangshan City. It thus can be seen that very different experiences exist between these two communities.

RESIDENT PERCEPTIONS

While Xidi has been criticized as indicated earlier, it needs to be noted that the community has retained control of tourism and the manifestations of tourism as noted are those acceptable to a Village Committee that reflects the views of the local community. Generally there is strong support for tourism as a means by which per capita income has significantly increased to the point where it is now four times the average for Yixian County and was approximately 7,000 RMB per person in 2006. This represents significant and comparatively rapid growth in per capita income in less than a decade, illustrating the quite dramatic impact that tourism can have. However, much of this growth only occurred after the Village Committee turned to outside sources of finance as suggested by Park and Minggao Shen (2001). As Ying and Zhou (2007: 105) comment: 'From the case of Xidi, it can be found that without funding from the outside, especially in the initial period of the tourism development process, the local community experienced a time of hardship with the lack of self-owned funds and the difficulties in obtaining loans or financial support from banks or government.' The success of Xidi was based on initial volunteer action in cleaning up the village and a willingness to plough back initial earnings into the Village Committee's tourism initiatives because a longer term perspective was adopted by the villagers. As noted earlier, the end-result has not only been increases in income for individual families but also an enhancement of community assets that include roads and new school buildings and a local television service. The programme is now primarily self-funding, but a key factor in this was the ability, and perseverance, of the Committee in the early 1990s to acquire bank borrowing. Total visitor spending in the village is now estimated at over 7.5 million RMB per annum.

The interesting aspect is that in Hongcun, although it has failed to achieve anything like the same success as Xidi in attracting resources, the community still supports tourism. In a survey carried out by the authors in 2006 with a sample of 418 residents, only 22 per cent 'agreed', 'strongly agreed' or 'very strongly agreed' that tourism was growing too fast for the community to cope. Of the total, 67 per cent 'agreed', 'strongly agreed' or 'very strongly agreed' that tourism boosted economic development. Similarly, 60 per cent felt tourism encouraged and created a market for handicrafts, and 29 per cent felt that 'tourists interfere with residents' lives'. A total of 63 per cent 'agreed', 'strongly agreed' or 'very strongly agreed' with the view that it was a good idea to attract more tourists. Gu and Ryan (2007) argued that the issue at Hongcun was not, from the residents' perspectives, one of tourism *per se*, but of the community wanting to better access the benefits generated by tourism. The current *modus operandi* was one whereby many operated tourism enterprises illegally, thereby incurring the displeasure of those who had sought licences; and it was the legal and operational framework that created these forms of tensions within the local community.

DISCUSSION

The case studies illustrate the importance of the role of community as a participant in the development of rural-based tourism when associated with local heritage, architecture and history. However, what the case studies also indicate are the importance of local government and the way in which it operates. Similarly both villages display the dominant nature of the economic imperative, and development has seemingly paid little heed to concepts of carrying capacities and the goals appear to be ones of increasing visitor numbers and hence expenditures. Marketing is still based upon numbers, especially in the case of Hongcun where the main revenue for the commercial partner is derived from tour and ticket sales, and where that partner has little direct interest in the community *per se*.

The physical development of Xidi is such that it has become a main street characterized by the souvenir shops and other paraphernalia of tourism including restaurants and some small hotels. The car-park is also lined with stalls selling souvenirs, and visitors arriving there are directed to the ticket office. At the office one buys a pass that provides entry to some of the historic houses. The scene that meets the tourist is similar to the picture of Pingyao in Chapter 15, namely, a loss of a sense of an agriculturally based community for one that, arguably, resembles some theme park of Chinese rural life. The overseas visitor, as in many places in China, will be the subject of calls from vendors who seek to engage interest in their wares. Hongcun produces a different perspective due to its geography. Residents have turned some of their rooms into small retail areas, but they remain relatively private when compared to Xidi. The narrow alleys and the sound of running water permeate the experience, until one runs into a tour group conducted by a guide with a microphone telling stories of the village and its history! The fact that there is no immediate commercial accommodation or apparent open restaurants inhibits overnight stays—certainly by tour groups, although independent travellers might stay in the farmhouse-style accommodation that can be found there on the perimeter of the village. Thus spatial patterns also have a role to play in the generation of the visitor experience. Both villages are subject to visitation from Huangshan City, and for many visitors the trip to the villages is a side trip from either the city or Huangshan National Park. Geography thus has a role in shaping tourism patterns and experiences and, when combined with a lack of village-based accommodation, reduces visitor spending. On the other hand, in the evenings the villagers can again reclaim their village and its social groups from the touring parties accompanied by microphone-wielding tour guides.

Both villages have very high visitor–host ratios which appear to get higher with each passing year. From a western perspective the villages can often appear crowded, especially when one encounters several tour groups of domestic Chinese visitors. However, as in the case of the hutong described in Chapter 20 and as just noted, pressures on the local community are eased by seasonal and daily patterns where for the most part, the villages empty of tourists. An additional factor is that both still retain agricultural occupations and thus it is possible if an individual wishes to work away from the crowds. And, given the population size and densities of China, many Chinese are habituated to higher densities of population than are their western counterparts.

Possibly what the case study also illustrates is the deficiency that exists within local Chinese communities of an experience of tourism planning and an awareness of the problems that tourism can create. While both are accredited with UNESCO status, the monitoring appears to be done with a light hand, and the authors were regaled with stories of how illegal businesses simply shut up shop during any period when UNESCO staff might visit the sites. As in other instances, a conundrum exists in a system once attributed with having strong central controls now appearing to generate systems of competing structures at national, provincial, municipal and local levels interleafed with differing levels of private sector entrepreneurial activity. Somewhere in this mêlée a system of structured tourism planning seems to go awry, perhaps because of (a) the economic imperative of a need for income generation, (b) little or no experience of how tourism in other parts of the world has created negative social and environmental impacts, (c) all parties being committed to tourism growth for reasons of securing prestige, income and political status and (d) the fact that behind all of this lies the shadow of corruption. In 2006, four judges, including Zhang Zimin, former president of the Intermediate People's Court in Juyang, Anhui, stood trial on charges of accepting payments of 1.3 million RMB. Indeed, to counter these practices, from January 2008 a process of swapping judges and provinces will commence to attempt to ensure that judges cannot have too close a tie with local officials. Corruption remains a festering sore as was recognized in public at the 17th National Congress. All of these factors seem to add weight to the thesis suggested in Chapter 8 that China is currently character-ized by a dismantling of its past statism.

ACKNOWLEDGEMENTS

The authors wish to acknowledge research grants from Beijing International Studies University and The University of Waikato Management School that made possible the research on which this chapter is based. Photographs are reproduced by the courtesy of Chris Ryan.

17 Community Involvement in Rural Tourism Development
Evidence from Pinggu, Yanqing and Miyun Districts, Beijing Municipality

Wen Zhang, An Yanyan and Jumei Liu

INTRODUCTION

As noted in previous chapters, the rapid growth of tourism in China has been not only characterized by benefits, but increasingly by unexpected negative economic, socio-cultural and ecological impacts. In recent years, much attention has been given to these issues, and the topics of a community approach and the sustainable development of tourist destinations have increasingly become pre-eminent in that country. Earlier studies of these issues in China included (a) the introduction of international research on both community-based tourism and the sustainable development of tourist destinations (e.g. He 2002; Zhang and Wang 2003); (b) the importance of community in destination development (e.g. Jiang 2003; Yang 2001); (c) sustainable development of both the tourism industry and the community (Hu and Zhang 2002); (d) planning and developmental modes of community tourism (Wang 2004) and (e) an analysis of the contents of community-based tourism development such as: management and operation, decision-making, distribution of income, protection of the environment, related education and training etc. (Wang 2004; Liu 2000). Empirical studies have indicated that community-based tourism development in China has generally contributed to the social and economic development of a locality, but there remain great differences in the degree of residents' involvement and the distribution of benefits while difficulties remain in implementing the community approach (Sun and Bao 2005; Sun 2005; Fan, Wall and Mitchell 2008; Gu and Ryan 2008). Consequently, the objectives of the research reported in this chapter are to examine: To what degree are residents of a tourism destination involved in tourism planning, to what extent are their culture and traditions respected, whose interests are considered in tourism development and what is it that residents perceive as the core of the local tourism development? Additional issues involve answering the question: To what extent is 'community-based tourism' actually community owned, operated and managed sustainably? Finally, how closely is local residents' involvement in tourism correlated with the development of the community?

The context of the research is rural tourism development around Beijing. This was chosen as an example because it is one of the earliest areas in China that developed rural tourism and is both typically market oriented and community based. The data for this study are obtained from a completed questionnaire and in-depth interviews conducted in the period from September to November 2006 in six villages in the rural parts of the municipality of Beijing. Altogether 300 questionnaires were handed to the village residents, and 81 per cent (n = 243) were collected and found complete. In addition, a total of 21 local residents or village administrators were interviewed.

COMMUNITY APPROACH AND SUSTAINABLE DEVELOPMENT

Tourism may be regarded as consisting of tourists, a business and an environment or a community in which this industry operates (Williams and Lawson 2001). As noted in the other chapters that form this segment of the book, a sense of 'community' has become increasingly important when considering tourism development for many reasons. There is a concern that communities should remain economically viable to sustain social relationships that in turn sustain differences of culture and human traditions, but changing economic circumstances create their own change. Further, Gu and Ryan (2008) refer to theories of place attachment and place identity (Breakwell 1992) to show how psychologically important attachment to place is to establishing self-identity, self-esteem and senses of self-efficacy at times of community change due to tourism impacts. Community involvement is also thought important in the development of regional economic planning (Getz and Jamal 1994). Equally, community involvement is considered essential for the longer term success of tourism product. In order to make tourism development more sustainable, Murphy (1983, 1985) was among the first to propose the community approach, arguing that the product and image that intermediaries package and sell is a destination experience, and as such creates an industry that is highly dependent on the goodwill and co-operation of host communities. It is, therefore, for these reasons that in recent years it has become widely recognized that planners and entrepreneurs must take the views of the host community into account if the industry is to be sustainable in the long term (Allen, Long, Perdue and Kieselbach 1988; Ap and Crompton 1998; Belisle and Hoy 1980; Doxey 1975; Maddox 1985; Murphy 1983). Indeed the terms 'sustainable tourism' and 'community development', increasingly seen in various combinations, became the 'buzzwords' of tourism development of the 1990s (Joppe 1996). In his article 'Community driven tourism development', Murphy (1988) examined the link between tourism and community development. He argued that the tourism industry had a responsibility to the community, in that it 'uses the community as a resource, sells it as a product and in the process affects the lives of everyone'

(1988: 97). To ensure that both the industry and community survive and prosper over the longer term it thus becomes necessary to develop at a scale and pace appropriate to local conditions. In this way tourism can become a true community enterprise, one that possesses mutually beneficial possibilities and synergism. The concept, 'community workshop', is proposed in Murphy's article as a means to obtain the mutual sustainable development of the tourism industry and the local community. Pearce (1980) also believes that the more local residents get involved, the greater are the benefits they can obtain from tourism development. Similarly Brown (1985) believes tourism development is a stimulus for community population growth and that the community plays an important role in providing tourism facilities. In China, many scholars have also studied the relationship between tourism and community development including Tang (1998), Liu (2000) and Lu (2001). Zhou and Ma (2008) examine this relationship in another chapter in this book, and argue that community involvement is also important in establishing an authenticity in rural tourism product that otherwise becomes simply yet another commodified reproduction lacking uniqueness even if based on local culture and tradition.

Past research of the nature cited earlier can thus be summarized as indicating that the support of the local community is crucial for tourism development. Indeed, the two components of community support and tourism development are reciprocal and supplementary. Further, the favourable development of a tourist destination will promote the healthy growth of the tourism industry. Consequently, a virtuous developmental circle can be envisaged whereby local support for tourism permits further tourism development that bestows benefits for the community, who in turn are reinforced in their support for tourism, leading to yet more developments and more benefits. Additionally, claims may be made for community sustainability (Zhou and Ma 2008). However, several caveats exist that effectively militate against such an optimistic scenario. Ryan (2003) summarizes social impact research and notes that local communities are far from homogenous, and that several psychographic studies (e.g. Lindberg and Johnson 1997; Lawson *et al.* 1998) indicate the simultaneous existence within a community of different attitudes towards tourism that can range from strong support to strong opposition. In a series of studies relating to Alberta, Getz and Jamal (1994) and Jamal and Getz (1995, 1999) have shown that community processes are expensive in time and effort, may delay commercially oriented decision-making and may in fact produce little real return for the effort expended. Butler (2006b, 2006c), in two edited volumes relating to the destination life-cycle, has indicated that diminishing returns of benefit to local communities can arise, and indeed it can be argued that further development of tourism shifts the focus of resource control away from local communities to external marketing organizations and sources of finance (Ryan 2003). Yet even here two further complicating factors arise. Fan, Wall and Mitchell

(2008) have argued that, in the Chinese situation, the role of government is ever present and thus one can argue that limitations can be imposed on tourism development through governmental control, while Ryan and Gu (2008) argue that the needs for distinctiveness in destination planning and marketing requires a process of 'glocalization' whereby both government and external business will involve the local to provide destination uniqueness. The question, however, is to what degree do local communities actually benefit from these processes?

CASE STUDY

Introduction to the Case Study

It is in the context of these contradictory views that the current research sought to assess what was effectively occurring in the wider rural parts of Beijing Municipality in the early years of the twenty-first century. Following the opening of the Chinese economy and the emergence of both domestic and international tourism, rural-based tourism has developed rapidly. From a wider perspective of economic and regional planning, this is quite important as it is a means of addressing the widening gap between urban development and rural poverty, and of offering employment and economic diversification to rural areas that will inhibit the high degrees of migration from west to east, and from rural to urban areas (e.g. see the description of red–green tourism by Gu, Ryan and Zhang 2007). Partly because of the proximity to a rapidly emergent affluent urban-based middle class, and an ability to attract at least part of the international tourism market travelling to Beijing, the rural parts of Beijing Municipality were among the first in China to seek benefits from tourism. Thus, since the 1980s, the suburban/rural areas of Beijing, especially those in the mountains, began to develop rural tourism based on local landscape, ecosystems and traditional customs. Consequently by the early 2000s a series of licensed accommodations were being made available for both domestic and international markets including farm-stay operations (Gu and Yu 2005). This research project thus sought to examine the success achieved by such tourism.

The Research

Design

The research reported in this chapter sought to examine the following issues:

1. Is rural tourism favourable towards the all-round development of the local community?

2. What are the attitudes of local residents towards tourism development?
3. Are the wishes, culture and traditions of local people respected?
4. Are the interests of residents placed at the centre of tourism development?
5. Are any negative tourism impacts perceived, and if so, what are they?

Consequently, this study proposed three hypotheses:

Hypothesis 1: The development of rural tourism around Beijing is community based.
Hypothesis 2: The positive impacts exceed the negative impacts brought by rural tourism.
Hypothesis 3: Local residents' involvement in tourism contributes to the local community's sustainable development.

The questionnaire was based upon the literature review cited earlier and included questions pertaining to perceived degrees of control consistent with the earlier work of Murphy (1985), and is also consistent with the subsequent utilization of place attachment theory noted by Gu and Ryan (2008). This section of the questionnaire posed statements and respondents were asked to express degrees of agreement on a five-point Likert-type scale ranging from (1) strongly disagree to (5) strongly agree. Additionally open-ended questions were also used to elicit more detailed responses and to allow respondents to give reasons for their answers. The structured questions included items on residence, degrees of community involvement and degrees of perceived sustainable community development. A further section asked for socio-demographic data. The open-ended questions mainly concerned the factors that enabled or inhibited a community approach to tourism policy implementation. A draft questionnaire was initially distributed to 20 graduate students undertaking tourism management studies at Beijing International Studies University and their responses led to some modification of the questionnaire. Subsequently, several colleagues in tourism were consulted about the modified questionnaire before the final form of the questionnaire was used in the field.

The questionnaire was used in six villages of three suburban/rural counties of the municipality of Beijing (Pinggu, Yanqing and Miyun), and these were supported with in-depth interviews and observational field studies over a one-month period in 2006. A total of 300 questionnaires were distributed of which 243 were fully completed, providing an effective rate of return of 81 per cent. The 21 in-depth interviews were of villagers, village officials and tourist company workers. The Kaiser-Meyer-Olkin statistic of sampling adequacy was found to be 0.77, indicating that further analysis of the quantitative data was appropriate.

RESULTS

Local Participation in Decision-Making

Table 17.1 indicates the mean scores and standard deviations for items used in the questionnaire. Place attachment theory implies that, if community members are to have a voice in tourism development, they must perceive that they have not only the means to voice an opinion, but also that such opinions are respected and heeded (Gu and Ryan 2008). This item asked respondents the degree to which they felt they could effectively voice opinions that were listened to. According to social exchange theory, residents are

Table 17.1 Perceptions of Community Control over Tourism Development

Item	Total Mean Std. Dev	Pinggu Mean Std. Dev	Yanqing Mean Std. Dev	Miyun Mean Std. Dev
The local residents have decision-making powers in community tourism development.	3.72 1.12	3.57 1.25	3.85 0.96	3.82 1.04
Community tourism planning is primarily done by the government.	4.29 0.82	4.37 0.86	4.63 0.49	3.86 0.85
You or your family's income has seen a rapid increase since tourism development commenced.	3.91 1.13	3.94 1.09	3.95 1.31	3.82 1.30
You have participated in tourism training courses organized by local government or non-government associations, etc.	3.84 1.31	4.06 1.12	3.88 1.31	3.48 1.50
You know where and how to voice problems in tourism development.	3.65 1.18	3.89 1.03	3.83 1.12	3.11 1.28
The problems you voice and the suggestions you propose are given attention.	3.29 1.15	3.51 1.00	3.51 0.98	2.75 1.33
It is your responsibility to maintain the tourist destination's image.	4.53 0.73	4.58 0.68	4.56 0.55	4.41 0.92
Co-operation with travel agencies will be a more effective way to promote your products.	4.06 0.93	4.15 0.92	4.22 0.79	3.77 1.01

unlikely to be concerned about tourism unless they perceive that benefits are being derived (Ap 1990, 1992; Ap and Crompton 1993), and given the role of tourism in Chinese rural development for economic purposes as discussed earlier, respondents were asked whether they or their families had achieved higher incomes because of tourism development. To be effective communicators and participants in community and tourism development, respondents were also asked to what degree they had received any training in tourism. The results are shown in Table 17.1.

Overall there was a perception that community planning was primarily undertaken by the government (mean = 4.29), but there was some agreement that economic benefits had been achieved by families (mean = 3.91). There was a strong sense of community responsibility in helping to sustain destination image (mean = 4.53) and that it was equally necessary to co-operate with travel agencies (mean = 4.06). Of interest is also the distribution of scores noted in Table 17.1 by showing the standard deviations. The finding that there is a perception that government is primarily responsible for tourism planning, and communities have a responsibility to support this through sustaining destination image is reinforced by standard deviations having values of less than one. These perceptions exist even where not everyone gains equally from tourism development in terms of additions to income. Of interest are the scores relating to assessments of community involvement in tourism planning. The overall mean scores imply comparatively lukewarm levels of agreement that individuals are involved, but also the table shows higher standard deviations on these scores, implying that while some feel comparatively disenfranchised, others feel much more involved in the process. Additionally, the residents of Miyun feel significantly less involved than do those of Pinggu.

The following conclusions might be drawn from these findings, and in arriving at these conclusions, data derived from the open-ended questions and interviews are also used.

Resident Involvement in Decision-Making Is Perceived as Being Insufficient. The Main Obstacles to Such Involvement Are Limited Local Democratic Participation and Insufficient Education Level

The scores on knowing where to voice opinions and having the expression of opinions respected are 3.65 and 3.29 respectively, indicating neutral responses to these items. The standard deviations imply that a proportion voiced active disagreement with the statements, and from open-ended questions it is also known that 22.8 per cent of residents thought that 'the degree of local democracy is not sufficient' to permit such exchanges of ideas, while, additionally, 25 per cent of the respondents believed that low levels of education acted as a constraint on residents' involvement in tourism decision-making.

Resident Involvement in Tourism Operations Is Active, Although Different Opinions Exist on the Levels of Income Increase Associated with Tourism Development

Of those surveyed, 69 per cent of respondents or their family members participated in tourism operations in some way, and 40 per cent directly worked for tourism enterprises or home-stays. A significant minority now derived substantial parts of their total income from tourism. Of the respondents, 30 per cent stated that income derived from tourism accounted for between 41 to 70 per cent of their total family income and a further 14.7 per cent stated that over 70 per cent of family income was derived from that source. This confirms the responses made to the item about family income as recorded in Table 17.1. Interview data also confirmed these wide differentials in additions to family income being derived from tourism. A number of factors can explain these findings and include differing degrees of entrepreneurially minded family members, different degrees of need to look at tourism as a potential source of income and different degrees of accessibility to tourist flows—these differences partially depend on existing resources that include degrees of education, type of home and language abilities. What also emerged from the data was an impression that residents are reconstructing their operational concepts and have begun to realize the effectiveness of 'cooperation' with tourism agencies and 'win–win' methods.

Some Training Programmes Are Provided for Residents by Governments or Tourism Enterprises; However, Not All Residents Have Access to the Opportunities

From the open-ended questions, it appeared that 91.5 per cent of residents perceived the role of government as including the provision of significant support for and in tourism training. However, many residents stated that not all can get access to the training programmes. The scores in Table 17.1 also reflect this issue.

Residents Possess a High Sense of Responsibility Towards Tourism

Table 17.1 showed that residents have quite a strong sense of 'maintaining the tourist destination's image'. From the open-ended questions, 90 per cent of residents surveyed believed it was necessary to establish a tourism industry association, 31 per cent thought that such an association could play a significant role in tourism development and 45 per cent thought it would have importance.

From the previous discussions, one can conclude that evidence exists to indicate support for rural tourism to be community based. It can be shown that a significant proportion of local people feel they have some decision-making power in tourism; are involved in the operation of tourism and have been provided with training and education programmes. More generally, respondents show a strong sense of responsibility for sustainable tourism development within their communities. However, the adoption of a community approach as understood and advocated by commentators like Murphy (1983, 1985, 1987) is still at an initial stage. This could be clearly seen from the relatively low level of democratic participation at the destinations, the unequal opportunities entitled to residents and the lack of a comprehensive supervision and feedback system, etc. To further develop community approaches to tourism requires that these obstacles must be overcome. However, as noted by reference to the work of Ap (1990, 1992) if communities are to be motivated to play a role in tourism development, then both opportunities (and threats) need to exist. What are the attitudes of respondents to tourism and its potential impacts?

Perceptions of the Impacts of Tourism

Consequently, as seen in Table 17.2, items used in this part of the questionnaire related to perceptions of local community infrastructure, intrusion impacts and harmonious social relationships. Respondents recognized the benefits of tourism in terms of improvements of local infrastructure, and this would appear to be a common finding in the current state of Chinese tourism. Gu and Ryan (2008), Ryan and Chou (2008) and Zhou and Ma (2008) found similar sentiments expressed by residents in Shi Cha Hai in Beijing, Hangzhou and the Baiyang Lakes, while Wall (2001) argues that tourism planning in China should be oriented to the development of benefits for local communities. The findings of this study show that there is disagreement that tourism is having an adverse impact on the local natural environment, and that too is consistent with other findings in China where the need to ensure scenic and aesthetic values to attract tourism turns attention to the improvement of local environments (Sofield and Li 1998). Possibly the example *par excellence* are the attempts to improve natural and urban landscapes in Beijing for the 2008 Olympic Games. The economic benefits are evidenced by wider consumer choices and experiences. Equally, at least in 2006, the volumes of tourists in the study area have no overall negative intrusion in residents' daily lives, but again the value of the standard deviation and the frequency distribution of responses indicate that a sizable minority are beginning to feel some unnecessary intrusion impacts.

Factor analysis indicated that three underlying dimensions existed on this perceptions scale. The first factor is premised on psychological and value impacts, and in that sense is consistent with the views expressed by Lindberg and Johnson (1997) and Ryan, Scotland and Montgomery (1998) that the core of assessments and perceptions of the impact of tourism rest upon value systems of what is important about respondents' communities. The items that form this dimension include the consideration that tourism

Table 17.2 Perceptions of the Impacts of Tourism on Local Communities

Item	Mean	Std. Dev
Infrastructure in the local community has improved a lot.	4.37	0.92
Tourism development is responsible for the destruction of local natural vegetation.	2.69	1.28
The life of local people has improved a lot since the development of tourism.	4.23	0.84
Local people's consumption has become more diversified as a result of the introduction of tourism.	4.19	0.92
Tourism is good for the development of local agriculture and other local industries.	3.65	1.06
The daily life of local people is disturbed by tourists.	2.83	1.34
Relationships in the local community are more harmonious since the development of tourism.	3.84	0.99
Traditional culture has been protected and passed on because of the development of tourism.	3.77	1.02
Contacts with tourists have changed the values and concepts of local people.	4.13	0.81
Most residents get benefits from tourism development.	4.09	0.99
The tourism industry is developing too quickly for the local community.	3.43	1.15
Tourism is very important in the construction of an equitable, harmonious society.	4.21	0.72

contributes to a more harmonious set of intra-community relations, respects and encourages local traditions and cultures, and generates benefits even while causing changes in values and concepts of local people. Reference to Table 17.2 indicates that for the most part these items generate high levels of agreement, although a flaw in the research design on the item 'changes the values and concepts of local people' strictly means that people agree there are changes, but from this item alone it is not possible to say whether these changes are regarded positively or negatively. Given that the other items tend to the positive, there is a temptation to indicate that such changes might be perceived positively, but the responses to open-ended questions imply a need for some caveats to be noted. For example, 12 per cent of residents surveyed said conflicts would appear when residents competed for travellers. This problem, on the one hand, is a reflection of competition among home-stays, while on the other hand, it arguably reflects the immaturity of the local rural tourism industry in failing to achieve co-operative mechanisms.

The second dimension was concerned with industrial impacts and concerns over the speed of development. Referring to Table 17.2, it can be noted that the industrial impacts are generally considered beneficial but there is also a mild overall agreement that the speed of development may be too

Table 17.3 Factor Analysis of Dimensions of Attitudes towards Tourism Impacts

Item	Factor 1	Factor 2	Factor 3
The life of local people has improved a lot since the development of tourism.	**0.718**	0.289	0.010
Contacts with tourists have changed the values and concepts of local people.	**0.717**	0.013	0.064
Local people's consumption has become more diversified as a result of the introduction of tourism.	**0.706**	0.243	−0.010
Most residents get benefits from tourism development.	**0.631**	0.307	−0.002
Relationships in the local community are more harmonious since the development of tourism.	**0.549**	0.220	−0.057
Traditional culture has been protected and passed on because of the development of tourism.	**0.548**	−0.205	0.053
Tourism is very important in the construction of an equitable, harmonious society.	**0.495**	0.163	−0.024
The tourism industry is developing too quickly for the local community.	−0.012	**0.815**	0.153
Infrastructure in the local community has improved a lot.	0.328	**0.656**	−0.097
Tourism is good for the development of local agriculture and other local industries.	0.297	**0.526**	−0.184
Tourism development is responsible for the destruction of local natural vegetation.	−0.088	0.098	**0.838**
The daily life of local people is disturbed by tourists.	0.143	−0.149	**0.818**
Percentage of total variance	24.98	14.64	12.07
Eigenvalue	3.56	1.47	1.17
Alpha Coefficient	0.76	0.56	0.57

fast, although again the ratio of the standard deviation to the mean is quite high. Again responses to open-ended questions confirm these overall results and the distribution scores. For example, evidence existed that residents appreciate governments' efforts on road construction, house renovation and water-supply systems. Of interest are the responses to item 19 on the questionnaire that was concerned with speed of development of tourism. Results show that most residents ratify the pace of development and believe tourism development has not had too many negative impacts on their lives. Indeed, some even think the process should be pushed faster. Yet, on the other hand, many residents have recognized the shortcomings of the locality in developing tourism, such as a limitation of natural and cultural resources, problems of seasonality and the relative lack of entertainment and operational space that has arisen as a result of capital and land policy restrictions. Again, such findings are consistent with previously published literature that identifies different psychographics among local community membership where some support and others question the speed of tourism development (Ap and Crompton 1993; Lawson *et al.* 1998).

The third dimension related to the negative social and environmental impacts of the developing tourism, and, as shown in Table 17.2, there tended to be disagreement that such impacts were being evidenced. On the other hand, the standard deviations tended to be relatively high, and examination of both the open-ended responses and the interview data revealed a small proportion of respondents not connected with the developing home-stay tourism who possessed quite strong opinions about tourism creating air pollution, excessive trash, litter and noise. Again, these findings are consistent with the literature in identifying the emergence of groups of residents not necessarily well disposed to continuing tourism at quite early stages of the tourist destination life-cycle. For example, Manente and Pechlaner (2006: 244) specifically identify 'residents' awareness of the role of tourists' among the early warning signs of a diminution in destination quality. This issue remains one of the major problems in any developing tourism zone—how to reduce levels of disturbance and environmental damage associated with tourism and minimize negative impacts on local residents. One aspect that does apply within Beijing Municipality is that marked business seasonality exists, and as far back as 1985 Murphy was noting that such marked seasonality in rural areas provides opportunities for local communities to recover from the noise and bustle of a tourist season, for the industry to undertake refurbishment of premises without impinging on customers and indeed for a community to actually look forward with anticipation to the next season. In short, seasonality provides periods of community refreshment whereby communities can return to themselves. Additionally, zoning which permits places of usage primarily by residents is another means by which local residents can retain areas of privacy. Thus, as Ryan (2003) notes, destination planning that incorporates information by its very silence about places is a means of preserving non-tourist zones for residential use and local community purposes free from tourism.

To summarize, in general, despite some dissatisfaction, residents hold a relatively positive attitude towards the changes that tourism has brought into their physical and psychological life, which can be reflected by people's living standards, consumption patterns, tourism benefits and normative challenges, etc. From the perspective of social and industrial aspects, tourism has played an important role in the construction of new infrastructure and improvements in local community resources. While some negative tourism impacts on the natural environment and community are being discerned, the negative attitudes are not as strong as the positive perceptions. In conclusion, the positive impacts exceed the negative ones which validate Hypothesis 2; that is, the positive impacts exceed the negative impacts brought by rural tourism around Beijing.

Relationships between Tourist Development and Community Enhancement

The next issue related to the degree to which resident involvement contributed to the sustainable development of both tourism and the local community. For this, data were examined using canonical correlations analysis. This method is similar to MANOVA in that it is used when one has two sets of two or more variables each and one wants to examine how differences in one set of variables relate to differences in the other set of variables. It, however, possesses the advantage of not requiring knowledge of distributions other than mean and variance, and is also ad hoc in its modelling process. However, it should be noted that interpretation is problematic, as Wood and Erskine (1976: 864) note:

> One researcher's canonical loading becomes another's canonical weight; canonical dimension to one is a canonical variate to another; and, canonical correlation is the relationship between data sets for one, but only the relationship between variates for another.

The canonical correlations are the correlations between a linear combination of the Set 1 variables and a linear combination of the Set 2 variables. Here it is employed to study the correlations between the application of community approach (Set 1) and the local community's sustainable development (Set 2). A series of questions from the questionnaire gave rise to two sets: Set 1 reflected the involvement condition of local residents while Set 2 reflected the sustainable development situation of the tourist destination perceived by local residents. Thus, Set 1 = {x1 = decision power, x2 = planning power, x3 = income increase, x4 = training programmes, x5 = problem reflection, x6 = problem responses, x7 = image responsibility and x8 = co-operative promotion}, and Set 2 = {y1 = infrastructure improvement, y2 =

land destruction, y3 = life level, y4 = consumption ways, y5 = agri-industry stimulation, y6 = life disturbance, y7 = relationship harmony, y8 = culture heritage, y9 = value changes, y10 = equal benefits, y11 = developing speed and y12 = harmonious society}.

It was found that the bivariate correlations between each variable in Set 1 were low excepting that between x5 (problem reflection) and x6 (problem responding), where r = 0.62. This is understandable because as residents identify problems in tourism development, they would like the authorities to listen and pay attention to the issues raised. The correlation coefficients between each variable in Set 2 were also low, implying the lack of any common underlying dimension for the items. Consequently, little relationship was also found between Set 1 and Set 2. The Standardized Canonical Coefficients can be interpreted to show which items are weighted most heavily in the linear combination of variables for each set of variables. These weights are created so as to maximize the correlation between the two sets of variables. The first canonical correlation is 0.647; the second is 0.37. With both canonical correlations included, $X^2(96) = 190.828$, $p < .001$, and with the first removed, $X^2 (77) = 115.795$, p = .003.

It was also found that the correlation between community involvement and community sustainable development could be transformed to a study of correlations among canonical group 1 (as group 2's index coefficients did not show great differences with those of group 1). From the linear combination of standardized items, one can assess the contribution of each item to the canonical variable, that is, the higher the coefficients, the greater the importance of the item as an explanatory variable The analysis indicates that the correlations between Beijing suburban tourism community's involvement and the local community's sustainable development may be explained by the canonical variable group 1(U1 and V1) and the canonical variable group 2 (U2 and V2) (see Table 17.4). In variable group 1, U1 is mainly explained by items x1 (decision power), x3 (income increase), x6 (problem responding), x7 (image responsibility) and x8 (co-operative promotion), while V1 is mainly explained by item y1 (infrastructure improvement), y7 (relation harmony), y9 (value changes), y10 (equal benefits) and y11 (speed of development). This canonical correlation indicates that local residents' decision-making power, economic benefits supervision power and self-operation concept are positively related to the local community's infrastructure improvement, the harmonious neighbourhood, the time-paced values and the imbalance of tourism development. From this evidence it can be concluded that local residents' involvement in tourism contributes to the local community's sustainable development.

CONCLUSION AND IMPLICATIONS

It might be argued that until now human development has passed through stages of being object centred, society centred and people centred. This is the

Table 17.4 Canonical Correlation Analysis—The Relationship between Community Participation and Development

Canonical Groups and Variables
One
$U1 = 0.310x1 + 0.128x2 + 0.396x3 + 0.170x4 - 0.127x5 + 0.274x6 + 0.337x7 + 0.217x8$
$V1 = 0.296y1 + 0.107y2 - 0.009y3 - 0.174y4 + 0.052y5 + 0.011y6 + 0.381y7 + 0.050y8 + 0.323y9 + 0.434y10 - 0.289y11 + 0.242y12$
Two
$U2 = -0.806x1 + 0.262x2 + 0.410x3 - 0.176x4 + 0.263x5 + 0.065x6 - 0.100x7 - 0.156x8$
$V2 = -0.359y1 - 0.244y2 + 0.807y3 + 0.150y4 - 0.164y5 - 0.005y6 + 0.024y7 - 0.302y8 + 0.160y9 - 0.669y10 + 0.544y11 + 0.290y12$

Note: Set 1 = {x1 = decision power, x2 = planning power, x3 = income increase, x4 = training programmes, x5 = problem reflection, x6 = problem responses, x7 = image responsibility, x8 = co-operative promotion}, and Set 2 = {y1 = infrastructure improvement, y2 = land destruction, y3 = life level, y4 = consumption ways, y5 = agri-industry stimulation, y6 = life disturbance, y7 = relationship harmony, y8 = culture heritage, y9 = value changes, y10 = equal benefits, y11 = developing speed and y12 = harmonious society}.

same with tourism destinations and their development, where a scientific, comprehensive and sustainable way is continually sought. This chapter is premised on a notion that community is a unity that possesses a sustainable development function. Thus, to develop tourism from a community perspective is an effective way for tourism to achieve its own overall sustainability. Community tourism is a concept and mode that pursues the harmonious development of tourism together with that of a community's economy, culture and environment. This study provides a brief review of the community approach in tourism development. Using rural tourism around Beijing as a case, it also investigates the current status of the community approach in China. By providing an analysis of empirical data, the results indicate that rural tourism in suburban Beijing is community based and has positive relations with the sustainability of community development. Residents' attitudes suggest that the positive tourism impacts on the destinations exceed the negative ones.

However, some issues still exist in Beijing's rural tourism development. The full implementation of a community approach is restrained by factors such as a weak consciousness of democracy, low economic development level, low levels of knowledge and expertise and a poor understanding of the community approach by the residents and government officials alike. In order to enable the destination community to derive maximum benefits from tourism as well as to make its development more sustainable, the following are suggested for future policy:

— Provide more decision-making power to the local residents.
— Support local residents by enhancing training and education.
— Provide preferable loans policy and a more effective land policy.
— Establish a healthy supervision and feedback system.
— Enable residents to form a tourism association of their own.

It might be thought that such calls are inconsistent with a government-sponsored approach to tourism policies, but the authors would argue that such policies become a means of enhancing the achievement of tourism policies. Zhang Benfa (1999) saw tourism as a means of emancipating females from purely domestic roles and to enhance them as citizens, while Liu Zhaoping (1998) argued that the development of tourism not only kept people from migrating, but also attracted those seeking jobs and development opportunities. There is little doubt that tourism can be a catalyst for change, and some, such as Zhou Xiao (2003), are of the view that tourism brings with it cultural and social change. The need is to ensure that by enhancing the capabilities of local communities tourism becomes a means of establishing a harmonious way to future well-being as is perhaps illustrated in the case of Hebei Province by the work of Zhou and Ma (Chapter 19 in this collection of Chinese research).

ACKNOWLEDGEMENTS

We would like to express our sincere thanks to the residents and administrators of the six villages for their co-operation in data collecting. The six villages are Cao Jia Lu and Yao Qiao Yu of Miyun District, Diao Wo and Gua Jia Yu of Pinggu District, and Liu Gou and Zhuo Jia Ying of Yanqing District. Our gratitude would also go to graduate students Na Li, Ning Zhang, Jinhua Chen, Weinan Lv, Guipei He, Yuhui Zhang, Yingqiu Gao, Yuanping Zhao, Dan Zhao and Cuibin Zhao for their assistance in data collecting and analysing. Last but not least, Professor Chris Ryan deserves our sincere thanks for the proofreading and modification of the final chapter.

18 The Psychological Carrying Capacity of Tourists

An Analysis of Visitors' Perceptions of Tianjin and the Role of Friendliness

Liang Zhi and Wang Bihan

INTRODUCTION

The concept of carrying capacity has been well established in the tourism literature, even allowing for debate as to the difficulties of measurement and implementation (e.g. see Mathieson and Wall 1982). Both Chinese and foreign scholars are aware that long-term sustainability of any tourist destination or site is subject to constraints imposed by the tolerance of the natural and social environments for a given number of visitors, even though good destination management techniques may permit a site to absorb quite large numbers of visitors without observable detrimental impacts, and indeed revenues gained may well help not only conserve but re-establish past damaged terrain (e.g. see Ryan and Stewart 2008). However, most studies adopt the perspective of a supply-led view-point with reference to the actual resource and its nature, and visitors are, in a sense, the determined variable in that the objective is to assess what levels of visitation are consistent with the tolerated impacts arising from those visits. This chapter adopts a different perspective. Carrying capacity involves more than one factor, and is determined by *both* supply and demand, and the perceptions and subsequent behaviours of visitors are themselves a determining factor when ascertaining the carrying capacity of a site. Therefore, when conducting research to assess carrying capacity it is necessary to not only consider the resources of tourist facilities, infrastructure and the natural and social environment, but one should also pay close attention to the expectations, motives and psychological characteristics of the visitors and their interactions with residents at the destination. Only in this way might one arrive at a better and more comprehensive understanding of the carrying capacity of a given destination or site.

This chapter therefore studies the perceptions of visitors to Tianjin with reference to their perceptions of degrees of crowding and the environment. It is important to understand the degrees of tolerance of crowding that visitors might display, and these might well be different in Chinese society when compared to studies derived from western sources because, for many urban-based Chinese, the densities of population encountered in daily living may be greater than for many of their western counterparts. An understanding of this may better enable site and destination managers to plan pedestrian flow through their sites.

The chapter is therefore structured to first provide a brief review of the literature to indicate the conceptual underpinning of the study. Second, it briefly describes the context of the study, namely, the city of Tianjin, before indicating the nature of the sample. It then provides a description of the results before discussing the implications of these results.

LITERATURE REVIEW

The concept of carrying capacity makes an early appearance in the tourism academic literature, informed in large part by prior studies emerging from the management of National Parks, particularly in North America. For example, in 1982 Mathieson and Wall wrote of physical and psychological carrying capacities in one of the earliest tourism text-books. In the 1980s carrying capacity was the focus of many studies. Pearce (1989) cites earlier work by Cooke and D'Amore where carrying capacity is defined as that point in the growth of tourism where local residents perceive, on balance, an unacceptable level of social disbenefits from tourist development. Certain variables such as the number of visitors or types of visitors that constitute the tourism environment and the degree of development as measured by volume of economic activity and the changing patterns of land use, retail provision and infrastructure can have a direct bearing on the socio-economic conditions that generate daily living for residents in a tourist zone. King Chan (1997) holds that a harmonious relationship between the individual visitor and the tourism environment lies in both the motive for and expectation from the visit and the provision of a tourism environment capable of meeting those expectations, motives and visitor needs. The greater the congruence between expectation and actual visit experience, the more satisfied is the visitor and the more harmonious the relationship.

For their part, in a study of rural tourism in Pennsylvania, Po-Ju Chen and Kerstetter (1999) found that the image of a destination mainly depended on factors such as the behaviour and social demography of actual and potential visitors. They found that visitors from different countries, cultural backgrounds and regions experienced different levels of satisfaction, and concluded that the characteristics of the tourist-generating zone were important determinants of final levels of satisfaction. For example, visitors with lower levels of income, poorer levels of education and lower social status generally had lower expectations and perceptions of a destination's image than other groups.

Among others that emphasize pre-visit conditions are Dann (1977), Crompton (1979), Iso-Ahola (1982) and Uysal and Jurowski (1993)—all of whom suggest that the travel motives of actual or potential travellers are essentially determined by their perceived images of the destination. While later research reviewed by Ryan and Gu (2008) indicates more complex relationships at work in shaping perceptions of place, it nonetheless remains true that the perception of place as being able to meet generic 'push' motives for change, relaxation and difference are important in 'pulling' tourists to a given destination—and understanding the nature of these perceptions is of help in the determination of carrying capacities. Bo Jigang (1996) observes that distance also exerts a factor—with more distant destinations being less familiar and often being perceived as being more 'exotic'. Wang Jiajun (1997) concurs with other research that adopts a confirmation–disconfirmation paradigm by arguing that congruence between destination 'realities' and visitor expectations *and* preferences is important, and thus, by the same token, emphasizing differences between these factors becomes a means of reducing potential travel to a given place.

This observation leads to a consideration of psychological carrying capacity. Tourist psychological carrying capacity is the maximum degree to which a visitor can tolerate the environment of the destination. It consists of feelings, sentiments and evaluations of experience of the place. The psychological carrying capacity is thus affected by the quality of the tourism environment and the tourists' prior expectations, images, perceptions and evaluations of experiences that arise from their interactions with the physical nature of the place and with the people encountered at that place. Thus, the attitudes of local residents towards visitors and tourists are also a determinant of psychological carrying capacity. These relationships are illustrated in Figure 18.1.

A relationship exists between the tourist psychological carrying capacity and the destination life-cycles model proposed by Butler (1980). In the initial stages of destination development, the numbers of tourists are

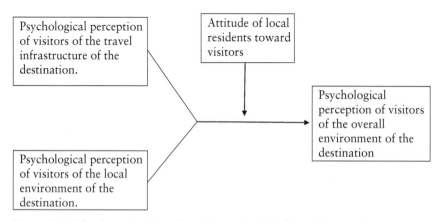

Figure 18.1 The determination of tourist psychological carrying capacity.

likely to be low. The local facilities and infrastructure are likely to be able to absorb the numbers easily, and given the psychographics of these early explorers (Cohen 1980), the tourist environment will be able to meet their needs and psychological demands, thereby generating high levels of satisfaction. As the numbers increase, and under conditions of short-term rigidities in tourism facility supply, and high elasticity of tourism demand, the increase in tourism demand will exceed the supply capacity of the destination. Therefore, visitors will face problems such as (a) increasing prices, (b) increased densities of visitors and overcrowding in key scenic areas or places of specific interest, (c) congestion and (d) an inadequate supply of tourism facilities and service personnel. As a result visitors may feel there is a decline in the standard of the tourism environment and quality of experience.

As noted, the attitude of the local residents can influence the psychological carrying capacity significantly. In the initial stages of the destination's development the social impact of small numbers of visitors may be small and the economic benefits of their spending and the income and employment opportunities generated by the visitors will be welcomed. As a result the local community is likely to be hospitable, and this will positively promote the tourist destination in the eyes of the visitors. However, as described by Doxey (1976), increasing numbers of visitors begin to cause social stress with heavy pressures emerging on local transportation systems, prices, public order, tourism facilities and tourist services. The constantly increasing numbers of visitors will also increasingly impinge on the residents' daily lives, causing dissatisfaction among the local residents who will begin to be less tolerant of tourists and the problems they are perceived as bringing. In more extreme cases a sense of hostility towards tourists might emerge, but generally tourists will view the community as less welcoming and evaluations of the overall destination will become less favourable. Should strong levels of dissatisfaction emerge, they will begin to view experiences as being negative, and begin to switch to alternative, competing destinations that now appear more attractive.

Building on this scenario, the authors argue that as the number of visitors increase, then a measure of the psychological carrying capacity will become larger than zero, whereas when the number of visitors decline, a measure of the psychological carrying capacity will become less than zero, with zero representing a state of homeostasis or balance. This can be represented mathematically as:

$$C = f(\text{PI, Pc, Ad})$$

Where: C = psychological carrying capacity of visitors—for which a proxy measure is visitor evaluation of the destination; PI = Visitors' perception of the tourism environment (facilities etc.); Pc = Visitors' perception of the local environment; Ad = Attitude of the local community towards visitors.

In turn, as noted from the literature, visitors' perceptions are also determined by age, sex, education and income of the visitor. Thus:

PI = f(A, S, E. Y)

Where: A = Age of the visitor; S = Gender of the visitor; E = Educational attainment of the visitor; Y = Income level of the visitors.

Following the proposed relationship it is possible to propose the following hypotheses:

Hypothesis 1: The psychological carrying capacity of a visitor positively correlates with the attitude of the local community towards visitors.

Hypothesis 2: The psychological carrying capacity possesses a U-shaped relationship with the visitor's age.

Hypothesis 3: The psychological carrying capacity is gender determined.

Hypothesis 4: The psychological carrying capacity is positively correlated with a visitor's income level.

THE CONTEXT OF THE STUDY

The study was undertaken in Tianjin, China's third most densely populated urban area with a population of 10.24 million people and 859 people per square kilometre. It is primarily a manufacturing centre while the wider municipality contains oil reserves and also thermal energy sources. Given its location, Tianjin has several claims to history and it dates as a trading centre from the completion of the Grand Canal of China in the Sui Dynasty (AD 581–618). It was the city in which the treaties ending the second Opium Wars were signed, it witnessed uprisings against the French in 1870, saw turbulent times in the Boxer Revolution and was largely occupied by the Japanese in the second Sino-Japanese War, and then riots in 1947 forced occupying US troops to withdraw. The Communist troops took control of the city in 1949 after a battle of some 29 hours. Given such a turbulent history among the sites that can be visited by tourists are Luzutang (Boxer Rebellion Museum), the Wanghailou Church—site of the 1870 Tianjin Massacre, while an older culture can be captured in Guwenhua Jie (Ancient Culture Street), Guwan Shichang (Antique market), the Confucius Temple—Wen Miao—and other sites, including that of the Shi Family residence, Shijia Dayuan. Nonetheless, apart from tourism associated with business travel and visiting friends and relatives, Tianjin is not ranked as one of China's leading destinations, although great strides are now being made. In 2008, Tianjin was selected as one of the venues for the Olympic Games Football Tournament. Like many Chinese industrial centres it experiences

Figure 18.2 The location of Tianjin.

poor air quality and has undergone significant transformation through new construction in recent years—but again, like many of its counterparts, new attempts are being made to improve the built and natural environment. While many international travellers are not aware of Tianjin, for those seeking to better understand the process of change that China has experienced Tianjin offers an excellent introduction—and has the added advantage of a reputation that its inhabitant are amongst those in China with a strong sense of humour.

THE RESEARCH

Reliability Measures

A total of 200 questionnaires were distributed to visitors during a period from November 2005 to July 2006, and a total of 146 usable surveys were returned. The questionnaire comprised four parts. The first was a scale measuring perceptions of tourist facilities, the second required an assessment of Tianjin's environment, the third was a measure of perceptions of the local community and their responses towards tourists and the final section consisted of questions to elicit socio-demographic data. Five-point Likert-type scales were used where 1 equalled 'strongly disagree' to 5 being 'strongly agree' with the mid-point, 3, being 'neither agree nor disagree'. The exception to this was a reversed scale for the item 'The overall environment of Tianjin is very crowded' where 5 was 'strongly disagree'. Each of the scales was then tested for reliability. In the case of the first series of items measuring perceptions of Tianjin's environment the item-to-item measures revealed correlations ranging from 0.44 to 0.61. The lowest scoring item was that where the Likert scale descriptors had been reversed, and in retrospect this decision was probably a poor one. Removing the item 'The overall environment of Tianjin is very crowded' improved the reliability of the scale and, with values in excess of 0.5 and in some instances above 0.6, while not 'good', the scores indicated a reasonable level of reliability to permit further examination with caution. The remaining scales scored better item-to-item correlations and Cronbach alpha coefficients were in excess of 0.51, this being the lowest score.

The Sample

The sample comprised 64 males and 82 females. Approximately 60 per cent of the sample was between the ages of 25 and 40, 43 per cent indicated they had professional qualifications and 17 per cent had graduate degrees. The sample thus displays the skew towards a younger and better educated respondent that is often characteristic of surveys of Chinese domestic tourism and which Ryan and Gu (2007) explain by reference to the emergence of an affluent newly emergent middle class able to travel, but who tend to be younger and better educated. Nonetheless, when compared to studies of, say outbound Chinese tourists, the income levels of this sample reflected the location being away from the major tourism centres such as Beijing, Shanghai, Hangzhou, Shenzhen and Guangzhou, with approximately 37 per cent of the sample earning between 1,500 to 3,000 RMB per month. Only about 20 per cent earned more than 3,000 RMB per month, and of the total sample, only 5.5 per cent earned more than 5,000 RMB per month. Of the respondents, 38 per cent earned less than 1,500 RMB per month, but these data were skewed in the direction of younger respondents.

Drawing upon data derived from other sources and comparing the socio-demographics of the sample with that data it was concluded that the sample, while small, was representative of the type of visitors being attracted to Tianjin.

The Findings

Given the nature of the sample, the full results are not provided here, although the wider implications of all the results are discussed, but it is evident that the small numbers of respondents impact on an ability to generalize from the results. Calculating the mean scores and standard deviations for the overall scores measuring visitors' perceptions of Tianjin's tourism environment, the city environment and overall perceptions generated the results shown in Table 18.1. In the table it can be seen that perceptions of the tourism environment tended to be low, as was the perception of the city

Table 18.1 Overall Mean Scores of Perception and Evaluation

Perception of Tourism Environment		Perception of City Environment		Overall Evaluation	
Mean	*Std Dev*	*Mean*	*Std Dev*	*Mean*	*Std Dev*
2.55	0.44	2.97	0.55	2.76	0.78

environment. Given the current nature of Tianjin's tourism and the nature of the city as described above, such low scores are not inconsistent with the current stage of tourism development in the city. They are inconsistent however with the literature based upon Butler's life-cycle and the Doxey Irridex, and this can be possibly explained by reference to these concepts being based on a model of tourism taking place in initially small communities that then grew with tourism. Transferring that model to tourism in Chinese towns and cities with their large populations and alternative sources of employment and income seems to be inappropriate, even if the models might have legitimacy in the context of China's rural communities as discussed elsewhere in this book.

The next stage was to test the various hypotheses relating to the role of socio-demographic factors.

The Role of Perceived Resident Attitudes Towards Visitors and Visitors' Evaluations of Tianjin

Using correlation analysis a Pearson coefficient of correlation was found to be 0.425 ($p < 0.001$), indicating a positive but relative weak relationship between the two variables.

The Role of Age

No correlation was found between age with $r^2 = 0.086$, $p = 0.30$.

The Role of Educational Attainment

Again no correlation was found with r2 = -0.015, $p = 0.855$.

The Role of Income

Again no correlation was found with r2 = 0.188, $p = 0.025$.

Having examined the data, the authors reached the following conclusions. First, when local residents seem to be well disposed towards visitors, the visitors reciprocate, and vice versa. That is, for example, when visitors have unfavourable feelings towards the local residents, in turn, they (the residents) tend to be unfriendly towards the visitors. The relationship between resident and visitor is thus interactive. Gender tends to play some small role here, with males tending to be better disposed towards visitors, and thus tending to elicit more positive results. However, why this might be the case is not known from the available data. It does appear that psychological carrying capacities are in part determined by resident attitudes, and these attitudes interactively reinforce and confirm a prevailing predisposition.

Second, income is also a factor in that higher income groups tend to be more tolerant of difference, possibly because they are more experienced travellers. Third, age and education are not determining factors. Finally, generally visitors were well disposed to Tianjin and enjoyed their visits.

The main finding is that there is reciprocity between visitors and residents. As such the findings fit loosely into a small stream within the wider literature. With reference to the first editor's past work, it has been argued that holidaymakers, being relaxed themselves, tend to be friendlier and thus evince more friendly responses from residents, thereby creating more satisfactory holiday experiences (Ryan 1995). From the view-point of intimacy theory (Trauer and Ryan 2005), among the significant others that shape the holiday experience, one can point to residents and other holidaymakers. However there are significant caveats to the research. The sample is small, but arguably its composition is not too dissimilar to the socio-demographics of other samples of holidaymakers in China—that is, they tend to be younger and of higher income and education levels than the general Chinese population. The finding is, as just indicated, arguably consistent with the wider literature. In terms of psychological carrying capacity the findings point to an inherent truth in our understanding of that capacity—namely, it is not fixed and is potentially variable for it is based on attitudes and perceptions that can be changed. Smiles can generate smiling responses, and it is perhaps easier to smile when on holiday and when one has an above-average income that permits holidays. Gender interactions also possess some importance. None of these are in themselves overly surprising findings, but it is perhaps their very normality that has importance. Tourism is often stated to be about experiences of places, people and interactions at places other than the home. That experience can be shaped by friendliness, and while this is possibly a statement of the obvious, the fact of being obvious does not diminish its importance.

19 Maintaining the Authenticity of Rural Tourism Experiences through Community Participation

The Case of Two Baiyang Lake Island Villages

Yong-Guang Zhou and Emily Ma

INTRODUCTION

Just as other contributors to this book have noted, tourism has experienced tremendous growth in China in the period commencing in the early 1980s as the benefits of a more open economic policy began to be felt. With this growth have come opportunities for many regions previously marginal in social, economic and political terms, and potential beneficiaries of these policies have been rural areas that are becoming enriched through the encouragement of tourism by central and provincial administrations. While actuality has sometimes fallen short of expectation, such economic and social enhancement of the previously marginalized remains an objective of Chinese regional and tourism policies. For a discussion of these problems in the Chinese context see, for example, Gao (1997) and Ding (2004). Like the chapter by Zhang, An and Liu in this book, this contribution identifies problems associated with tourism in rural settings. However, while they sought to establish a link between tourism and community development, this chapter differs by seeking to illustrate how community participation aids the enhancement of tourism experiences for visitors by retaining a sense of the 'authentic', while at the same time it describes two different approaches to community-based tourism that are being put into practice in two villages. It is well recorded in both western and Chinese academic literature that tourism offers potential negatives as well as opportunities, such as a loss of natural environments, threats to local traditions and culture, a loss of a sense of both 'rurality' and 'authenticity' for both host and visitor (e.g. see Bramwell and Lane 1994; Oppermann 1996; Gao 1997; Zhou 2002; Zhong 2004). Additionally, within the Chinese context there exist issues pertaining to service quality and resource (for example, concerns over sanitation; see Gu and Yu 2005) and the problems described by Ryan and Gu (2008) with reference to unbridled competition in Beijing, which also have their counterparts in rural settings.

Be this as it may, the authors nonetheless subscribe to a view that the maintenance of a sense of 'rurality' and 'authenticity' remain critical determinants of rural tourism's success, and that a means of achieving this is through effective community participation. The goal is for tourists to gain better experiences of the rural through rural residents' participation, while rural residents can benefit from increased income and improved infrastructures due to tourism development. This approach can improve the living conditions and economic independence of rural areas, leading to the sustainable development of rural communities and tourism.

The context of this study is two island villages in Baiyang Lake, Hebei Province. Both are popular tourist destinations and their adoption of a community-based approach towards tourism development attracted the authors' attention in 2002 and 2003. As such, they provide insights into China's rural tourism development and community advancement. Consequently, the study involved on-site interviews and an investigation of

1. features of the two approaches towards tourism development adopted by the two villages;
2. a comparison of the similarities and differences of the two villages in their usage of the concepts of 'authenticity' and 'rurality';
3. the subsequent provision of recommendations for the future development of Baiyang Lake's island villages.

LITERATURE REVIEW

The concepts of 'rurality' and 'authenticity' are entwined. A sense of the rural refers to traditions and culture that arise from close connections to land and the senses of seasonal change that working with the land bring. The rural exists as a counterpoint to the urban, which seeks to protect and shelter humans from the weather and create independence from the seasonal changes. Further urbanization orients humans to production- and service-based economies divorced from considerations of naturally imposed constraints of soil and weather. Consequently, rural traditions are based upon patterns of seasonality such as harvest and planting, and rituals associated with fertility, family and generational change. Local village residents thus fulfil roles of participants, guardians of tradition, and interpreters and gatekeepers as to what is explained and not explained to non-residents. Tourists thus become formalized guests, operating within constrained periods of time, but often, it is assumed, wishing to participate in the non-urban as a source of change, nostalgia, education or entertainment. From one perspective the exchange of authenticity is an unequal one. The visitor knows only that the tradition is not one with which he or she is familiar, but any 'performance' is filtered through expectations of what constitutes the authentic. However, residents will have a knowledge of what is the 'authentic' and the issue as Ryan and Aicken (2005) point out in the context of indigenous-based tourism, is possibly not so much 'authenticity' but 'authorization'—who is it that authorizes and gains from the portrayal of the authentic? Additionally, other issues can be explored—as Ryan (2005) notes—not all traditions are necessarily to be valued in a contemporary age given its current different values as to, for example, gender roles. Thus, from the perspective of this study, 'authenticity' rests in the authorization by the host community of those aspects of their culture, traditions and way of life that are presented to tourists for their consumption. It is a form of 'approved' commodification—although to state that it is approved by a host community is to assume that the community is homogeneous in its agreements and is able to enforce its interpretation—something not always present, as discussed by Ryan (2003).

For its part, 'rural tourism' is defined as tourism activities that occur in rural areas (by the European Union [EU] and the Organization for Economic Co-operation and Development [OECD] in 1994), where 'rurality' lies at its core. Lane (1994) proposed several characteristics of rural tourism, including: (a) it is located in rural areas; (b) tourism activities are closely related to nature and/or the traditional culture of the rural area; (c) it is small scale; (d) it often involves host families of the rural area concerned and (e) it is diversified in product because of the complexity of the rural areas. Academic research in rural tourism may be traced back into the 1950s. It should also be noted that 'rural tourism' has overlapping connotations with nature-based tourism, but distinctions can be made between tourism associated with a rural life-style and tourism that uses natural settings as a backdrop but which are not wholly consistent with the traditional life-styles of a given area. For example, adventure tourism products like jet boating or white water rafting take place in natural settings, but may be inconsistent with traditional life-styles, although here again distinctions can be fuzzy as traditions evolve and take advantage of new technologies.

One of the first articles that dealt with tourism in mountainous areas (Ager 1958), found that tourism could bring additional income, support the livelihood of farmers and reduce outmigration of the younger generation by providing additional jobs. Thus rural tourism can generate many benefits such as generating income for agricultural or other rural households; reducing outmigration by providing job alternatives; bringing a transfer of ideas from urban to rural areas; providing urban people with an experience of rural living; diversifying the rural economy, especially when it is linked to other sectors such as arts and crafts; and making the provision of additional and improved infrastructure, e.g. roads, arts centres etc. (Demoi 1991; Oppermann 1996). Problems also exist, such as high seasonality (Lehle 1982; Schijppner 1988), which makes the additional income generated by rural tourists less significant (Oppermann 1996), but which, as noted by Murphy (1985), generates periods in which communities can rest from the pressures of the tourism season.

China's rural tourism product offerings have grown rapidly since 1985. In the 1980s, when modern lodging facilities were more noted for their absence and inadequacy in rural locations, tourists had to eat and live with local rural families at very cheap prices. This served as an introduction to rural tourism in China. Rural family hostels first came into being to supplement other tourism attractions, and served as an accommodation base in the tourism zone. In mountain areas (those above 800 to 1500 metres), rural family hostels can be a good choice as an escape from the summer heat of the cities and for leisure and recreational pursuits such as walking in the mountains and enjoying the scenery. However, this form of tourism was almost incidental and was not planned or promoted in any meaningful way until comparatively recently. It was in the 1990s that rural tourism became more popular in farms and villages that surrounded cities, and it could be divided into two categories. First was the experience of eating and staying overnight with a rural host family, which, translating from the Chinese, was a form of rural cultural tourism called 'Fun with Rural Families'. The second classification is based on agriculture, which is now known as *agri-tourism*. However, initially they provided relatively simple products with limited activities, while the lodging facilities often lacked traditional rural features being modern blocks cheaply constructed and designed to provide urban-style plumbing facilities wherever possible. Rural villagers who participated in rural tourism often sought to offer an urban-style form of accommodation and dining removed from their traditional roots because (a) this was thought to meet client needs and (b) it represented a notion of progress of their own well-being. Arguably the better rural tourism product emphasizes differences between cities and rural areas, which not only include living and dining with rural host families, but which also requires tourists' participation in rural daily activities. Without this participation by visitors, tourists cannot gain a real insight into a rural life-style and the rural nature of the location simply becomes a backdrop for relaxation, which could be equally well gained at city or beach resorts.

In terms of how to operationalize 'rurality' in rural tourism development, there are two major views in China. Li (2000), Wu (2003) and others state that the design of tourism products should be based on tourists' needs, and one should increase the degree of involvement of tourists in rural-based activities in order to give them more fun. The emphasis thus becomes one based on 'difference'—differences, however, that filter perceptions of the rural through the tourist gaze of their own needs as perceived by the host. However, this approach may lead to a loss of 'rurality' as well as authenticity. Authenticity is a term borrowed from museum studies that has been extended to tourism (Trilling 1972; Wang 1999). Authenticity connotes 'traditional culture and origin' with a sense of the genuine (Sharpley 1994: 130). Wang (1999) classified authenticity issues in tourism from two perspectives. First, authenticity in tourism can be differentiated into two separate issues: that of tourist experiences (or authentic experiences) and that of toured objects. Second, the complex nature of authenticity in tourism is exhibited in the fact that it can be further classified into objective, constructive and existential authenticities. Therefore, if the design of tourism products solely considers tourists' needs for entertainment and fun, it may remove itself from a sense of authenticity, which will lead to a loss of 'rurality' and potentially the sustainable development of a rural community in terms of the community holding to its past traditions and culture. The demand for entertainment may, for example, require capital resources in excess of those available to a rural community, and product based on entertainment alone may not successfully compete with those available at more financially endowed locations. The alternative of attracting major financial resources to the rural area then runs the risk of changing the countryside into simply yet another urban complex, thereby threatening the life-style that tourism development was supposed to sustain!

Other researchers such as Liu (2000), Zhong (2004) and Zheng (2004) believe that rural tourism development should be based on the community's situation, and primarily consider the local community's needs. Until now the former view has dominated rural tourism practices in China. Historically, China is an agriculturally dominated county, and 70 to 80 per cent of its population lived in rural areas until very recently. People resident in China's countryside usually have lower incomes and living standards, and improving their living standards continues to be a major goal for China's modernization programmes. Many researchers believe that rural tourism can be an effective way to help rural areas escape poverty (Gao 1997; Liu and Yang 2002; Zhou 2002), and the China National Tourism Administration (CNTA) also formally approved tourism's role in rural area development in 1996 and again in 2006, as discussed in the introduction to this section and elsewhere in this book. However, some commentators argue that currently most benefits go to investors from outside the rural communities and local residents only enjoy limited benefits from tourism development. For example, Ding (2004) argues that only more local community participation in rural tourism development will benefit more local residents in more direct and effective ways. However, there is limited English language research published that is derived from the perspective of local communities, and thus this chapter contributes to that literature from the Chinese view-point.

METHODOLOGY

In order to possess a systematic understanding of tourism development in Baiyang Lake's island villages, a qualitative methodology was employed as being well suited to the study of a social process over time (Babbie 2001; Ying and Zhou 2007). In addition to reviewing past research, the study was generally conducted in the form of non-participation observation and open-ended on-site interviews in the year 2005 with several follow-up telephone interviews in 2006.

Non-participation observation was adopted by the authors for a number of reasons. As native residents of Baiyang Lake and aided by previous research findings, the authors could identify the degrees of rurality and authenticity represented in the tourism activities of the two villages. However, no longer being resident there for long periods of time, the authors wished to be 'phenomenologically distant' in an attempt to adopt a more objective or post-positivistic view of development—a perspective born of practical issues of time as much as preference for a given research paradigm. In order to obtain a richer as well as comparative understanding of the two villages' tourism activities, a subsample of 10 carefully chosen local people was selected for interview. These interviewees included two county level governors in the tourism department of Anxin County, two presidents of the two villages and three host families and three non-host families from each of the two villages. These interviewees were believed to represent different stakeholders able to provide valuable insights for this study.

THE RESEARCH

Context of the Research

Baiyang Lake Island Villages Tourism Development

As the largest lake in the Huabei Plain, Baiyang Lake has an area of 366 square metres, and 85 per cent of its area falls into the administration of Anxin County. There are thousands of islands and residents live on several of the bigger islands, thus forming a specific village, an island village. Currently, there are 36 island villages with a total of 100,000 residents, and 62 peninsula villages with a further 120,000 residents. Together these villages form a community based on similar life-styles (Zhe 2000). Indeed, the island village is an independent community both in terms of natural environment and administration. The island villages are separated from the outside world by water, which provides a relatively quiet and isolated environment for the local community. The majority of the island villages are independent administrative units with integrated administrative and education organizations, infrastructures (such as water and electricity supply) and transportation. Additionally, the island villages have unique social and cultural characteristics because of a lengthy isolation from the outside world. They have a relatively low pace of economic development, low levels of population movement with close genetic relationships within the local people, and many unique customs have formed from this special way of living over time. From this analysis, it is clear that the island village has preserved a highly integrated communal characteristic and can be regarded as a typical exemplar of a village community in China.

As the *Pearl of Huanbei*, Baiyang Lake has become a famous attraction for many, including past emperors since the Jin Dynasty. Historical records show that Emperor Kangxi of the Qing Dynasty visited Baiyang Lake on more than 40 occasions. So too did Emperor Qianlong. Baiyang Lake was an important battlefield from 1937 to 1945, when Japanese armies broke into the Huabei Plain. The defiance of local people was written into stories by the famed writer, Sunli, whose work is widely read in China. The area was also the location for the popular 1963 film *Little Soldier ZhangGa* that featured the heroism of an island village boy during World War II and the resistance to the Japanese (Xu 2007). A more recent cartoon animation of the story has also been released. The image of Baiyang Lake as a quiet place with a beautiful landscape and brave people is deeply rooted in Chinese culture and it is this that has provided several opportunities for its tourism development in recent years. The area's scenic features also include Baiyang waterfall and its architectural heritage included Houxi Bridge and Minghe, a town whose history extends for 1000 years. The area is also close to the thermal springs of Xiong County, and the county town is currently the subject of a proposal to create a 'no smoking town'—that is, a town wholly dependent for thermal district heating and thus not dependent on coal fired power stations (Wang Li and Wang Ning 2000). In addition Baiyang Lake Hot Spring Town is the largest tourism project in Xiong County with an investment of 600 million RMB (Wang Li and Wang Ning 2000). Apart from its natural assets, from a marketing perspective, tourism in the Baiyang Lakes benefits from its relatively near access to the cities of Beijing and Tianjing Baoding—each representing significant markets, as shown in Figure 19.1.

Tourism Development

Tourism in Baiyang Lake started in the 1980s. During the majority of the past 20 years, the Anxin County government has been the only initiator and implementer of tourism strategies and policies. However, at the beginning of the twenty-first century, several enterprises from both the local county and outside participated in additional investment and operation of newer tourism attractions in the locale of the lakes, and tourism has become one of the major economic drivers of Anxin County. Many local residents now participate in the service sectors of tourism; a situation different to the period prior to 2000 when island villagers, due to their previously isolated location, rarely benefited from tourism. As resources improved, the Baiyang Lake region was reappraised by CNTA and in 2001 was certified as an 'AAAA' level tourism attraction.

Wangjiazhai and Dongtianzhuang Island Villages

Seeing the benefits being attracted by tourism development, some villages have attempted to participate in building tourism attractions in the islands,

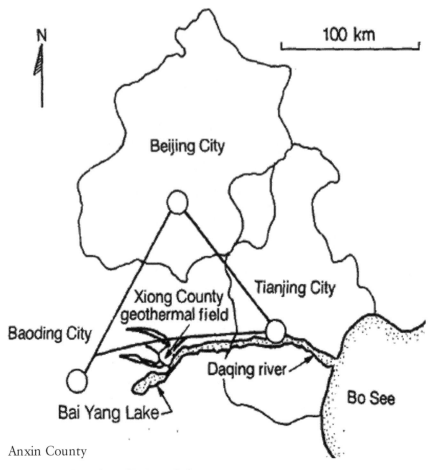

Figure 19.1 Location of Baiyang Lakes.

Table 19.1 Tourist Arrivals and Income in Baiyang Lakes

Year	Tourist Arrivals	Income from Tourism (RMB)	Multiplier Effect (RMB)
2001	510,000	10,400,000	150,000,000
2002	520,000	10,520,000	160,000,000
2003	740,000	10,700,000	220,000,000
2004	860,000	27,800,000	280,000,000
2005	890,000	42,600,000	560,000,000
2006	970,000	47,000,000	580,000,000

Source: Baiyang Lake Tourism Bureau (2007).

and this led to a new form of tourism attraction, the Island Cultural Village. The events at two villages, Wangjiazhai (Wang) Village and Dongtianzhuang (Dong) Village provide the basis for this case study. Wang is 2.5 km to the east of Anxin town and has 6,000 inhabitants. At the beginning of 2002, the administrative representatives of this village decided to build a cultural village as a tourism attraction at a remote island near their village. The county administration agreed to provide necessary support infrastructure, such as approach roads, while encouraging the villagers to both raise required funds and to help build the village. Each villager or family who wanted to participate needed to invest 150,000 RMB, and in return, each of the investors could get a house and ownership of the land on which the house was built to an area of 200 square metres. Seventeen families participated in the project, and built 17 village houses of the same traditional design consistent with the features of Baiyang Lake and local culture. The Island Cultural Village is 2.5 square kilometres in size and is surrounded by a good environment with clean water, water plants and trees, which is important given that some parts of the lake have suffered from environmental degradation (Xu 2004; Xin 2004). Equally, some upgrading of the local villages nearby has taken place. Seasonality is very obvious in Baiyang Lake's tourism, with the period from April to late October being the 'golden' season of tourism development and the rest of the year being the low season.

Partly to ensure that the houses would be properly maintained over the winter months, the County and Village Administrations decided that these houses could be designated as being primarily for the villagers' own use, but the residents have responsibilities to maintain the houses, keep them tidy and clean and to ensure that they are consistent with the standards required for tourism. The Wang Village Administration has also enacted policies and rules to ensure the service quality provided by this cultural village meet industry standards. The policies cover, for example, fees and food safety regulations. The village administration co-operated with a local travel agency in Anxin town to ensure that there would be a flow of tourists, and the Village Administration assigns groups of tourists to different houses on a quota system that provides equal opportunities for the 17 families. The fees (in 2005) were 100 RMB per tourist, which included one night's stay and bed and food for three meals, a tour guide and some traditional recreational activities. The Village Administration retains 10 to 30 per cent of the income as administration levies and to maintain the infrastructure, with the remaining 70 per cent being income for the villagers. The distribution of income benefits the participating villagers directly, and consequently more villagers are now becoming involved in this project. By April 2005, the number of families participating in the programme had increased to 50 and the project had expanded to two small islands linked by a long bridge. Average income of each family in the golden season is between 15,000 and 20,000 RMB. Observation indicates that tourists seem to enjoy this new attraction, and this is supported by the increasing numbers of returning visitors. The Village

Administration undertakes informal surveys periodically, and the president of the village proudly told the authors that 97 per cent of visitors are satisfied with this attraction.

Dong village is 5 km to the east of Anxin town, with an area of 20 square kilometres and 3,000 residents on the island. As it is located in the centre of Baiyang Lake, the environment is better preserved, and the traditional life of island people has been largely unchanged over the years. The water quality is extraordinarily good, with the largest natural grouping of lotus flowers being clustered here together in Baiyang Lake. However, the longer term quality of the reed-beds does depend upon water levels, and the region is scattered with reservoirs as part of a water-supply system for the region and its expanding industrialization. Xu (2004) and Xin (2004) both note, therefore, the importance of careful water management to maintain the health of the reed-beds and overall water quality. Women mainly spend their time at home, taking care of the children and older family members, and making a special kind of cloth made of reed, which is a typical natural product of Baiyang Lake. Men mainly spend their time in fishing or breeding, such as duck, crab and shrimp. As it is completely surrounded by water which is generally undisturbed by outsiders, people feel very relaxed and safe in the island. They still generally keep their doors unlocked during the whole day and night and will welcome visitors at any time. The village itself is very clean, with broad streets and green trees.

The approach taken towards tourism development in Dong is a little different from that of Wang. Instead of building a new village, it regards the original village as a tourism attraction. Based on existing facilities, the Dong Village Administration has invested money to improve the infrastructure, to educate villagers about the importance of tourism and how to treat tourists and market itself to the outside world. This is also related to the economic situation of this village, where most families had limited incomes and could not afford to build a new village. The Administration has selected several families who have homes thought appropriate (including a requirement of a harmonious family atmosphere), and has certificated them as 'Family Hostels'. Consequently, these families have the right as well as responsibility to provide hospitality to tourists. A fee of 60 RMB is charged for every tourist, and this includes one night stay with a bed and three meals. There are no specially designed recreational products. However, tourists can enjoy the existing sports and recreational facilities. They can chat with the villagers if they like, or visit the morning market. They can either dine with the host family or go to the local restaurant to enjoy more professional cuisine. They can go fishing, swimming or just appreciate the beauty of lotus flowers. They can learn the skills of reed weaving from local women, or listen to the stories of past conflicts and heroism from the elders. Dong started its form of rural tourism in 2003, when eight families participated. In 2003 and 2004 the average family income was about 3,000 RMB. Return visitation is now becoming very common. Seeing the potential market opportunity, the

Village Administration decided to expand the number of 'Family Hostels' to 20 in 2005.

At the same time, in 2005, the Village Administration began plans to attract outside investors to build more attractions surrounding the village. Currently, both villages have retained the essence of rurality. First, the design of buildings and landscapes are consistent with the original style of the local residents. Second, tourism activities are developed from villagers' daily life. Third, tourism is small scale, with shared activity spaces and close relationships and interactions between tourists and host community. Both villages have involved community participation, even while the extent and format are different. Wang is actually a built commercialized/commodified tourism attraction; while at Dong the original community (village) became an attraction, which the authors define as an 'attractionalized community'. In Wang, the attraction operator must be its villagers, and the participation of the local community is encouraged. The attraction was built on an island that was originally uninhabited, and the tourism products are centrally designed, which ensures consistency of product standards and makes it easier for administration. However, only a minority of villagers can be really involved in the tourism activities and enjoy the improved facilities and infrastructures of the built attraction. Therefore, community participation is partial.

In Dong, every villager can enjoy the improved infrastructures, and have more chances to have contact with tourists, either economically or socially. Although the community may not be deeply involved in tourism activity, it is widely involved. The daily life of local residents is still relatively undisturbed while the improved infrastructure in the village itself enhances the quality of local life. The educational efforts of the Village Administration have helped to ensure villagers' sense of ownership, participation and a feeling of responsibility for the local environment and culture.

A Comparison of Practices

As indicated earlier, Wang is actually a newly built tourism attraction that involves the local community. It is a built 'authenticity' and responds to the view that tourism products are designed according to tourists' needs and expectations. Thus, here the activities are designed to meet tourists' expectations of what constitutes local people's daily life in the island while offering them a high level of participation in that life during the period of their visit. Administration members, employees and host families are from the local community. All of this is consistent with the requirements of a 'built' or 'replicated authenticity', that is, it adheres to local custom but is literally a newly built resource with no prior history as a place of local residence.

Dong's tourism presents a truer construct of island daily life given that tourists experience exactly the way the village exists. Further, tourism and village development are engaged in the same process and occupy the same physical space. From the perspective of the Dong Village Administration, a key objective is the improvement of the lives of local residents. Therefore, both the convenience of villagers and tourists are considered while building extra facilities. Additionally, protection of the environment and the preservation of cultural heritage consistent with the notion of village improvement are incorporated into the development process. It is therefore suggested that Dong is based on an 'Objective Authenticity'. Tourists are satisfied with the authenticity experienced here, and it has been found easier to generate return visitors, thereby aiding the possibility of economically sustainable tourism development.

Wang possesses more investment capital that has meant the development of new facilities (including new integrated reception facilities) and this has helped to make marketing easier. Access to this capital resource and the development it has permitted has brought more income for local residents. It is, however, characterized by a complex relationship where the villagers in the cultural village are operators of an attraction. A complex social and commercial set of relationships occur whereby the community possesses within its own ranks its own buyers and sellers of resources, who in turn interact with relationships between villagers and tourists, who again represent, respectively, sellers and buyers. Thus, at least a two-tiered set of commercial transactions arise with potentially unequal results. The personal investments of 150,000 RMD also generate a strong motive to produce immediate rates of return, and one reaction has been the sale of agricultural products, but some have been of questionable quality—and 'authenticity'. This practice is consistent with the observation made by Zhong (2004) that modernization and increased commercial needs together emasculate the original authenticity based on the premodern and the non-commercial. Together these forces may easily destroy the 'replicated authenticity', simply creating yet another attraction, and tourists may lose interest unless marketing, pricing and promotion can sustain interest. The authors suggest that Wang will experience rapid growth at the commencement of the project, but its very commercial nature makes it easier for competitors to replicate the product, and its growth will slow down as similar competitors come into being. Besides, such authenticity is easily copied and it may be difficult to impress tourists as they become habituated to the product. The opportunity to generate consumer involvement is important in the generation of repeat visitation based on place identification (see Hwang, Lee and Chen 2005). A lack of a sense of authenticity undermines consumer involvement by increasing the possibility of substitute products, therefore making it more difficult to generate long-term returns.

Dong too might possess potential longer term problems. In Dong, tourism is a partial substitute for the traditional island industry. Host families have their own jobs, and the reception of tourists is a supplement to their daily work during the golden season. Part of the income will go to the administration of the village and serve as the funding for public fairs and improving infrastructure. At the beginning, there will not be many tourists because of limited investment and marketing. On the other hand, the relationship between tourists and host families is more like guests and hosts, rather than buyers and sellers. In this situation, tourists can catch glimpses of the back-stage life of the island village (MacCannell 1999), thereby potentially obtaining a better sense of the authentic. Easier communication between guest and host provides a possibility that they can form relationships with each other, which friendship may become a determinant of future repeat visitation. Thus, the authors predict that while Dong will experience an initial low rate of increase in terms of tourism arrivals and income at the beginning, the number of tourists will gradually increase later on, and return tourists and positive word of mouth will help to broaden its market appeal. The problems that Dong will face are those of assessing and adhering to given carrying capacities and perhaps resisting temptations to become just another commercial tourist entity based on bed and breakfast accommodation.

DISCUSSION AND CONCLUSIONS

Baiyang Lake has both beautiful scenery and a unique culture. There are 100,000 residents living in island villages. They have been isolated by water from the outside world, and preserve a traditional way of life that has slowed down past economic development. The living standards and income levels of the island residents are much lower than in other outside areas of Anxin County. Tourism development of the island villages has also lagged far behind the whole system of Baiyang Lake development. The island villagers have, in the past, worked as boatmen, tour guides, in hostels and restaurants, or sold traditional products at some tourism attractions. Although the role played by Wang and Dong is still not significant in the totality of Baiyang Lake's tourism system, the initiatives described here have fundamentally changed the situation. Island villagers have become the designers, investors and operators of tourism attraction, and not simply employees. They participate widely in tourism activities. They directly benefit from tourism as never before. This benefit is not restricted to economics, but also enhances personal self-esteem, living standards and the environment. The two islands have been successful and serve as reference points for other island villages and the Anxin government.

Rural tourism development in China has experienced two stages, both of which concentrated on tourists' need without paying enough attention to local community's interests. The uniqueness of Baiyang Lake island village tourism development is that it fully considered the interests of the local community, therefore making tourism a means by which to improve local residents' lives. This is especially true in Dong's case, in which not only are the local community's interests carefully considered, but tourism and community development becomes an integrated process. This could be a sustainable approach to sustain future well-being of the local community.

The authors suggest that this approach is important for the future direction of rural tourism in China, as shown in Table 19.2.

Opportunities and challenges still exist in Baiyang Lake's island tourism development. With the advent of the 2008 Beijing Olympic Games, it needs to be noted that Baiyang Lake is but two hours away from Beijing by highway, while within Beijing it is gaining increased recognition and reputation as a tourist location. It is certain that Baiyang Lake will attract more long-haul domestic and even international tourists. However, is Baiyang Lake well prepared for that? How about the island villages? The authors believe that they both need more training and skill enhancement, not only in building and operating attractions, but also in educating human resources to have the required skills and attitudes for the tourism and hospitality industry. It is also necessary to pay more attention to the promotion of Baiyang Lake and the island villages to international tourists. Improvements in infrastructure, especially in Internet technology, provide a good opportunity for that. Baiyang Lake should use this effective yet relatively low-cost means to take advantage of the opportunities brought by the 2008 Beijing Olympics and their aftermath to promote Baiyang Lake and its island villages to the international market. However, above all, the projects need to continually assess community interests and monitor carrying capacities to ensure that they remain consistent with village objectives. This, in turn, within the Chinese political context, requires sustained expressions of those interests at the level of Anxin government to ensure that, as signalled by the local government reforms of 1987, the relationship genuinely possesses liaison and not a

Table 19.2 Rural Tourism Development Approaches in China

	Rural Family Stay	*Cultural Village*	*Wang*	*Dong*
Decision-Maker	Villager, Government	Government, Corporation	Villagers, County Government	Villagers
Investment	Limited	Large	Large	Limited
Income	Major	Major	Major	Minor
Community Involvement	Food and Bed	Isolated from Community	Partial Involvement	Total Involvement
Authenticity	Low	Replicated, Commercial Authenticity	Replicated Authenticity	Objective, Experiential Authenticity
Rurality	Lost	Partial	Good	Sustained
Sustainability	Low	Low	Medium	High

top-down directive of perceived wider interests of commercial development. This would be consistent with the objectives of those government reforms as expressed by Peng Zhen. As noted by Worden, Savada and Dolan (1987):

> In May 1984 Peng Zhen described the relationship between the NPC Standing Committee and the standing committees at lower levels as 'one of liaison, not of leadership.' Further, he stressed that the institution of standing committees was aimed at transferring power to lower levels so as to tap the initiative of the localities for the modernization drive.

Thus the realities of Chinese rural tourism, and the nature of community involvement, remains an evolving programme as demonstrated by this case study and the work of Zhang, An and Liu and Gu and Ryan in this book. The next chapter now considers a very different context, that of a hutong in the centre of Beijing itself.

20 Place Attachment, Identity and Community Impacts of Tourism
The Case of a Beijing Hutong

Gu Huimin and Chris Ryan

INTRODUCTION

The purpose of this chapter is to examine place attachment and attitudes of residents of Shi Cha Hai hutong to the growth of tourism as visitors are drawn to this part of Beijing, China. The chapter will describe the location where the study was undertaken, indicate the research methods used and describe the results. The description of the results has two main sections: (a) descriptive statistics reporting the scores derived from a quantitative survey, followed by (b) multivariate analysis. The research design was informed by two sets of literature discussed in the following. These are (a) past studies of the impacts of tourism published in the tourism literature, and (b) the psychological theories of place attachment associated with Breakwell's (1986, 1992) identity process model (and see Breakwell and Canter 1993). However, given that the data were derived from a Chinese cultural and heritage setting, the article commences with a description of the research locale.

THE CONTEXT OF THE STUDY—SHI CHA HAI HUTONG

Shi Cha Hai is an area of Beijing renowned for its architectural and historical heritage. In 2005, 25 historical and cultural protection zones in Beijing were gazetted, and of these Shi Cha Hai is the largest. Tourism development is generally thought to have commenced in 1994 with the formation of the Beijing Hutong Tour Agency by Xü Yong, a photographer familiar with the area (N. Wang 1997). Certainly the hutongs of Chinese cities represent a disappearing pattern of life. Comprising small streets and alley-ways, they link homes of varying sizes through a maze of lanes in which traditionally family life took place in the public and semi-public places of the alleys and courtyards. These districts were not solely the preserve of one social group and this is perhaps most evident in the hutong district of Shi Cha Hai, which lies just north of the Forbidden City. Being located near to the seat of imperial power, this specific hutong not only possessed the homes of lower income groups, but also houses belonging to court officials and members of the royal family. Walking through the district today one can find doorways marked with the two, three or four bosses (see Figure 20.1) or stone lions that marked the homes of courtiers and officials of the palace. Additionally, near the Shi Cha lakes the tourist can visit Prince Gong's Palace, home of various members of the royal family of the Qing dynasty, and the childhood home of the last Emperor.

In the modernization of Beijing, as in other major Chinese cities, hutong have been demolished to make way for skyscrapers or high-rise apartments. It should not be thought that such destruction has been without benefits. For many living in the hutong the reality of their daily lives was one of homes without hot and cold running water, inadequate sewage disposal and small living spaces. Indeed, one of the major public building programmes that the Chinese government has undertaken in Shi Cha Hai has been the provision of modern public toilet blocks for local residents. While it is possible to modernize the hutong, the costs of modernization that retain the character of these areas are high, and might be higher than the costs of total reconstruction of an area. Nonetheless, there is a growing recognition that the hutong represent something that is quintessentially 'urban Chinese' and thus protecting the hutong is a means of retaining 'difference' that will help sustain a sense of heritage and a tourist attraction (N. Wang 1997; Le 2006).

Figure 20.1 Door markings in the hutong.

Consequently Shi Cha Hai represents a potentially important tourist resource of historical value—one already being recognized as former homes of the wealthier members of Chinese imperial society are being converted into heritage hotels and restaurants. At the time of this study in September 2006, Shi Cha Hai remained a vibrant community of Beijing families maintaining a pattern of life not unlike that followed for decades. Wandering through the back alleys of the hutong it is still possible to see families chatting, playing mahjong or eating their meals. However, the growing impact of tourism is being felt, not only in terms of the refurbishment of the streets as shown in Figure 20.2, but also, to some degree, in terms of increasing property values and costs of living. Indeed, of the residents surveyed in this study, 29.5 per cent 'very strongly agreed' with the statement that 'the hutong is becoming too expensive for me to live in because of tourism'. Such a perception does not necessarily mean that increasing house prices and rents are wholly due to tourism. Economic growth and increasing incomes in Beijing are of themselves a cause of increasing house prices, but there appears to be evidence that in Shi Cha Hai a premium exists over and above the normal increase. Between 2004 and 2005 the value of commercial house sales in Beijing climbed from 1,249.1 million yuan to 1,758.8 million yuan and the

Figure 20.2 Development in the hutong—June 2006.

price of houses increased by over 20 per cent to a mean of 6,274 yuan per square metre (Beijing Statistical Information Net 2006). In 2003, hutong tourism in Xicheng was officially worth 303,500 yuan (Beijing Statistical Information Net 2006). While Xicheng district (which includes the Shi Cha Hai area and its hutong) contains a population of 831,000, its population if anything has declined slightly, and so the pressure on property prices has primarily stemmed from economic growth. The area is primarily one of service industries (which account for 89 per cent of its earnings) and in recent years finance, advertising and the real estate have been economic drivers along with tourism.

As China is still a planned economy the government remains heavily involved in the planning process. The municipal government focuses on general planning, while the Beijing Tourism Administration is responsible for tourism planning. In 2005, the government started infrastructure enhancement following the rapid development of tourism in this area. Remedying dilapidated houses in Shi Cha Hai is an important part of this plan. In early 2007 there were almost 20 blocks under reconstruction and refurbishment. The government is also concerned with conserving the architectural heritage. For example, in 2006 a major project commenced to protect Smokebag Slanting Street, including road repairs. An island on the lake was also under construction and a new viewing spot was completed in May 2007 (http://www.bjsch.net 2006).

With the development of Shi Cha Hai tourism, visiting hutongs by pedicabs is increasingly popular. Since 1999, 17 tour companies and almost 1,000 pedicabs have become engaged in hutong tourism. This led to disorganized competition and congestion. Consequently the government created a special office, the Shi Cha Hai Administration Office, to supervise pedicab operations and services. This office has taken many measures, including unifying the design and appearance of the pedicabs, periodically training the pedicab drivers and regulating the routes of pedicab tours to improve the image of hutong tourism and reduce the negative impacts on the residents' lives. The local government also invests to beautify the surrounding environment (http://www.bjsch.net 2005) (see Figure 20.3).

An additional feature is the Shi Cha Hai Tourism Culture Festival annually promoted by the government to promote and market the district. It has also built the web site for Shi Cha Hai, and co-operates with hutong tour companies to improve awareness of Shi Cha Hai. A key issue facing the authorities is how to balance the development and conservation of the hutong. The Beijing municipal government plans to build quadrangle hotels characterized by the traditional Beijing architectural style to replace the current sub-standard inns in Dazhalan area. By 2008 these quadrangle hotels themselves will be part of a new urban scenery (www.bjta.gov.cn 2006; www.bjsch.net 2006; http://big5.china.com.cn/news 2006).

Figure 20.3 Banner about tourism found in the hutong—July 2006.
Note: The banner reads 'The Reconstruction of Luoguxiang Street needs your support'.

LITERATURE REVIEW

It might be fair to characterize the subject matter of resident perceptions of the impacts of tourism as existing within the mainstream of the tourism academic literature. This is a rightful recognition of the important impacts that tourism can have on communities, affecting as it might not only incomes but also the daily patterns of lives as tourists can throng the streets, use shops and induce retail outlets to change to meet a tourist rather than local demand. Additionally, it has been argued tourists' actions may be the source of demonstration effects whereby some parts of the local community may seek to change their own behaviours.

Attitudes might be said to be an enduring predisposition towards places, people and behaviours, and are usually stated as possessing three components: the cognitive, what is believed to be factual; the affective, the emotive response to that perceived knowledge; and the conative, a predisposition to various forms of behaviour resulting from the evaluation of the perceived factual. Various observations can be made as to such a formulation. First, knowledge of the conative does not necessarily permit accurate prediction as to actual behaviour. Intervening variables may inhibit (or reinforce) a predisposition. Thus, for example, a lack of income may inhibit purchase of a desired object. Second, the role of salience, importance and determinance needs to be considered. For example, Ryan, Scotland and Montgomery (1998) argue that altruistic attitudes towards tourism based upon the desire to see others economically benefit can be eroded if nuisance values grow; thereby bringing into conflict a series of values (e.g. the desire for community good against personal comfort). To some extent, this is an echo of the earlier concept of the Irridex where Doxey (1975) postulated a growing resistance to tourism when the numbers of tourists began to negatively impact upon accepted patterns of daily life. For his part, Ap (1990, 1992) suggested that trade-offs existed within the concept of social exchange theory, that is tourism is tolerated so long as 'benefits' outweighed the disadvantages. Thus a third observation might be made of the earlier formulation of attitude, namely, it presumes a lack of fuzziness in attitude formulation and implies a consistency which may be absent because of imperfect degrees of perceived knowledge, value conflict and intervening circumstance.

Given this, it is not surprising that while researchers have been able to suggest a number of factors that can impact on resident attitudes, not all factors have been replicated in subsequent studies. Factors identified as possessing importance have been duration of residency within a place, occupation (especially if employed in the tourism industry), place of residence and proximity to the main tourist zones, age, gender, ethnicity, value systems and sub-segments of population and their sense of place attachment and stage of destination development within the tourism life-cycle. (For a review and discussion of such factors one can refer to, among others, Long, Perdue and Allen 1990; Pearce, Moscardo and Ross 1996; Waitt 2003; Ryan and Cooper 2004; Carmichael 2006). What might be concluded is that individual research findings may be time and place specific. Consequently there are reasons to expect a lack of consistency on the part of both individuals and communities towards tourism as the industry waxes and wanes, and patterns of residency change over time.

Therefore, it might be said a study of the Shi Cha Hai hutong possesses interest as representing a site in the relatively early stages of what is envisaged to be a rapid expansion of tourism in Beijing, which itself is experiencing rapid economic development. At the same time, the concept of tourism is, itself, relatively new for many in China, and indeed, as Hsu, Cai and Wong (2007) indicate, it is a concept marked by significant intergenerational differences as older generations may still perceive holidays as acts of irresponsibility even whilst younger generations anticipate holidays as both a right and period of self-fulfilment. The study can therefore be justified on the grounds noted by Fan, Wall and Mitchell (2008) that there is still little empirical evidence of the way that tourism impacts on communities in China.

Another aspect that has been overlooked in the western tourism academic literature is the role of place attachment and the contribution of place identity to self-identity on the part of residents. Ironically, concepts of place attachment and involvement have been used by researchers in studies of repeat visitation. For Hwang, Lee and Chen (2005) place attachment potentially explains repeat visitation, and site recreation research in terms of how involvement theory explains levels of commitment to given activities. Similarly Alegre and Juaneda (2006: 686) argue, 'When tourists visit a place they develop emotional links with it, and this is important in understanding their behaviour'. Accepting this to be potentially true, then the same process can be applied to residents. It can be argued that communities can possess strong senses of co-operative and communal identities based on networks of extended family relationships. Given this assumption, it can be further argued that the impact of tourism upon a vibrant community in a Chinese setting may have impacts upon such a sense of place attachment given the potential changes that can occur.

Hogg and Abrams (1988: 325) argue that social identification is 'identity contingent self-descriptions deriving from membership in social categories (nationality, sex, race, occupation, sports teams . . .)' to which can be added neighbourhoods. Thus, the social and physical attributes of place can be internally subsumed to help create a sense of being. Breakwell (1986, 1992) postulates four ways in which this process occurs. This is illustrated in Figure 20.4 with reference to place.

Breakwell (1986) suggests that the first principle of identity is the establishment of a sense of personal uniqueness. Feldman (1990) and Hummon (1986) examined the nature of self-concept of being a rural- or urban-based person and found evidence that type of location contributes to

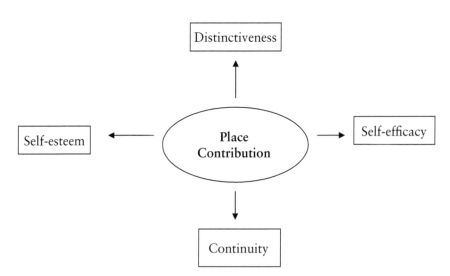

Figure 20.4 Place and identity formation.

self-perception. Similarly Twigger-Ross and Uzzell (1996) present evidence of how membership of a London Docks neighbourhood had a strong impact on not only sense of place but sense of identity. It can be postulated that membership of a hutong can either be a source of pride as representing membership of a traditional society that is disappearing, or alternatively, a source of dissatisfaction as being a member of something not perceived as being 'modern'. In short, the evaluation of place membership impacts upon self-esteem, but equally the nature of close family networks can reinforce processes of positive evaluation of place and thus identity. Uzzell (1995) found evidence that living in a historic town generated a sense of pride through association while Lalli (1992) concluded that spatial–physical environments were unequivocally important for human identity. The remaining two components of Figure 20.4 thus assume importance when considering locations faced with change. A degree of stability of self-identity is arguably important for healthy self-esteem in that a lack of continuity implies a loss of self-esteem, while equally degrees of permitted change are required for personal development and growth. The self needs to be in a state of flux, but equally too much change or stability can be psychologically harmful. Place referent continuity has been discussed by many authors in terms of place-identity relationships (e.g. Korpela 1989; Giulani 1991; Lalli 1992; Twigger-Ross and Uzzell 1996; Hidalgo and Hernández 2001) with the conclusion that changing place and relocation can create opportunities for self-development, but a key issue is the degree of control that people feel they possess over the nature and rate of change of place. Consequently, a sense of continuity with place needs to go hand in hand with self-efficacy, that is, a sense of belief in one's own abilities to meet and cope with changing circumstances. Manageable environments have both a psychological and physical manifestation that are inextricably linked (Twigger-Ross and Uzzell 1996; Hidalgo and Hernández 2001). Thus, the latter authors have established that place attachment comprises both physical and social bonding that impacts on place identification and sense of self.

Given this approach, the concepts of place attachment and identity progress the nature of the debate on the impacts of tourism towards senses of self and perceptions of change, and consequently attitudes towards tourism are not solely explained in terms of social exchange theory, but in terms of personal value systems as suggested by Ryan, Scotland and Montgomery (1998). If, following Brown and Raymond (2007: 90), 'Place dependence refers to connections based specifically on activities that take place in a setting', the change of activities that are due to tourism may well reflect changes in place dependency, place identification, social interaction and self-identity that help explain assimilation and demonstration effects. Data derived from surveys that include items on landscape assessment, employment patterns, income generation, changing retail usage and the items used by researchers of tourism's social and community impacts thus also offer measures of place attachment.

Hypothetically it becomes possible to construct a series of hypotheses, for example, longer stay residents will have stronger place attachment, but problems exist with such an approach when using exploratory research techniques. Place attachment is a social psychological construct, and various potential alternative scenarios can be easily envisaged. For example, is place attachment equated with a perception of past patterns of hutong life, or is there pride in perceived progress? Other studies of community and tourism in China have emphasized strong senses of community responsibility and participation in tourism promotion (e.g. Zhang, An and Liu 2008; Zhou and Ma 2008), but such studies have been of rural communities. Two further complicating factors might also exist in the Chinese environment. First, the role of the local administrative town (*xiang*) government, while subject to reforms in 1982 and 1987 to sustain grass-root participation, is still dominated by top-down initiated liaisons (Worden, Savada and Dolan 1987). Second, the very population size of Chinese communities creates communication problems (a Chinese village may comprise a population of many tens of thousands). Population densities affect concepts of place attachment and self-efficacy in playing a role in community politics. Such issues have been rarely explored in the English language tourism academic literature on Chinese destinations and tourism policies and impacts.

THE RESEARCH

Characteristics of the Samples

This chapter reports findings from the third stage of a research project that involved two prior stages of qualitative research based on interviews with residents and local tourism business people. The results described in the following are also informed by the findings of these other studies, some of which are reported by Ryan and Gu (2007). A convenience sample total of 400 Beijing residents were interviewed in the hutong, of whom 69 per cent were normally resident in the Shi Cha Hai hutong. The remaining respondents were all resident in Beijing, and had close family or business associations with the hutong, and thus also possessed a sense of place identification. Of the sample, 55 per cent were male, and 45 per cent female. With reference to age, 54 per cent were between the ages of 19 and 45. With reference to income levels, 48 per cent described themselves as possessing 'average income', 17 per cent stated they had 'above-average income' while the remainder stated they had below-average levels of income.

Residents of the Shi Cha Hai hutong were asked how long they had been resident in the hutong. The mean number was 31 years, the maximum being 70 years. The standard deviation was 19.3 years and the median was 34 years. Given this mean duration of residency it can be noted that many respondents provided additional comments and expressed interest in the survey and having an opportunity to express a view. The questionnaire and interviews were all conducted in Chinese (Mandarin) and no names and addresses were collected—meaning that respondents retained privacy.

Questionnaire Construction

The questionnaire comprised three parts. The first was a filter to assess the nature of their relationship with the hutong. Only those resident or otherwise having close relationships with the hutong were included in the study. The second section was the main part of the survey and listed a series of statements. Degree of agreement with the statements were elicited through the use of a seven-point Likert-type scale with 1 being the lowest score ('very strongly disagree') to 7 ('very strongly agree') with a non-response option being provided. Examination of line diagrams indicated widespread use of the full range of options and in this case there did not appear to be a reluctance to use the 'extreme' scores of '1' and '7.' There was little evidence of any specific response set operating. The items sought comments about congestion, changing retail patterns, likelihood of petty crime, income generation, attitudes towards the management of tourism and the nature of the tourists themselves (for example, were they friendly or intrusive), and reactions to the changes induced by tourism. A third section of the questionnaire permitted open-ended responses to questions and invited any additional comments. A final section collected socio-economic data.

With reference to sampling adequacy, the Kaiser-Meyer-Olkin statistic was 0.85. Different split half reliability measures were in excess of 0.77 while alpha coefficients were in excess of 0.80, thereby indicating internal reliability in the data. Tukey's test of non-additivity also possessed appropriate values with $F=61.3$, $p<0.001$. These statistics indicate the data met the usual tests for internal rigor, sample adequacy and suitability for further analysis.

Results

This section of the chapter reports overall results, patterns of determination and interpretative comments. As indicated in the literature review, a common finding is that residents feel that economic gains from tourism are both likely and important. In this study, however, cultural and heritage concerns feature more strongly, implying the potential importance of landscape to place attachment as indicated by Hidalgo and Hernández (2001). Both residents of Shi Cha Hai and other areas of Beijing rate the retention of the hutong and its architecture as being the most important item, with a mean score of 6.24. When the same principle was stated in an opposite manner, that the hutong should be demolished in favour of more modern housing, the mean score for the total sample was but 2.95, with Shi Cha Hai residents scoring even less with a mean score of 2.88. One interpretation is the importance of landscape in image formation of place, of place dependency and that tourism has value in retaining architectural heritage as a means of reinforcing place attachment. However, while residents agree that the hutong is cleaner and smarter in appearance because of tourism (mean = 4.66) they do not agree that tourism can improve the quality of life in Shi Cha Hai (mean = 3.29). Neither do they strongly perceive that tourism is necessarily a means of protecting the architectural character of the hutong (mean = 4.40) or that tourism is a long-term means of ensuring the survival of the district (mean = 2.73). As discussed in the following, these findings are of interest in terms of place attachment and identity.

The major disadvantage identified by respondents was traffic congestion (overall mean score = 6.63) with hutong residents feeling this more strongly (mean = 5.86) than other Beijing residents (mean = 5.10) where $t = 3.11$, $df = 369$, $p = 0.002$. A significant difference between the two groups was found in those items relating to the impact of visitors upon the daily life of hutong residents. Shi Cha Hai residents certainly were more aware that the presence of tourists could impact upon their lives, attributing a score of 5.50 to the item 'Local residents are the people who mainly suffer from living in a tourist area' compared to 4.00 by other Beijing residents ($t = 5.70$, $df = 370$, $p < 0.001$). Given that localized activities are a key determinant of place attachment (Twigger-Ross and Uzzell 1996), potential disturbance of those activities has potential importance. However, at the time of the study (September 2006), these impacts appeared to be manageable. For example, hutong residents scored the item 'I think visitors are far too intrusive in our everyday lives' at 4.94, representing a point of agreement, but not too strong a level of agreement on the scale being used. Again, with reference to the concept of self-efficacy as a component of self-identity and place attachment, this finding has importance in terms of degrees of self-control over daily life. The main forms of intrusion appear to be potential traffic congestion, as noted earlier, and noise (mean = 5.13). It should also be noted that 39 per cent of the hutong's residents 'strongly agreed' that there were already things that they did not do because of the intrusion of tourists. There is a recognition by hutong residents that their environment is cleaner than before, with 47.2 per cent strongly or very strongly agreeing with the statement that 'the hutong is a lot smarter and cleaner because of tourism'. On the other hand, 48 per cent 'strongly agreed' with the statement that 'I feel tourism is growing too fast for the hutong to cope with', over half of hutong residents strongly or very strongly agreed that 'the tourism authorities should restrict the growth of tourism in the hutong', nearly 30 per cent strongly agreed that the hutong was becoming too expensive to live in and 17 per cent strongly agreed that 'with the changes I would like to move out of the hutong.' One implication, therefore, is that place attachment is, for a minority, being strongly eroded by tourism's impacts. Further evidence would suggest that self-efficacy is undermined with possible deleterious impacts on self-identity if place attachment is weakened due to a perceived loss of control on local environment (e.g. see Twigger-Ross and Uzzell 1996). One can suggest that while, in September

2006, there was tolerance of the presence of tourists, there might not continue to be such tolerance on the part of a growing minority of residents if numbers increased further, as might be expected with the promotion of the 2008 Olympics. Scores on items relating to increasing the promotion of the hutong as a tourist asset are consistent with these findings. For example, the mean score on the item 'I think that attracting more tourists to the hutong is a good idea' was but 4.35, and 18 per cent of residents 'strongly disagreed' with this statement compared to 9 per cent who strongly agreed. One implication is the sense of community becomes threatened as opinions become potentially polarized, and for some, senses of place attachment and identity based on place may be weakened as the place visibly changes. On the other hand, others may take pride in the place changes enhancing place attachment, yet some with strong attachments to a construction of place based upon past patterns may become more vociferous in challenging change.

The literature has indicated that various determinants of attitudes towards tourism might exist. For example, socio-demographic variables might have a role, as might duration of residency. Thus it might be hypothesized that younger people, being more used to a world of change, may be more tolerant of tourism and its impacts, while those who have been long-term residents might not be more resistant to change solely because of older age, but because they may be more emotionally wedded to a status quo of social and communal networks that they see as being threatened by tourism's growth and consequent impact. Consequently it can be hypothesized that longer stay residents may have stronger attachments to past patterns of place. The data set permitted some examination of these issues.

With respect to the age of the Shi Cha Hai residents, age was a determinant of attitude when tested by ANOVA in 13 of the 44 items. At first sight this implies that age was perhaps not overly important, but the patterns of agreement and difference between age groups are significant. For example, those under the age of 25 years scored twice the score of those aged 36 to 50 years on the item that tourism created jobs—but these scores were still only 4.3 as against 2.3—that is, the younger age group tended to neutrality as to whether tourism created jobs, while those aged 36 to 50 years tended to disagree with the statement. Those over the age of 50 years also tended to disagree with the statement, but not as strongly as their 36 to 50 year old counterparts ($F = 4.93$, $p < 0.001$, Scheffé test significant for those aged 36 to 50 years at $p < 0.05$). There was almost a direct correlation between age and the perception that tourists were intrusive in daily lives, with the scores increasing from 3.8 for those under the age of 25 years to 5.8 (almost 'strong agreement') for those between 60 to 65 years ($F = 3.22$, $p = 0.005$). Similarly those under the age of 36 years stated higher degrees of agreement (mean = 5.2) with the statement that attracting tourists to the hutong was a good idea, while older age groups did not ($F = 2.71$, $p = 0.014$).

What possess interest are those items where differences are not statistically significant. All age groups have very similar scores that tourism will add to congestion and disagree that tourism is a means by which the hutong can be protected. All expressed a view that the planning authorities could do better with reference to the hutong and tourism. Equally of interest is that while not statistically significant, the younger age groups tend to be slightly more favourably disposed than their older peers to more invest-ment in tourism and that, albeit at low levels, their incomes are better as a result of tourism (the test for homogeneity of variance, Levene statistic, is here significant at $p = 0.015$). With the data described, a picture emerges of a tolerance of current levels of tourism but concerns across all age groups about future impacts and growth, with younger age groups tending to be a little more accepting of tourism—while all recognize that tourism is but one factor within a context of change.

Turning to the duration of residency in Shi Cha Hai, this variable appears to explain difference in attitudes towards the development of tourism. The variable appears to be significantly independent of 'age'. By this it is meant that someone aged 40 years with 30 years of residency in the hutong is much more likely to have views in common with someone 20 years older with a similar period of residency than with someone else also of 40 years of age but with a shorter period of residency. Of the 44 items in the questionnaire, 22 show statistical differences by residency when using ANOVA. The Levene statistic for measuring homogeneity of variance also shows significant varia-tion within the sample on this variable. Interestingly, there are no statistical differences with reference to job opportunities arising from tourism, perhaps because tourism is not perceived as a generator of many jobs. On the other hand, long-term residents do perceive tourism as a generator of cultural opportunities (scoring a 6 on the 7-point scale, $F = 3.92$, $p = 0.001$), but these long-term residents are very strongly opposed to tourists staying over-night in the hutong (it should be noted that all residents tend to oppose this). In many ways this denial of overnight stays in the hutong is a form of tempo-ral zoning that permits favourable tourism development in terms of sustain-ing the architecture and heritage of the hutong but allows residents to retain their social space over the evenings and nights when arguably a traditional communal life is at its most evident. The lack of overnight-stay tourists in the hutong is also a potentially important means of retaining the social life that creates strong feelings of place identity and attachment and the formation of personal identity as a hutong resident. Longer term residents are also statisti-cally significantly different in their statements that there are things that they no longer do because of tourism, and that tourists are intrusive in daily life, scoring levels of agreement on the former statement that exceed a value of 5.80 (agree). These scores are consistent with a view that those of less dura-tion of residency will not have established patterns of life in a 'tourist-free' time and thus will have become more habituated to the presence of tourists, but even in the case of those of less than five years duration of residency the scores exceed the mid-point of 4.0; implying no disagreement with the state-ment that they now do things differently because of tourism.

Gender is generally not a determinant of attitude with t-tests showing significant differences in only two items. Here males tend to score a little higher on the economic benefits of tourism—but as the mean scores are approximately 3.0, disagreements with the statements are still high. Past studies have shown that employment in the tourism industry is often a factor that serves to distinguish between subsamples of a given population. In this instance, given that employment in the industry was low, a proxy question was asked—to what extent did respondents agree that they would wish to have a job in the tourism industry? The responses to this question were recoded into a threefold classification, those who disagreed, those who were indifferent and those who stated they would like a tourism-related occupation. This proved to be a significant variable in a number of instances. Foremost among these was that while those who were not attracted by a job in tourism, or were indifferent to the occupational choice, scored above 5.1 on the item about tourists being intrusive; those expressing an interest in a tourism job scored significantly lower at 4.08 ($F = 7.17, p < 0.001$). In short, this group expressed scores that were more favourable towards tourism, subject to the caveat that they nonetheless often perceived problems with the industry. For example, they did agree that tourism would create congestion (5.56 as against 6.0 for the other groups and thus were not significantly different from those groups, $F = 1.03, p = 0.385$). It is therefore of interest to note that a quarter of Shi Cha Hai residents expressed favourable attitudes towards having a job in tourism.

On this premise it became possible to develop a psychographic profile of hutong residents using cluster analysis. Four clusters were discerned, the first comprising 23 per cent of the sample, the second, 37 per cent, the third 17 per cent and the fourth the remainder. The first were generally the most favourably inclined towards tourism. Nonetheless, this cluster still agreed with statements that tourists had an intrusive impact on the hutong. The second cluster were actually less critical of tourist behaviours, but were oriented towards tourism on the cultural opportunities it offered, including meeting tourists who were generally considered to be friendly, and they did therefore have the highest score on the item that tourism was the best means of preserving the hutong, albeit only at 3.71. Of interest was that of this cluster nearly 50 per cent wished to have a job in tourism—by far the highest proportion (it ranged from 17 per cent in cluster one to less than 7 per cent in the remaining two clusters). The third cluster was the most critical of tourism and of the planning mechanisms, while the final psychographic grouping were the strongest in their recognition of the hutong as a special place of heritage and history (scoring a mean of 6.73 on this item), again being critical of tourism, but expressing a more positive view about planning mechanisms than the third group. Using discriminate analysis to assess the 'correct allocation' of respondents to clusters it was found that in clusters one and four 94 per cent of cases were 'correctly' allocated, whereas in clusters two and three 100 per cent of cases were correctly classified.

It has often been commented that any discussion of tourism in China is never far removed from a consideration of government, and this also true in this instance. From the work of Zhou and Ma (2008) and others, place attachment can be connected with participation in local administration. There is evidence in this set of data that key determinants of attitudes towards tourism impacts and the role of the Beijing planning authorities are determined by a few key variables that include duration of residency in the hutong, favourable attitude towards tourism as measured by proxies such as attractiveness of a tourism-related job, the degree to which tourism can protect the houses in the hutong, degrees to which tourists are perceived as intrusive, and attitude towards the hutong as heritage and historical assets. These relationships are indicated in Figure 20.5. It is possible to test such a relationship through structural equation modelling.

The results of regression modelling and structural equation modelling using these variables reinforced the role of job creation as a discriminatory variable. Given the previous comments about self-efficacy and relationship with political processes possessing importance in the Chinese context,

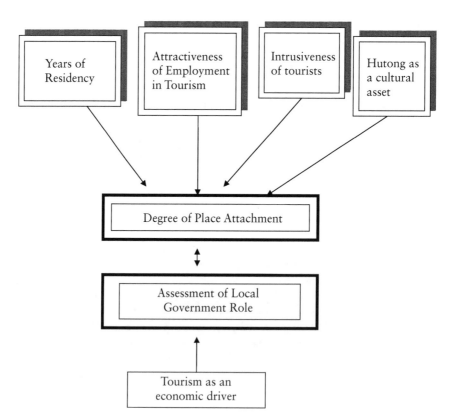

Figure 20.5 Determinants of perceptions of the role of local government.

regression analysis used as the dependent variable the item that the planning authorities were undertaking an excellent job in balancing the needs of local residents with the desire to increase tourism. The results showed that the independent items accounted for 16.1 per cent of the variance (adjusted R^2 = 16.1) with β = .340 ($p < 0.001$) on the item 'I would like to have a job in tourism' and $\beta = 0.165$ ($p = 0.006$) for the item 'the tourism industry is good for the hutong's economy'. Taking Figure 20.5 as the basis for modelling the variables using the AMOS V program and the maximum likelihood option resulted in goodness of fit criteria failing the normally accepted criteria (e.g. CFI = 0.67, NFI = 0.62, RMSEA = 0.101 as against a preferred value in excess of CFI = 0.8, NFI = 0.9 and RMSEA < 0.08). Nonetheless, both Hair Jr *et al.* (1998) and Kline (2005) utter cautionary words about data interpretation and indicate that SEM is a useful means of conceptual exploration even where the data fail to fit rigorous criteria of fit. An examination of the regression weights and intercepts identified three key elements—intrusion effects, the protection of hutong heritage and tourism being generally good for the hutong's economy as the main determinants by which residents assessed local government performance. From the qualitative evidence these three items were commonly mentioned and although difficult to assess, these items appear important in social constructs of place and place attachment.

DISCUSSION

In terms of destination development the findings represent the attitudes of residents towards tourism and place at a time when Shi Cha Hai hutong is moving from an involvement into consolidation stage of the tourism destination life-cycle (e.g. see N. Wang 1997) and facing potential further tourism development prior to the 2008 Beijing Olympic Games. The findings show a current tolerance of tourism accompanied by a growing concern about levels of intrusion into daily life, and some scepticism about the ability of tourism to create jobs and help preserve the nature of the hutong. In terms of the debate about place identification, the respondents distinguish between the hutong as a collection of buildings and as a social entity. Many note that the preservation of buildings and built heritage does not equate to the maintenance of a way of life, yet it is the architectural heritage that attracts tourism and tourism that gives commercial value to that architecture, thereby preserving the place markers for the future. The qualitative data shows that a compromise between tourism development and the maintenance of social patterns emerges through zoning (Ryan and Gu 2007). First, there is temporal daily zoning when tourists may see the hutong during the day, but are largely absent during the evenings and nights and when, certainly in the warmer months of the year, the alley-ways become a centre of social interaction for local people. Thus, the current comparative absence of hotel building is a means of preserving the hutong as a social entity. Second, there is seasonality when levels of tourism fall in winter. Temporal zoning thus aids place attachment by generating increased tolerance for tourism by providing periods of respite from the pressures induced by tourism.

From the perspective of the wider literature, the study shows how a strong sense of place and living heritage informs responses to tourism. In this specific context this understanding of the hutong as a special location of architecture and social interaction had more of an impact than the issue of economic gain. This is contrary to many findings derived from the English-speaking world and is shown by the low overall mean scores allocated to the items measuring economic impacts. Yet there is an important caveat arising from the distribution of scores on these items. Attitudes towards economic impacts did play a significant role as a discriminating variable in attitudes towards further tourism development. One reason for this was the differences displayed between two subgroups, namely, a difference partly based upon age and, second, the perceived attractiveness of a job within the tourism industry. In that sense the asking of a question about the attractiveness of such employment proved to be a pertinent item, and for other researchers it may well be worth asking this in addition to asking whether the respondent is actually employed in the industry. The importance of age as a variable has, it is suggested, something to do with the nature of place attachment as conceptualized in this chapter. Younger people, especially during recent decades in China, have been socialized at a period of rapid economic and social change where change is often deemed to be progress. For them, senses of place attachment are not necessarily formed in a context of lack of infrastructure change or social evolution. Indeed, change may be a prerequisite of forming place attachment, for change is an attraction and not a threat. These considerations might account for the failure of the data to produce estimates of good fit when applied to the structural equation modelling of Figure 20.5. Equally the discussion of place attachment can be related to theories of personality. Kelly (1963: 177) refers to group expectancies and the manner in which the young adopt 'negativistic' roles to be contrary to expected behaviours but wind up 'using the very same dimensional system his [sic] parents use.' Thus, the younger person may express dissatisfaction with the status quo of the hutong, yet within their life experience of Beijing at this date, their experiences have been wholly experiences of change. Consequently, perceptions of change of the hutong and place attachments may well be expressed in statements of ambiguity and inconsistency. In the qualitative study not reported here, examples of inconsistent attitudes and at times careful delineations of positives and negatives were found. For example, younger people would enjoy the night atmosphere of restaurants at the lakes, but might complain of an inability to spend money there—so thereby enjoying a place attachment of being 'fashionable' but equally feel in part excluded other than in a 'hanger-on role'. But 'hanging on' in itself contributes to the summer sense of activity.

As noted earlier, the high levels of importance attributed to the architectural heritage of the hutong emerges as a strong contributor to place attachment and the sense of being special by living in a special place. This finding also emerged from the qualitative data, but it also needs to be stated that identity efficacy and place attachment are complex in the understandings of Chinese communal political processes. Examples were found where some respondents clearly stated that these issues were only matters for the government, and not for them, and the quantitative data clearly showed that some had reservations about future policy. As Worden, Savada and Dolan (1987), Xie (2001), H. Wang (2003) and Fan, Wall and Mitchell (2008), among others, all comment, in China the role of government in the planning processes of tourism is wholly pervasive, thereby potentially complicating the role of self-efficacy as a contributor to self-image as understood in the western academic literature. Future research that wishes to engage in place attachment and place identity as part of an assessment of the impacts of tourism on Chinese communities may need to consider this more explicitly. In this sense future research should seek to clarify more specifically the relationships of (a) perceptions of the role of local government (b) change induced by tourism and (c) place attachment and self-efficacy within the framework suggested by Breakwell (1986, 1992).

In short, as this research project evolved, various limitations in the project became evident. Questions were primarily related to tourism-induced change, but the hutong is experiencing change as Beijing also changes around it. The hutong is linked to the wider Beijing economy and the social, economic and political changes that are occurring are not solely due to tourism. This seems to imply a failing in not only the current research, but with past literature because by concentrating on tourism-induced change only research has failed to link tourism to the wider socio-economic changes that might also be occurring. Ryan and Gu (2007) argue that hutong-based tourism has become linked into a process of globalization, and possibly of glocalization too. In short, the original design of this research may be limited by the very focus it created on assessing resident identification with the hutong and reactions to tourism-induced change. Perhaps the project should have been contextualized within a wider framework, and this need may be particularly true in contemporary China. It is therefore suggested that future research into the impacts of tourism within China need to take account of these wider social and economic issues.

ACKNOWLEDGEMENTS

This chapter is part of the result of the project sponsored by the Beijing municipal government namely the '21st Century Beijing Creative Group'. In addition, a travel grant from the University of Waikato Management School

21 The Context of Chinese Tourism
An Overview and
Implications for Research

Chris Ryan, Gu Huimin and Zhang Wei

The first chapter sought to establish a context for the study of Chinese tourism. In this final chapter we wish to return to contextual issues to highlight issues in the understanding of Chinese tourism, the policies that accompany them and the implications for research. It has become increasingly clear to us in our own past research and in the reviewing of the chapters and other materials for this book, that simply transferring concepts from western academic literature to inform research in China is, at least for the present, not wholly appropriate. It is also evident that we are not alone in this opinion. For example, in their chapter Li and Sofield explicitly questioned at length the appropriateness of science-led western paradigms for the interpretation and classification of National Parks and natural areas in China, arguing that such approaches misunderstand the cultural significance and meanings for Chinese visitors. Similarly, Wall (2005) has advised against adopting western notions of community participation in tourism without questioning those notions in a system. Rather he argued the key test is participation in *benefits*, not involvement in a *process* as is understood in the West. The challenge for Chinese tourism planning is to generate such benefits from what is still often a top-down mode of planning. While in the future, as China changes, it may become more appropriate to use western concepts for tourism and destination planning, at present key factors are still at work in China that inhibit such easy transfers of ideas.

While it is true that some imperatives of tourism remain, such as the need to meet tourists' demands, and to generate satisfying experiences for visitors that will engender repeat visitation where required by economic imperatives, the wider socio-economic, political framework within which Chinese tourism planning operates is very different to that existing in the West. Among the factors that we feel researchers must take into account when considering tourism in China are the following:

MAN–NATURE RELATIONSHIPS

Western and Chinese cultures have different views regarding the relationship between man and nature. In western culture, there exists one perspective derived from Biblical tradition that man has dominion over the natural world, which is to be utilized, controlled and plundered (Meng 2004). On the other hand, traditional Chinese culture embraces the concept of 'harmony between people and nature'.

A typical example is the Chinese garden, which not only reflects natural beauty, but integrates several forms of art, including landscape painting, calligraphy, architecture and sculpture, which together represent reflections of a philosophical nature (Fan, Shen and Zheng 2002). For example, Taoist ideas are commonly expressed in a garden. In a typical Chinese garden there are winding paths between bamboos and trees, small pavilions hidden behind rocks or on the top of hills and long corridors with traditional colour paintings. Taoism places man in nature. In doing so the harmonious balance is achieved, where man is present but does not dominate (Panich 2008). This is also reflected in classical Chinese literature. For example, Chapter 17 of Cáo Xuěqin's novel, *The Story of the Stone*, describes the process of selecting calligraphy and sayings for the gardens designed by 'Horticultural Hu' where the process of the selection of sayings is dominated by setting and literary allusion that the reader is able to locate.

Thus, as both Li and Sofield and Ma, Ryan and Bao independently noted in their chapters, the Chinese concept of nature (*da-ziran*) represents everything coming together. It is a holistic concept and mankind is part of this process where man and nature are engaged in a mutually supportive or harmonious relationship. Nature without man does not represent an ideal; rather it is an incomplete state. Similarly, consideration of the nature of man without reference to the natural is also incomplete. Therefore, the Chinese perspective will both attribute meanings to natural features that become rich in symbolic significance apart from any botanical or geomorphic value, and such examples would include the pine trees at Huangshan that have been a feature of Chinese landscape painting for decades. But equally, given the importance of calligraphy as an expression of art and aspiration, it is appropriate that landscapes incorporate calligraphy, and that such calligraphy be maintained in a fresh state where the colours selected are inherent to the sense of the characters as much as the location in which they are painted. Thus, red is often used as a colour representing happiness or good fortune. The same perspective lends itself to the construction of buildings in National Parks from at least three view-points: (a) as an aesthetic enhancement of natural settings, as illustrated in Figure 4.1, (b) as a means of access so that man can become part of and appreciate the landscape, (c) a form of economic regeneration that remains consistent with cultural norms. This is not to deny that within the West a new awareness of man–nature relationships exists within the environmental movement with its appreciation of a need for a harmonious coexistence. So too there exists a view of the idyllic derived from the Victorian Romantic movement, but the West does not possess quite the same nuanced understandings of, say, the role of calligraphy that itself is nuanced, and the importance of poetic signings as caught by Cáo Xuĕqin's work cited earlier.

At the same time, the view that a human presence in natural areas such as National Parks can be justified is being taken to extremes when one observes the degrees of urbanization associated with these parks. National Parks have suffered long periods of under-funding with the result that the state has been encouraging under licence the involvement of private sector capital investment to better provide accommodation and services to visitors in the pursuance of economic gains. For example, Wang and Bai (2002) provide the examples of Huang-long-dong and Bao-feng lake in Hunan Province where the operating rights of 300 tourism spots and National Parks have been transferred to authorized corporations. Again, at Wulingyuan World Heritage Site one of the major attractions is the One Hundred Dragon Lift that takes visitors to the summit of the mountain in less than two minutes, and as described by Zhang *et al.* (2008) this facility is abused by tour operators seeking to develop restricted itineraries. They conclude the main beneficiaries of the park are tour and the lift operators. Consequently, it can be maintained that profit motives and the need for an achievement of economic objectives are writ much larger for Chinese National Parks than is the case for their counterparts in, say, Europe. This creates an obvious tension between under-funded conservation efforts and the needs for maximizing returns on capital expenditure, where the rationale for the investment is that the generation of visitors creates revenue for conservation expenditure (Chen and Huang 2006; Guo and Zheng 2006). The potential problems for nature protection are obvious in terms of overcrowding and overuse in zones used for tourist sightseeing. Yet such crowding is also consistent with a culture where feelings of the collective group may rate higher than individualistic gazing, and arguably such sentiments ease park planning through means of zoning. While these issues are not unknown in the West, arguably they are much more intensely experienced by an under-funded and under-paid (by western standards) National Park work-force.

THE ROLE OF FACE, CONTRACTS AND PERSONAL RELATIONSHIPS

In spite of the Revolutionary period, Confucianism still heavily influences Chinese thinking about social relationships. The individualism of the West is thus significantly mitigated by a series of norms that sustain personal relationships. As in other societies, for example the Maori of New Zealand, individuals are members of families, and family relationships and sustenance of the family in an extended sense is very important, and is indeed an important component of identity. The *ying* and *yang* of personal–social goals in Confucianism is represented by the tension between *ren*, the relationship between individual people, and *li*, the social norms within which relationships are found acceptable. While as Weymes (2007) cites, Generation Y are showing different attitudes as revealed by various Gallup polls, middle and senior management is still governed on the whole by those adhering to more traditional schools that rejected Logicalism (S.H. Liu 1998). Prior to China's membership in the World Trade Organization, business relationships were still primarily determined by personal and family relationships rather than by formal legal rules of contract, and although international business requires adherence to international laws of contract, business with Chinese organizations is still characterized by the importance of relationship. The Chinese believe that establishing a more personal relationship with prospective business partners is foremost. If this is successful, transactions and business relationships will follow. It is a very different style of doing business from westerners, whose approach is to get 'straight down to business'. Rather, in the Chinese context, it is rare to mention business in the first meeting.

It is thus argued here that research that is purely quantitative in nature and dependent upon completion of Likert-type questions is poorly sited to capture the nuances of such important cultural norms, especially when dealing with the impacts of tourism. Ethnographic approaches are important, and thus change the nature of the researcher–researched relationships. Researchers must also then become part of the networks that establish *guanxi* by which people favour those within relationships in reciprocal patterns. The researcher must also be aware of ensuring that respondents do not lose 'face', that is questions do not place respondents in positions of implied or actual criticism, especially in public. Given also predispositions to adopt roles of humility within structures of inherently understood hierarchies, it may take several months if not years before a researcher can begin to access the social structures that effectively articulate the social discourse that lies behind tourism place planning and decision-making. The protection of face, the *mian zi* syndrome by which people project perceptions of wealth and power is well recognized within Chinese society, as is *jia chou bu ke wai yang*, which in a western sense can be equivalent to 'not washing your dirty laundry in public' although possibly a better translation is 'family disgrace should not be revealed to an outsider'. This applies often to organizations as well as the network of personal relationships to which an individual belongs. In short, these are collective concerns.

The facets have implications for management and service encounters. To take an extreme perspective, one can argue that in the market economies of the West, the client is 'king', but in the relationship-laden environment of a Chinese organization, one's superior is 'king'. Criticism of poor service in a society where relationships and 'face' is important, even where these concepts are changing, will be differently expressed and it is possible to query whether uncritical applications of confirmation–disconfirmation paradigms characteristic of service industry marketing research can be applied unmodified to a Chinese context. These issues have been explored in various works on management in China. For example, Selmer (1998), Chunman (2000) and others have explored the relationships between culture, state intervention and management practice, but it can be observed that as yet little of this debate has been reported in the academic English-language tourism journals.

THE PREDISPOSITION TOWARDS CLASSIFICATION, CATEGORIZATION AND TAXONOMIES

While these are social norms based upon cultural traditions and modes of thought that may seem at odds with the more detached western view, there is also a 'scientism' that exists within formal Chinese thinking and government pronouncements. This is evident in a Chinese predisposition to generate categorizations and classifications in taxonomic orders. As discussed in Chapter 2 of this book, this is evident in Chinese tourism planning when scenic areas are categorized as a '4 A' location, or where one is told that this is 'an international tourist city' where that categorization is applied (at least in 2007) to 40 cities in China. The sources of this approach are many but we would suggest that among them are the separate influences of Maoist and Confucian thought. Maoist Marxism emerges from a belief in the progress of society as evidenced by observable historical processes that lend themselves to scientific analysis. Mao often made reference to a scientific process. For example, in 1942 he wrote,

> In opposing subjectivism, sectarianism and stereotyped Party writing we must have in mind two purposes: first, 'learn from past mistakes to avoid future ones', and second, 'cure the sickness to save the patient'. The mistakes of the past must be exposed without sparing anyone's sensibilities; it is necessary to analyse and criticize what was bad in the past with a scientific attitude so that work in the future will be done more carefully and done better. This is what is meant by 'learn from past mistakes to avoid future ones'. (Mao 1942)

Progress was made on the basis of empirical observation, and it is in this tradition that taxonomies are created as patterns of observed, objective categories subject to testing by reference to various benchmarks. Taxonomies can also be subjected to hierarchical classification, and within a Confucian tradition of respect for received wisdoms from authoritative scholarly figures, such taxonomies are to be respected. Evidence for this perspective is relatively easy to find. For example, in the National Palace Museum in Taipei that displays over 650,000 artefacts from 'The Forbidden Palace' evidence is displayed from the Ming Dynasty of a centralization of control of ceramics, the modes of production to be used and the categories of design that were officially authenticated. This view-point is also consistent with a school of scientific management thinking in that, given a need to allocate resources across many competing areas, the notion of benchmarking and classifying aids decision-making. Indeed, by creating rules as to resource allocation it simplifies such decision-making as China moves away from centralized control. In short, past modes of centralized decision-making can be devolved by reference to commonly understood classifications, although in practice, at an anecdotal level at least, the implementation of such rules may be markedly different from one province or municipality to another. Certainly there is a rich literature on the impacts of culture on Chinese management practices. One such study, that of Ralston *et al.* (1996) and subsequent work, looked at the work values of a sample of 869 Chinese managers and noted significant generational differences in terms of adherence to older values and newer values of greater individualism. But what such studies also indicate are the long hours of work, the tensions that exist between differing sets of expectations across public–private sector negotiations, the role of the 'boss' in SMEs and adherence to authoritarian roles, tensions between centralization versus decentralization in larger organizations and the stress under such conditions felt by Chinese administrators and managers (Warner 2003). One consequence of this is that strategic planning suffers and holistic views of management are more noted by their absence than presence (Moore and Wen 2004 2007).

THE ECONOMIC IMPERATIVE

Previous chapters have alluded to the driving need to lift large sections of the population from poverty. Chapter 1 was one such chapter and it is a theme repeated elsewhere within this book. There is little point in replicating past arguments, but one can observe that many have been critical of Chinese economic growth policies because of the environmental costs. A case can be made that the incurring of these costs has, in the past, been acceptable to the Chinese government, but we are of the opinion that a new discourse is increasingly being voiced and acted upon. This can be seen in Shanghai's building of wind farms, and of other similar initiatives noted by the Chinese Renewable Energy Industries Association (CREIA) prior to the 2005 Beijing International Renewable Energy Conference. The web page, http://energy.sourceguides.com, in March 2008 listed over 314 Chinese manufacturers of solar panels and a total of 11,000 companies engaged in various forms of environmental engineering. Much of this is being prompted by government sponsorship and a growing awareness that the future well-being of the Chinese people cannot be purchased at the expense of environmental degradation. According to Xinhua News Agency (12 March 2008), China will henceforth check provincial-level governments' performances in conserving energy and reducing pollutant emissions, and these evaluations in turn feed into major indicators of administrative efficiency. About 1,000 key enterprises are also to be monitored by provincial-level governments, and the results are required to be publicized (see http://en.chinaelections.org/newsinfo.asp?newsid=16310). The Chinese government is also planning to start levying an environmental tax. These and other measures have emanated from the National People's Congress of March 2008, and will be of future significance. Additionally, the attempt to create an environment-friendly Olympics might be dismissed by some as propaganda, but in Chinese society where 'face' is so important, public governmental utterances do possess importance.

THE SOCIALIST MARKET SYSTEM
AND EROSION OF STATISM

Jia Qinglin, in his report to the 11th CPPCC made continued reference to the socialist responsibilities of private sector entrepreneurs. While a stereotypical image of the Chinese is that of an entrepreneurial nation, and in 2007, a reputation fostered in some quarters for sub-standard exports, the governmental reactions to this was not always fully reported in the western press. Again, it is important to remember Chinese culture, and these criticisms and problems were seriously regarded by governmental circles. Food quality was certainly an issue in the third China–US Strategic Economic Dialogue in December 2007. Prior to that there was significant tightening of regulations with the General Administration of Quality Supervision, Inspection and Quarantine being established; the founding of the State Food and Drug Administration; and the strengthening of law enforcement by government agricultural, health and market regulatory authorities. In August 2007, the government initiated a series of educative programmes to teach farmers the proper use of pesticides etc. According to a press release of the Chinese Embassy in Australia (http://au.china-embassy.org/eng/sgfyrth/t389454.htm 2007), a total of 676 large- and medium-sized cities involved in rural production were subjected to rigorous monitoring regimes. In short, there is an expectation that while the market system is expected to generate economic well-being, it is also expected to behave in ways not inimical to the wider social good. Failure to behave according to the appropriate social norms (*li*) thus merits a restoration of state control through an enforcement of given benchmarks and standards. Thus, the norms of Chinese society and 'scientifism' combine to permit entrepreneurship in a socialist market system that operates from a Chinese perspective. While perhaps to the western ear the terminology might appear convoluted and possibly even oxymoronic, that is not the case to the Chinese mind.

HARMONIOUS SOCIETY

The Chinese government hosted a news conference on the release of China's 11th Five-Year Plan in 2006. Minister Ma Kai and Vice Minister Zhu Zhixin of China's National Development and Reform Commission responded to questions and the former confirmed the plan had two strategic lines, 'the scientific concept of development and the goal of building a harmonious society'—which themes ran through the whole draft policies (Gov.cn 2006). The linking of these two themes of 'scientific development' and 'harmonious society' represent important components of official thinking, and it can be argued, represent the current ideological approach of a government that has moved far from its original doctrines of Maoist thought, even while being careful not to publicly repudiate the architect of modern China. An issue is therefore, what are the concepts that underlie this rhetoric? Jing Tiankui (2008) of the Chinese Academy of Social Science, in an interview published in *China Daily* on 8 March 2005, argued that the policies enacted under these policies were characterized by leaving space for, and the support of, rights of social groups, and an enhancement of their abilities to improve their situations. This was to be achieved with government support implemented through increased expenditure on public welfare and education. In these ways China has begun taking steps towards a 'harmonious society'. In other words, GDP will no longer be the only or most significant goal. The Chinese government is thus shifting its emphasis from the economic prosperity of country to one of 'peaceful living of people', though economic development is still important. The concept incorporates democracy, the rule of law, equity, justice, sincerity, amity and vitality—and its incorporation into official government thinking means that social advancement is now accorded equal status with economic advances. It also justifies campaigns against corruption as that trait threatens the relationship between people and government. Ieong (2007) has analysed the language and has argued that the symbolic importance possesses policy implications and a significant shift in governmental policies as the Community Party seeks to modernize China. Certainly a significant literature exists in the Chinese language journals on these issues.

With reference to tourism, researchers and policymakers (Liu and Sheng 2006; Tang and Sui 2006) agree that tourism is a significant driving force for building a harmonious society. Tourism development is thought to contribute to the formation of a harmonious society by reducing the economic gaps between regions. Many less developed regions possess high-quality and bountiful tourism resources. Effective tourism planning and development will make these potential resources available to the tourism market. In the course of tourism development, local residents also obtain benefits by being provided with more information, technology, funds and an exchange with the outside world made possible by infrastructure improvement and increased incomes. Tourism planning must thus incorporate notions of the harmonious society by, for example, respecting the rights of people, environmental protection, poverty elimination and sustainable development. The implications for tourism have thus already been identified in different chapters in this book. The most obvious examples relate to the recognition of minority groups and the different religions of China, and the use made of tourism to generate economic benefits for previously marginalized groups. But as also evidenced, such initiatives have brought to the fore well-rehearsed arguments relating to how tourism presents cultures, commodifies them, yet also creates opportunities previously not existing. Recognition of minority cultures for purposes of tourism alone ultimately fails to keep in check the aspirations for a recognition of degrees of autonomy within the Chinese state. This may well represent one of the growing challenges for internal Chinese policies in the maintenance of social harmony, for that harmony will require recognition of difference. Whether tourism can play a role as a bridge between the status quo and an uncertain future is itself a contributory factor to the uncertainty.

CONCLUSIONS

Within China, tourism has many advantages not present in other industries. Being an under-developed and indeed an almost ignored industry for so long, it does not have to grapple with the issues involved in state-owned enterprises to the same degree as in other areas of the economy. Yet there are sectors of the accommodation and ground transportation sectors that still exist within the public sector, and the problems of the authorization of managers as against the rights of the state as owner are found in these sectors. Associated with this, and of particular importance for tourism, is the release of assets from central, municipal and provincial control. In some more conservative local government administrations, there is a reluctance to open publicly owned assets to more competition from a licensed private sector because of high investment needs in the public sector, poor rates of demand and possibly reluctance to forgo an area of planning thought important. Personal observation would indicate this is more problematic in bus and coach operations in some provinces as one such example impacting on tourism. There is also the issue of destination management and promotion. Within western economies there is a copious literature that questions the degree to which destinations are managed as distinct from promoted, and how one manages a destination when so many of the assets remain within the private sector. Given that these are issues in advanced economies, it is not surprising that the same issues can be felt more sharply in an economy such as China's.

In short, there are many important differences of structure, policy, culture and nuanced understandings when comparing China to other economies. It must also be said that there is a wide understanding of these issues within Chinese academic circles, but they are not perhaps being articulated strongly enough in western circles, perhaps because of undue respect being given to western paradigms of research. Given what is often perceived as a past dominance of post-positivistic paradigms based upon quantitative methods in the West, this is arguably a mistake (e.g. see Phillimore and Goodson 2004). By the same token, the renewed interests in researcher reflexivity and the need for more immersion techniques to achieve better understandings of tourist and employee experiences within tourism organizations in western academic circles does bode well for a transformative approach to tourism research in the Chinese context as a dialogue commences between the members of the international academic community. The opening of China represents not simply an opportunity for China, but also represents a catalyst for western managers and academics to question their own approaches to tourism. China thus potentially offers many gifts to western researchers through forcing a revaluation and questioning of taken-for-granted assumptions about the research paradigms used and the cultural milieu within which they are formed. But the potential gift requires Chinese academics in tourism to find and then articulate a confident voice. We believe this is an evolving process, and the current stage is but embryonic as Chinese colleagues have yet to wholly articulate a distinct discourse. Yet we discern the beginnings of this new confidence. It has been a privilege for us to work with so many Chinese colleagues, and we thank them for sharing their ideas with us and for widening our own understandings of tourism in China. So too, we hope, this book will help others to better their own understandings of the issues facing destination planning in China.

Bibliography

Achabal, D.D., Heineke, J.M. and McIntyre, S.H. (1984) 'Issues and perspectives on retail productivity', *Journal of Retailing*, 60 (3): 107–27.

Agarwal, S. (1994) 'The resort cycle revisited: implications for resorts', in C.P. Cooper and A. Lockwood (eds) *Progress in Tourism, Recreation and Hospitality Management*, Chichester: Wiley.

—— (1997) 'The resort cycle and seaside tourism: an assessment of its applicability and validity', *Tourism Management*, 18 (2): 65–73.

Ager, T. (1958) 'Der Fremdenverkehr in seiner Bedeutung fiir die Gebirgsbevijllcerung und fur die Bergbauernbetriebe', *Agrarpolitische Revue*, 14: 455–68.

Albers, P. and James, W. (1988) 'Travel photography: a methodological approach', *Annals of Tourism Research*, 15: 134–58.

Alderson, W. and Low, S. (1995) *Interpretation of Historic Sites*, Nashville, TN: American Association for State and Local History.

Alegre, J. and Juaneda, C. (2006) 'Destination loyalty: consumers' economic behavior', *Annals of Tourism Research*. 33 (3): 684–706.

Allen, L.R., Long, P.T., Perdue, R.R. and Kieselbach, S. (1988) 'The impact of tourism development on residents' perceptions of community life', *Journal of Travel Research*, 27 (1): 16–21.

Ancient Town Luzhi (2004) Tourism brochure edited by Suzhou Luzhi Tourism Development Corporation.

Anderson, R.I., Fok, R. and Scott, J. (2000) 'Hotel industry efficiency: an advanced linear programming examination', *American Business Review*, 18 (1): 40–8.

Anderson, R.I., Lewis, D. and Parker, M.E. (1999) 'Another look at the efficiency of corporate travel management departments', *Journal of Travel Research*, 37 (3): 267–72.

Anhui Province (1987) *Plan for the Places of Scenic and Historic Interest in Huangshan*, Huangshan: Anhui Province Government Press.

Ansson, R.J. (1998) 'Our national parks—over crowded, under funded, and besieged with a myriad of vexing problems: how can we best fund our imperiled national park system?' *Journal of Land Use and Environmental Law*, 14 (1): 1–52.

Ao, R.J. and Wei, Y.S. (2006) 'A study on the regional tourism resources and the unbalanced development of the tourism industry in China', *Journal of Finance and Economics*, 32 (3): 32–43.

Ap, J. (1990) 'Residents' perceptions research on the social impacts of tourism', *Annals of Tourism Research*, 17 (4): 610–16.

—— (1990) 'Resident perception research of the social impacts of tourism', *Annals of Tourism Research*, 17 (7): 481–94.

—— (1992) 'Residents' perceptions of tourism impacts', *Annals of Tourism Research*, 19 (3): 665–90.

────── (1992) 'Residents' perceptions on tourism impacts', *Annals of Tourism Research*, 19 (4): 665–90.

Ap, J. and Crompton, J. (1993) 'Residents' strategies for responding to tourism impacts', *Journal of Travel Research*, 32 (1): 47–50.

────── (1998) 'Developing and testing a tourism impact scale', *Journal of Travel Research*, 37 (2):120–30.

Arlt, W. (2006) *China's Outbound Tourism Entering the Third Stage*, Online. Available HTTP: <http://www.china-outbound.com/Newsletter/2006_05_phases.html> (accessed 15 June 2006).

Arnould, E.J. and Price, L.L. (1993) 'River magic: extraordinary experience and the extended service encounter', *Journal of Consumer Research*, 20 (1): 24–45.

Ashworth, G.J., Graham, B. and Tunbridge, J.E. (2000) *A Geography of Heritage: power, culture and economy*, London: Arnold.

Au Loong-yu, Nan Shan and Zhang Ping. (2007) 'Women migrant workers under the Chinese social apartheid', *Committee for Asian Women* (May): 20.

Augustyn, M. (1998) 'National strategies for rural tourism development and sustainability: the Polish experience', *Journal of Sustainable Tourism*, 6 (3): 191–209.

AZNFP (Administration of Zhangjiajie National Forest Park) (2004) *Building Zhangjiajie Forest Park and Promoting its Tourism Industry*, Online. Available HTTP: <http://www.hnforestry.gov.cn/listinfo.aspx?ID=14001> (accessed 20 December 2006).

Babbie, E. (2001) *The Practice of Social Research*, 9th edn, Wadsworth: Thomson Learning, Inc.

Baedcharoen, I. (2000) 'Impacts of religious tourism in Thailand', unpublished thesis, University of Otago, New Zealand.

Bähre, H. (2007) 'Privatisation during a market economy transformation as a motor of development', in P.E. Burns and M. Novelli (eds) *Tourism and Politics: Global Frameworks and Local Realities*, Oxford: Pergamon.

Baker, M. and Riley, M. (1994) 'New perspectives on productivity in hotels: some advances and new directions', *International Journal of Hospitality Management*, 13 (4): 97–111.

Banff-Bow Valley Study (1996) *Banff-Bow Valley: at the crossroads. Technical Report of the Banff-Bow Valley Task Force*, R. Page, S. Bayley, J.D. Cook, J.E. Green and J.R.B. Ritchie (eds). Prepared for the Honourable Sheila Copps, Minister of Canadian Heritage. Ottawa: Minister of Supply and Services Canada.

Banker, R.D., Charnes, A. and Cooper, W.W. (1985) 'Some models for estimating technical and scale inefficiencies in Data Envelopment Analysis', *Management Science*, 30 (9): 1078–92.

Bao Jigang (1996) *Tourism Development Research: principles, methodology and practice*, Beijing: Science Press.

Bao, J. (1998) 'Tourism planning and tourist area lifecycle model', *Architect*, 12: 170–78.

Bao, J.G. and Chu, Y.F. (2001) *Tourism Geography*, Beijing: Higher Education Press.

Bao, J. and Zhang, Z. (2006) 'The TALC in China's tourism planning: case study of Danxia Mountain, Guangdong Province, PRC', in C. Cooper, C.M. Hall and D. Timothy (series eds) and R.W. Butler (vol. ed.) *The Tourism Area Life Cycle (Vol. 1): applications and modifications*, Clevedon: Channel View Publications.

Barros, C.P. (2005a) 'Evaluating the efficiency of a small hotel chain with a Malmquist productivity index', *International Journal of Tourism Research*, 7 (3): 173–84.

────── (2005b) 'Measuring efficiency in the hotel sector', *Annals of Tourism Research*, 32 (2): 456–77.

────── (2006) 'Analyzing the rate of technical change in the Portuguese hotel industry', *Tourism Economics*, 12 (3): 325–46.

Barros, C.P. and Matias, A. (2006) 'Assessing the efficiency of travel agencies with a stochastic cost frontier: a Portuguese case study', *International Journal of Tourism Research*, 8 (5): 367–79.

Bate, S. (1993) 'Middle Eastern promise', *Marketing Week*, 16 (10): 46–8.

Baum, T. (1998) 'Taking the exit route: extending the tourism area life cycle model', *Current Issues in Tourism*, 1 (2): 167–75.

—— (2006) 'Revisiting the TALC: is there an off-ramp?' in R.W. Butler (ed.) *The Tourism Area Life Cycle, Vol. 2: conceptual and theoretical issues*, Clevedon: Channel View Publications.

Beck, L. and Cable, T. (2002) *Interpretation for the 21st Century*, 2nd edn, Sagamore Publishing.

Beckwith, B. (1999) 'Visiting the land where Jesus was born', *St. Anthony Messenger*, 107: 18–23.

Beijing Statistical Information Net (2006) Online. Available HTTP: <www.bjstats. gov.cn> (accessed 13 April 2007).

Belisle, F.J. and Hoy, D.R. (1980) 'The perceived impact of tourism by residents: a case study in Santa Marta, Colombia', *Annals of Tourism Research*, 7: 83–101.

Berry, E.N. (2001) 'An application of Butler's (1980) tourist area life cycle theory to the Cairns Region, Australia 1876–1998', unpublished doctorial dissertation, School of Tropical Environment Studies and Geography, James Cook University of North Queensland, Australia.

Berry, T. (2006) 'The predictive potential of the TALC model', in R.W. Butler (ed.) *The Tourism Area Life Cycle, Vol. 2: conceptual and theoretical issues*, Clevedon: Channel View Publications.

Big5.china.com.cn (2006) *Centuries-old Commercial Center in Beijing to Take on New Look*, Online. Available HTTP: (accessed 13 October 2006).

Bishop, Peter (1989) *The Myth of Shangri-la. Tibet, Travel Writing and the Western Creation of Sacred Landscape*, Berkeley: University of California Press.

Bitner, M.J. (1992) 'Servicescapes: the impact of physical surroundings on customers and employees', *Journal of Marketing Research*, 56 (April): 57–71.

Bolton, R.N. and Drew, J.H. (1991) 'A multistage model of visitors' assessment of service quality and value', *Journal of Consumer Research*, 17: 375–84.

Boyd, S.W. (2006) 'The TALC model and its application to national parks: a Canadian example', in C. Cooper, C.M. Hall and D. Timothy (series eds) and R.W. Butler (vol. ed.) *The Tourism Area Life Cycle (Vol. 1): applications and modifications*, Clevedon: Channel View Publications.

Bramwell, B. and Lane, B. (1994) *Rural Tourism and Sustainable Rural Development*, Clevedon: Channel View Publications.

Bramwell, B. and Sharman, A. (2001) 'Approaches to sustainable tourism planning and community participation', in G. Richards and D. Hall (eds) *Tourism and Sustainable Community Development*, London: Routledge.

Breakwell, G.M. (1986) *Coping with Threatened Identity*, London: Methuen.

—— (1992) *Social Psychology of Identity and the Self Concept*, Guildford: Surrey University Press.

Breakwell, G.M. and Canter, D.V. (1993) *Empirical Approaches to Social Representations*, Oxford: Clarendon Press.

Briedenhann, J. and Wickens, E. (2004) 'Tourism routes as a tool for the economic development of rural areas-vibrant hope or impossible dream?' *Tourism Management*, 25 (1): 71–9.

Brown, Bryan J.H. (1985) 'Personal perception and community speculation: a British resort in the 19th century', *Annals of Tourism Research*, 12 (3): 355–69.

Brown, G. and Raymond, C. (2007) 'The relationship between place attachment and landscape values: toward mapping place attachment', *Applied Geography*, 27 (1): 89–111.

Bryman, A. (1988) *Quantity and Quality in Social Research*, London: Routledge.

Bu, X.L. (2006) 'Environmental protection for sustainable tourism in Zhangjiajie', *China Environment Newspaper* (20 September).

Bureau of the World Heritage Committee (1998) *Decisions of the Twenty-second Extraordinary Session of the Bureau of the World Heritage Committee (Kyoto, 28–29 November 1998) With Regard to the State of Conservation of Properties Inscribed on the World Heritage List*, Kyoto: World Heritage Centre.

Butler, R.W. (1980) 'The concept of a tourist area cycle of evolution: implications for management of resources', *Canadian Geographer*, 24: 5–12.

—— (1993) 'Tourism—an evolutionary perspective', in J.G. Nelson, R. Butler and G. Wall (eds) *Tourism and Sustainable Development: monitoring, planning, managing*, Waterloo: Heritage Resources Centre Joint Publication Number 1.

—— (2006a) 'The future and the TALC', in R.W. Butler (ed.) *The Tourism Area Life Cycle, Vol. 2: conceptual and theoretical issues*, Clevedon: Channel View Publications.

—— (2006b) *The Tourism Area Life Cycle, Vol. 1: application and modifications*, Clevedon: Channel View Publications.

—— (2006c) *The Tourism Area Life Cycle, Vol. 2: conceptual and theoretical issues*, Clevedon: Channel View Publications.

Butler, R. and Hinch, T. *Tourism and Indigenous Peoples: issues and implications*, Oxford: Butterworth-Heinemann.

Caffyn, A. and Lutz, J. (1999) 'Developing the heritage tourism product in multi-ethnic cities', *Tourism Management*, 20: 213–21.

Carmichael, B.A. (2006) 'Linking quality tourism experiences, residents' quality of life, and quality experiences for tourists', in G. Jennings and N.P. Nickerson (eds) *Quality Tourism Experiences*, Oxford: Elsevier Butterworth Heinemann.

Carr, A. (2004) 'Mountain places, cultural spaces: the interpretation of culturally significant landscapes', *Journal of Sustainable Tourism*, 12 (5): 432–59.

Chan, Wing-tsit (1969) *A Source Book of Chinese Philosophy*, New York: Colombia University Press.

Chang, J., Yang, B.T. and Yu, C.G. (2006) 'The moderating effect of salespersons' selling behaviour on shopping motivation and satisfaction: Taiwan tourists in China', *Tourism Management*, 27 (5): 934–42.

Charles, K. N., and Paul, S. (2001). Competition, privatization and productive efficiency: evidence from the airline industry. The Economic Journal, 111(473), 591–619.

Charnes, A., Cooper, W.W. and Rhodes, E. (1978) 'Measuring the efficiency of decision making units', *European Journal of Operational Research*, 2 (6):429–44.

—— (1981) 'Evaluating program and managerial efficiency: an application of Data Envelopment Analysis to program follow through', *Management Science*, 27 (6): 668–97.

Ch'en, K.K.S. (1973) *The Chinese Transformation of Buddhism*, Princeton: Princeton University Press.

Chen, C.K., Gao, Y.G. and Yu, K.J. (1990) 'A study on tourism development of the DanXia scenic district', *Acta Geographica Sinica*, 45 (3): 284–94.

Chen, G.L. and Chun, T. (2004) *The Investigation of Rural Chinese*, Beijing: The People's Literature Publishing.

Chen, H.Y. (2000) *The Institutional Transition of China's Township and Village Enterprises: market liberalization, contractual form innovation and privatization*, Ashgate Publishing.

Chen, Ping (1999) *Modern Chinese: History and Sociolinguistics*, Cambridge: Cambridge University Press.

Chen, W.J. and Huang, G.W. (2006) 'Discussion on the concessionary management of national parks', *Journal of Jiangxi Science and Technology Normal University*, 6: 34–7.

Cheung, T.M. (2001) *China's Entrepreneurial Army*, Oxford: Oxford University Press.

Chiang, W.E. (2006) 'A hotel performance evaluation of Taipei international tourist hotels—using data envelopment analysis', *Asia Pacific Journal of Tourism Research*, 11 (1): 29–42.

China National Tourism Administration (2005) China Tourism Industry Statistics Report.

—— (2006a) *China in Brief*, Online. Available HTTP: <www.cnta.org.cn> (accessed 25 June 2006).

—— (2006b) *A Summary of Tourism Industry Development in 2005*, Online. Available HTTP: <www.cnta.org.cn> (accessed 20 April 2006; original in Chinese).

China National Tourism Office (2007) *China Tourism Statistics*, Beijing: CNTO.

China Org. Cn. 2007b, Shenzhen, Korea's second hometown http://www.china.org.cn/english/china/226395.htm (accessed 2 February 2008).

China.org.cn (2007a) *Shenzhen to Have 2 More International Schools*, Online. Available HTTP: <http://www.china.org.cn/english/LivinginChina/228476.htm> (accessed 2 February 2008).

China.org.cn (2007b) *Shenzhen, Korean's Second Hometown*.

China Tibet Information Centre (2005) *Religion*, China Tibet Information Centre. Online. Available HTTP: <http://www.tibet.cn/english/zt/religion/200402004518134156.htm> (accessed 20 March 2005).

China Travel Net (2005) *Wutai Mountain*, Online. Available HTTP: <http://www.chinatravel1.com/english/destination/dashandachuan/wutai.htm> (accessed 1 July 2005).

Chinese History Museum (1992) *Illustrated History of China's 5000 Years*, Tianjin: Tianjin People's Arts Publication [in Chinese].

Chinese National Geographic (2005) 10.

Chon, K. (1989) 'Understanding recreational travelers' motivation, attitude and satisfaction', *The Tourist Review*, 44 (1): 3–7.

Clevenger, A.P., Wierzchowski, J., Chruszcz, B. and Gunson, K. (2002) 'GIS-generated, expert-based models for identifying wildlife habitat linkages and planning mitigation passages', *Conservation Biology*, 16 (2): 503–14.

Coelli, T.J. (1996) 'A guide to FRONTIER Version 4.1: a computer program for frontier production function estimation', CEPA working paper 96/07, Department of Econometrics, University of New England, Arm idale.

Cohen, E. (1976) 'In Haec Signa: pilgrim-badge trade in Southern France', *Journal of Medieval History*, 2 (3): 193–214.

—— (1988) 'Authenticity and commoditization in tourism', *Annals of Tourism Research*, 15 (3): 371–86.

Collins-Kreiner, N. and Kliot, N. (2000) 'Pilgrimage tourism in the Holy Land: the behavioral characteristics of Christian pilgrims', *GeoJournal*, 50 (1): 55–67.

Conti, G. and Perelli, C. (2007) 'Governing tourism monoculture', in G. Richards and D. Hall (eds) *Tourism and Sustainable Community Development*, London: Routledge.

Cooper, C.P. and Jackson, S. (1989) 'Destination life cycle: The Isle of Man case study', *Annals of Tourism Research*, 16 (3): 377–98.

Cope, J. (2003) *Exploring the Nature and Impact of Critical Experiences within Small Business Growth and Entrepreneurial Development*, working paper, Lancaster University Management School. Online. Available HTTP: <www.lums.lancs.ac.uk/publications/viewpdf/000193/> (accessed 10 January 2007).

Cottrell, S.P., Vaske, J.J. and Shen, F.J. (2007) 'Modeling resident perceptions of sustainable tourism development: applications in Holland and China', *China Tourism Research*, 3 (3): 219–34.

Cowen, T. (2002) *Creative Destruction. How Globalization is Changing the World's Cultures*, Princeton and Oxford: Princeton University Press.

Crain, M.M. (1996) 'Contested territories: the politics of tourism development at the shrine of El Rocio in southwestern Andalusia', in J. Boissevain (ed.) *Coping with Tourists—European Reaction to Mass Tourism*, Providence: Berghahn Books.

Crompton, J. (1979) 'Motivation for pleasure vacation', *Annals of Tourism Research*, 6 (4): 408–28.

Cruz, J. (1984) *Relics*, Huntington, IN: Our Sunday Visitor, Inc.

Czarda, R. (1997) 'Vereinigung und Systemtransformation als Governance Problem', in M. Corsten and H. Voelzkow (eds) *Transformation zwischen Markt, Staat und Dritten Sektor*, Marburg: Metropolis.

Dahms, F. (1991) 'Economic revitalization in St. Jacobs, Ontario: ingredients for transforming a dying village into a thriving tourist destination', *Small Town* (May–June): 12–18.

Daniels, S. (1989) 'Marxism, culture, and the duplicity of landscape', in R. Peet and N. Thrift (eds) *New Models in Geography*, London: Unwin Hyman.

Dann, G. (1977) 'Anomie, ego-enhancement and tourism', *Annals of Tourism Research*, 4 (4): 184–194.

—— (1996) 'The people of tourist brochures', in T. Selwyn (ed.) *The Tourist Image: myths and myth making in tourism*, Chichester: John Wiley.

Davis, J. (1994) 'Consumer response to corporate environmental advertising', *Journal of Consumer Marketing*, 11 (2): 25–37.

Debbage, K. (1990) 'Oligopoly and the resort cycle in the Bahamas', *Annals of Tourism Research*, 17: 513–27.

Deng, J., Bauer, T. and Huang, Y. (2003) 'Ecotourism, protected areas, and globalization: issues and prospects in China', *ASEAN Journal on Hospitality and Tourism*, 2 (1): 17–32.

Deng, R. (2006) *South Koreans: new lovers of Zhangjiajie*, Online. Available HTTP: <http://www.zjj-trip.com/news/NewsInfo.aspx?ID=3172andpageUp=220> (accessed 20 September 2006).

Deying, Z., Yanagida, J.F., Chakravorty, U. and Ping, S.L. (1997) 'Estimating economic impacts from tourism', *Annals of Tourism Research*, 24 (1): 76–89.

Dezan Shira and Associates (2005) *China's New Regulations for Establishing Foreign-Owned Trading Companies*, Online. Available HTTP: <http://www.global sources.com/TNTLIST/2005/0805/DZRA_TRADINGCOS.HTM> (accessed 12 February 2008).

Dietvorst, A.G.J. and Ashworth, G.J. (1995) 'Tourism transformations: an introduction', in G.J. Ashworth and A.G.J. Dietvorst (eds) *Tourism and Spatial Transformations: implications for policy and planning*, Wallingford: CABI.

Ding, H. (2004) 'Critical review of research on "tourism alleviating poverty" in China', *Tourism Studies*, 19 (3).

Ding Yu-lian, Lu Lin, Huang Liang 2006, A study on the symbol of cultural leisure tourism. Tourism Tribune 21(7): 12–16.

Dodin, Thierry and Raether, Heinz (2001) *Imagining Tibet. Perceptions, Projections and Fantasies*, Boston: Wisdom Publications.

Dogan, H.Z. (1989) 'Forms of adjustment: sociocultural impacts of tourism', *Annals of Tourism Research*, 16: 216–36.

Domberger, S. and Jensen, P. (1997) 'Contracting out by the public sector: theory, evidence, prospects', *Oxford Review of Economic Policy*, 13 (4): 67–78.

Dong, J., Yang, D.Y., Wang, P. and Jiang, X.W. (2006) 'Discussion on the meaning and value of protection of national parks near city—taking Zhongshan national parks in Nanjing as an example', *Economic Geography*, 26 (1): 137–9.

Douglas, N. (1997) 'Applying the life cycle model to Melanesia', *Annals of Tourism Research*, 24 (1): 1–22.

Doxey, G.V. (1975) 'A causation theory of visitor-resident irritants, methodology and research inferences', *The Impact of Tourism. Sixth Annual Conference Proceedings of the Travel and Tourism Research Association*, 1: 195–8.

—— (1976) 'When enough's enough: the natives are restless in Old Niagara', *Heritage Canada*, 2 (2): 26–7.

Dubois, G. (2001) 'Codes of conduct, charters of ethics and international declarations for a sustainable development of tourism: ethical content and implementation of voluntary initiatives in the tourism sector', in N. Moisey, N. Nickerson and K. Andereck (eds) *Proceedings of the 32nd Annual Travel and Tourism Research Association Conference*.

Dumreicher, H. (2008) *Chinese Villages and their Sustainable Future: the European*.

Durbarry, R. (2004) 'Tourism and economic growth: the case of Mauritius', *Tourism Economics*, 10 (4): 389–401.

Dwyer, L. and Forsyth, P. (1998) 'Estimating the employment impacts of tourism to a nation', *Tourism Recreation Research*, 23 (2): 1–12.

Dwyer, L., Forsyth, P., Spurr, R. and Vanho, T. (2003) 'Tourism's contribution to a state economy: a multi-regional general equilibrium analysis', *Tourism Economics*, 9 (4): 431–48.

Ebrey, Patricia Buckley (ed.) (2003) *A Visual Sourcebook of Chinese Civilization*, New York: Washington University. Online. Available HTTP: <http://depts.washington.edu/chinaciv> (accessed 10 December 2004).

Edgell, Sr, D.L., Allen, M.D., Smith, G. and Swanson, J.R. (2008) *Tourism Policy and Planning: yesterday, today and tomorrow*, Oxford: Butterworth Heinemann.

Edwards, J., Fernandes, C., Fox, J. and Vaughan, R. (2000) 'Tourism brand attributes of the Alto Minho, Portugal', in G. Richards and D. Hall (eds) *Tourism and Sustainable Community Development*, London: Routledge.

Elvin, M. (1973) *The Pattern of the Chinese Past*, Stamford: Stamford University Press.

Fan Yezheng and Hu Qingping (2006) *The Development of Tourism Planning and Research in China. Tourism Tribune, Chinese Tourism Research Annual, 2001–2004*, Beijing: Social Sciences Academic Press.

Fan, L.Y., Shen, S.G. and Zheng, H. (2002) 'On analysis of the philosophical origin of the art in Chinese classical gardens', *Journal of University of Science and Technology of Suzhou*, (2).

Fan, C., Wall, G. and Mitchell, C.J.A. (2008) 'Creative destruction and the water town of Luzhi, China', *Tourism Management*.

Fang, K. (2000) *Dangdai Beijing Jiucheng Gengxin: Diaocha, Yanjiu, Tansuo [Contemporary Redevelopment in the inner city of Beijing: survey, analysis and investigation]*, Beijing: China Construction Industry Publishing House.

Faulkner, H.W. (2001) 'Towards a framework for tourism disaster management', *Tourism Management*, 22 (2): 135–47.

Feinberg V.M. (1995) 'Pilgrims and Profits', *Eretz* (January to February): 32–42.

Feldman, R.M. (1990) 'Settlement identity: psychological bonds with home places in a mobile society', *Environment and Behavior*, 22: 183–229.

Feng Qun Wei (2006) 'The development of tourism and real estate', *City Planning Review* (July): 83

Fisher and Sharone, P. (1994) 'Braving the violence', *Advertising Age*, 65 (10): 8–9.

Foucault, Michel (1980) *Power-Knowledge: selected interviews and other writings, 1972–1977*, edited by Colin Gordon, Brighton, Sussex: Harvester Press.

Frank, T. (1920) *An Economic History of Rome: to the end of the republic*, John Hopkins Press.

Fuchs, M. (2004) 'Strategy development in tourism destinations: a DEA approach', *The Poznan University of Economics Review*, 4 (1): 52–73.

Gallarza, M.G., Saura, I.G. and García, H.C. (2002) 'Destination image: towards a conceptual framework', *Annals of Tourism Research*, 29 (1): 56–78.

Gao, F. and Jiao, J. (2004) *Wutai Mountain Plans to Apply for WHS*, People's Daily Online. Available HTTP: <http://www.people.com.cn/GB/paper39/11133/1008498.html> (accessed 15 March 2005).

Gao, S. (1997) 'Preliminary attempts on tourism's role in alleviating poverty', *China Administration*, 7: 22–4.

Gaubatz, P. (1999) 'China's urban transformation: patterns and processes of morphological change in Beijing, Shanghai and Guangzhou', *Urban Studies*, 36 (9): 1495–1521.

Getz, D. (1992) 'Tourism planning and the destinations life cycle', *Annals of Tourism Research*, 19 (4): 752–70.

—— (1993) 'Tourist shopping villages: development and planning strategies', *Tourism Management*, 14: 15–26.

Getz, D. and Jamal, T. (1994) 'The environment-community symbiosis: a case for collaborative tourism planning', *Journal of Sustainable Tourism*, 2 (3): 152–73.

Ghimire, K.B. (1997) 'Conservation and social development: an assessment of Wolong and other panda reserves in China', in K.B. Ghimire and M.P. Pimbert (eds) *Social Change and Conservation*, Earthscan.

Ghimire, K.B. and Pimbert, M.P. (1997) 'Social change and conservation: an overview of issues and concepts', in K.B. Ghimire and M.P. Pimbert (eds) *Social Change and Conservation*, Earthscan.

Gladstone, D.L. (2005) *From Pilgrimage to Package Tour: travel and tourism in the third world*, New York: Routledge.

Go, F. and Ritchie, J.R.B. (1990) 'Tourism and transnationalism', *Tourism Management*, 11: 287–90.

Golany, B. and Roll, Y.A. (1994) 'Incorporating standards via DEA', in A. Charnes, W.W. Cooper, A.Y. Lewin and L.M. Seiford (eds) *Data Envelopment Analysis: theory, methodology and applications*, Norwell: Kluwer Academic Publishers.

Goodwin, H. and Francis, J. (2003) 'Ethical and responsible tourism: consumer trends in the UK', *Journal of Vacation Marketing*, 9 (3): 271–84.

Gov.cn (2006) *Ten Features in China's 11th Five-Year Plan*, Online. Available HTTP: <http://english.gov.cn/2006-03/08/content_246945.htm> (accessed 11 March 2008).

Graefe, A.R. and Vaske, J.J. (1987) 'A framework for managing quality in the tourist experience', *Annals of Tourism Research*, 14 (3): 390–404.

Greenley, D.A., Walsh, R.G. and Young, R.A. (1981) 'Option value: empirical evidence from a case study of recreation and water quality', *The Quarterly Journal of Economics*, 96 (November): 657–72.

Greenwood, D.J. (1989) 'Culture by the pound', in V. Smith (ed.) *Hosts and Guests: the anthropology of tourism*, 2nd edn, Philadelphia: University of Pennsylvania Press.

Gregoriou, G.N., Rouah, F., Satchell, S. and Diz, F. (2005) 'Simple and cross efficiency of CTAs using data envelopment analysis', *The European Journal of Finance*, 11 (5): 393–409.

Griffin, J. (1994) 'Order of service', *Leisure Management*, 14: 30–2.

Gu, C. and Zhong, Y. (2005) *Zhangjiajie: a story of 2000 years*, Beijing: Five-Continent Dissemination Press.

Gu H. (2008) *Tourism Crisis Management*, Beijing: Beijing Tourism Press.

Gu, H. and Ryan, C. (2007a) 'Chinese clientele at Chinese hotels—evaluations and importance attribution—a comparison between different star ratings', 13th APTA Conference, Beijing International Studies University.

Gu H. and Ryan, C. (2007b) *Hongcun: Residents and Communities—Tourism and the Diminution of Place Attachment*. 4th China Tourism Forum—Yunnan University, United Nations World Tourism Organisation, Hong Kong Polytechnic University, Kunming, China, 15–16 December.

Gu, H. and Ryan, C. (2008) 'Place attachment, identity and community impacts of tourism—the case of a Beijing hutong', *Tourism Management*, 29 (4): 637–47.

Gu, H., Ryan, C. and Zhang, W. (2007) 'Jinggangshan Mountain: a paradigm of China's red tourism', in C. Ryan (ed.) *Battlefield Tourism: history, place and interpretation*, Oxford: Pergamon.

Gu, H. and Yu, L. (2005) *Homestay Inn in Rural Beijing: an exploratory study. Proceedings of the 12th Asia Pacific Tourism Association Conference*, Taiwan: Hualien.

Guiliani, M.V. (1991) 'Towards an analysis of mental representations of attachment to the home', *Journal of Architecture and Planning Research*, 8 (Summer): 133–46.

Gunn, C.A. (1988) *Tourism Planning*, 2nd edn, New York: Taylor and Francis.

Gunn, C.A. and Var, T. (2002) *Tourism Planning: basics, concepts, cases*, 4th edn, New York: Routledge Taylor and Francis Group.

Guo Y.Y. and Zheng, J.X. (2006) 'To abate the paradox between the protection and development of Chinese national park—taking Luo Fu Mountain as an example', *Chinese Construction*, (9): 69–71.

Gustafsson, B. and Zhong, W. (2000) 'How and why has poverty in China changed? A study based on microdata for 1988 and 1995', *The China Quarterly*, 164 (December): 983–1006.

Hair, Jr, J.F., Anderson, R.E., Tatham, R.L. and Black, W.C. (1998) *Multivariate Data Analysis*, 5th edn, Upper Saddle River, NJ: Prentice Hall.

Hall, C.M. (2006) 'Buddhism, tourism and the middle way', in D.J. Timothy and D.H. Olsen (eds) *Tourism, Religion and Spiritual Journeys*, London: Routledge.

—— (2008) *Tourism Planning: policies, processes and relationships*, 2nd edn, Harlow: Pearson Prentice Hall.

Hall, C.M. and Page, S.J. (2002) *The Geography of Tourism and Recreation: environment, place and space*, London: Routledge.

Halme, M., Joro, T., Korhonen, P., Salo, S. and Wallenius, J. (1999) 'A value efficiency approach to incorporating preference information in data envelopment analysis', *Management Science*, 45 (1): 103–15.

Ham, S. H. (1992) *Environmental interpretation—A Practical Guide for People with Big Ideas and Small Budgets*, North America Press.

Han, N. and Zhuge, R. (2001) 'Ecotourism in China's nature reserves: opportunities and challenges', *Journal of Sustainable Tourism*, 9 (3): 228–42.

Harrill, R. (2004) 'Residents' attitudes toward tourism development: a literature review with implications for tourism planning', *Journal of Planning Literature*, 18 (3): 251–66.

Harrison, D. (1995) 'Development of tourism in Swaziland', *Annals of Tourism Research*, 22 (1): 135–56.

Harrison, L. and Husbands, W. (1996) *Practicing Responsible Tourism: international case studies in tourism planning, policy and development*, Chichester: John Wiley and Sons Ltd.

Harvey, D. (1987). 'Flexible accumulation through urbanization: reflections on postmodernism in the American city', *Antipode*, 19 (3): 260–86.

Harvey D, 1988, The geographical and geopolitical consequences of the transition from ... 'Fordist to flexible accumulation', in B. Sternlieb and J. Hughes (eds) *America's New Market Geographies: nation, region and metropolis*, Rutgers, NJ: Center for Urban Policy Research.

—— (1988) The geographical and geopolitical consequences of the transition from Fordist to flexible accumulation, in B. Sternlieb and J. Hughes (eds.) America's New Market Geographies: nation, region and metropolis, Rutgers, N.J. Centre for Urban Policy Research.

—— (1989a) *The Condition of Postmodernity: an enquiry into the origins of cultural change*, Oxford: Blackwell.

—— (1989b) 'From managerialism to entrepreneurialism: the transformation in urban governance in late capitalism', *Geografiska Annaler*, 71B (1): 3–17.

Haywood, M.K. (1986) 'Can the tourist area life cycle be made operational?' *Tourism Management*, 7 (3): 154–67.

He, Y.L. (2002) 'How to develop community-based eco-tourism (CBET)?—experiences learned from Ban Huay Hee, Thailand', *Tourism Tribune*, 6: 57–60.

Hendee, John C. and Stankey, George H. (1973) 'Biocentricity in wilderness management', *BioScience*, 23 (9): 535–8.

Hendee, John C., Stankey, George H. and Lucas, Robert (1990) *Wilderness Management*, 2nd edn, North America Press, Golden.

Hickman, L. (2007) 'The new gold rush', *The Guardian*, 22 September, Online. Available HTTP: <http://www.guardian.co.uk/travel/2007/sep/22/saturday.green 1?page=all> (accessed 17 January 2008).

Hidalgo, M.C. and Hernández, B. (2001) 'Place attachment: conceptual and empirical questions', *Journal of Environmental Psychology*, 21: 273–81.

Higgins-Desbiolles, F. (2007) 'Taming tourism: indigenous rights as a check to unbridled tourism', in P.E. Burns and M. Novelli (eds) *Tourism and Politics: global frameworks and local realities*, Oxford: Pergamon.

Hilton, James (1933) *Lost Horizon*, New York: Macmillan.

Hjalager, A.M. (2006) 'Stages in the economic globalization of tourism', *Annals of Tourism Research*, 34 (2): 434–57.

Hogg, M.A. and Abrams, D. (1988) *Social Identifications: a social psychology of intergroup relations and group processes*, London: Routledge.

Holloway, C. (2004) *Marketing for Tourism*, 4th edn, London: Pearson Higher Education.

Houlihan, M. (2000) 'Souvenirs with soul: 800 years of pilgrimage to Santiago De Compostela', in M. Hitchcock and K. Teague (eds) *Souvenirs: the material culture of tourism*, Aldershot: Ashgate.

Hovinen, G. (1981) 'A tourist cycle in Lancaster County, Pennsylvania', *Canadian Geographer*, 15 (3): 283–6.

—— (2002) 'Revisiting the destination lifecycle model', *Annals of Tourism Research*, 29 (1): 209–30.

Hsu, C.H.C., Cai, L.A. and Wong, K.K.F. (2007) 'A model of senior tourism motivations—anecdotes from Beijing and Shanghai', *Tourism Management*, 28 (3): 1262–1273.

China. org. cn 2007, Shenzhen, Korea's second hometown. http://www.china.org.cn/english/China/226395.htm (accessed 2 February 2008).

Hu, Z.Y. and Hang, Z.G. (2002) 'Community involvement and sustainable development of tourism industry', *Human Geography*, 2: 38–41

Huang, R. (2004). 'The experience of Chinese international students in the UK: a tourism perspective', unpublished thesis, University of Derby.

—— (2006) 'The influence of culture on the travelling behaviour of Chinese international students in the UK', paper presented at the Tourism in Asia Conference, Leeds Metropolitan University, June.

Hummon, D. (1986) 'City mouse, country mouse: the persistence of community identity', *Qualitative Sociology*, 9 (1): 3–25.

Hunter, C. (1997) 'Sustainable tourism as an adaptive paradigm', *Annals of Tourism Research*, 24 (4): 850–67.

Hwang, S.N. and Chang, T.Y. (2003) 'Using data envelopment analysis to measure hotel managerial efficiency change in Taiwan', *Tourism Management*, 24 (4): 357–69.

Hwang, S.N., Lee, C. and Chen, H.J. (2005) 'The relationship among tourists' involvement, place attachment and interpretation satisfaction in Taiwan's national parks', *Tourism Management*, 26 (2): 143–56.

Ieong, S.L. (2007) 'The construction of a harmonious society in China: the concept, language and implications', *International Journal of the Humanities*, 4 (7): 9–18.

Inskeep, E. (1991) *Tourism Planning: an integrated and sustainable development approach*, New York: Van Nostrand Reinhold.

International Council of Monuments and Sites (ICOMOS) (1990) *World Heritage Nomination—ICOMOS Summary 547 Mount Huangshan Scenic Beauty and Historic Site (China)*, Online. Available HTTP: <http://whc.unesco.org/archive/advisory_body_evaluation/547> (accessed 2 February 2005).

International Council of Monuments and Sites (ICOMOS) (1999) *International Cultural Tourism Charter (Managing Tourism at Places of Heritage Significance)*, Online. Available HTTP: <http://www.international.icomos.org/charters/tourism_e.htm> (accessed 27 January 2008).

International Ecotourism Society (2004) *Consumer Demand and Operator Support for Socially and Environmentally Responsible Tourism*, Online. Available HTTP: <www.ecotourism.org> (accessed on 20 November 2005).

Ioannides, D. (1992) 'Tourism development agents: the Cypriot resort cycle', *Annals of Tourism Research*, 19 (4): 711–31.

Isen, A., (1984) 'Toward understanding the role of affect in cognition', in T. Srull and R. Wyer (eds) *Handbook of Social Cognition*, Hillsdale, NJ: Lawrence Erlbaum Associates.

Iso-Ahola, S. (1982) 'Toward a social psychological theory of Tourist Motivation—a rejoinder', *Annals of Tourism Research*, 9 (2): 256–62.

IUCN (1990) *World Heritage Nomination—IUCN Summary 547 Mount Huangshan Scenic Beauty and Historic Site (China). Summary Prepared by IUCN (April 1990) Based on the Original Nomination Submitted by the Peoples Republic of China*, Online. Available HTTP: <http://whc.unesco.org/archive/advisory_body_evaluation/547> (accessed 2 February 2005).

Jackowski, A. (2000) 'Religious tourism—problems with terminology', in A. Jackowski (ed.) *Peregrinus Cracoviensis*, Cracow: Publishing Unit, Institute of Geography, Jagiellonian University.

Jafari, J. (1990) 'Research and scholarship: the basis of tourism education', *Journal of Tourism Studies*, 1 (1): 33–41.

—— (2005) 'Bridging out, nesting afield: powering a new platform', *The Journal of Tourism Studies—Special Issue on Scholarship*, 16 (2): 1–5.

Jamal, T. and Getz, D. (1995) 'Collaboration theory and community tourism planning', *Annals of Tourism Research*, 22 (1): 186–204.

—— (1999) 'Community roundtables for tourism related conflicts: the dialectics of consensus and process structures', *Journal of Sustainable Tourism*, 7 (3/4): 290–313.

Jayaram, V. (2000) *Chinese Buddhism*, Online. Available HTTP: <http://hinduwebsite.com/buddhism/chinese_buddhism.htm> (accessed 2 March 2005).

Jeong, J. and Lee, K. (2006) 'The physical environment in museums and its effects on visitors' satisfaction', *Building and Environment*, 41 (7): 963–9.

Ji Li and Fang Yongqing (forthcoming) 'The inter-organisation learning process of China's township enterprises', unpublished paper, National University of Singapore and Nanyang Technological University.

Jiang, Y.T. (1998). 'TVEs should not forget farmers (Xiang-zhen Tuonong Quingxiang Yaobude)', *Economic News (Jingli Ribao)* (6 September): 7.

—— (2003) 'A study on community participation in tourism development in underdeveloped areas', unpublished dissertation, Xiamen: Huaqiao University.

Jin, G. and Ding, D.S. (1998) 'Primary study on indices system of tourism operating, managing and controlling of national parks', *Journal of NanJing University (Philosophy, Humanities and Social Sciences)*, 35 (4): 126–9.

Johnson, J.D. and Snepenger, D.J. (1993) 'Application of the tourism life cycle concept in the Greater Yellowstone Region', *Society and Natural Resources*, 6 (2): 127–48.

—— (2006) 'Residents' perceptions of tourism development over the early stages of the TALC', in C. Cooper, C.M. Hall and D. Timothy (series eds) and R.W. Butler (vol. ed.) *The Tourism Area Life Cycle (Vol. 1): applications and modifications*, Clevedon: Channel View Publications.

Johnston, R.J., Gregory, D. and Smith, D.M. (1988) *The Dictionary of Human Geography*, Oxford: Blackwell.

Joppe, M. (1996) 'Sustainable community tourism development revisited', *Tourism Management*, 17 (7): 475–9.

Joseph, C. and Kavoori, A. (2001) 'Mediated resistance: tourism and the host community', *Annals of Tourism Research*, 28 (4): 998–1009.

Keitumetse, S. (2007) 'Celebrating or marketing the indigenous? International rights organisations, national governments and tourism creation', in P.E. Burns and M. Novelli (eds) *Tourism and Politics: global frameworks and local realities*, Oxford: Pergamon.

Keller, P.C. (1987) 'Stages of peripheral tourism development—Canada's Northwest Territories', *Tourism Management*, 8 (1): 20–32.

Kelly, G.A. (1963) *A Theory of Personality: the psychology of personal constructs*, New York: W.W. Norton and Company.

Kieschnick, J. (2003) *The Impact of Buddhism on Chinese Material Culture*, Princeton: Princeton University Press.

Kim, H.J., Chen, M.H. and Jang, S. (2006) 'Tourism expansion and economic development: the case of Taiwan', *Tourism Management*, 27 (5): 925–33.

King Chan (1997) 'Tourist satisfaction in relation to a holiday in Thailand', in *The Development and Evaluation of Activity Schedules for Tourists on a One Day Commercial Reef Trip*, James Cook University, Queensland: Co-operative Research Centre for Ecologically Sustainable Development of the Great Barrier Reef.

Kline, R.B. (2005) *Principles and Practice of Structural Equation Modeling*, 2nd edn, New York: The Guildford Press.

Köksal, C.D. and Aksu, A.A. (2007) 'Efficiency evaluation of A-group travel agencies with data envelopment analysis (DEA): a case study in the Antalya region, Turkey', *Tourism Management*, 28 (3): 830–4.

Kolas, Ashild and Thowsen, Monika P. (2005) *On the Margins of Tibet. Cultural Survival on the Sino-Tibetan Frontier*, Seattle: University of Washington Press.

Korpela, K.M. (1989) 'Place identity as a product of environmental self regulation', *Journal of Environmental Psychology*, 9: 241–56.

Kotler, P., Bowen, J. and Makens, J. (2006) *Marketing for Hospitality and Tourism*, 4th edn, New York: Prentice Hall.

Krippendorff, K. (1980) *Content Analysis: an introduction to its methodology*, Newbury Park, California: Sage.

Kuznets, S. (1955) 'Economic growth and income inequality', *American Economic.*

La Tour, S.A. and Peat, N.C. (1979) 'Conceptual and methodological issues in consumer satisfaction research', *Advances in Consumer Research*, 6: 431–7.

Lagiewski, R.M. (2006) 'The application of the TALC model: a literature survey', in R.W. Butler (ed.) *The Tourism Area Life Cycle (Vol. 1): applications and modifications*, Clevedon: Channel View Publications.

Lai, K., Li, Y. and Feng, X. (1996) 'Gap between tourism planning and implementation: a case of China', *Tourism Management*, 27 (6): 1171–80.

Lai, Y. (2001) *An Encyclopaedia of Chinese Buddhism*, Shanghai: Xinhua Publications.

Lalli, M. (1992) 'Urban related identity: theory, measurement and empirical findings', *Journal of Environmental Psychology*, 12: 285–303.

Laushway, E. (2000) 'Santiage de compostela the pilgrims' choice', *Europe*, 396 (1): 33–5.

Laws, E. (2002) *The Design Analysis and Improvement of Tourist Services*, Champaign, IL: Sagamore.

Lawson, R.W., Williams, J., Young, T. and Cossens, J.A. (1998) 'Comparison of residents' attitudes towards tourism in 10 New Zealand destinations', *Tourism Management*, 19 (3): 247–56.

Le, Z. (2006) 'A discussion on a city's charm and brand', paper presented at the Conference—International Tourist Cities under the Globalisation Context: Shenzhen, China. Peking University, Shenzhen's People Municipal Government and Guangdong Provincial Tourist Administration.

Leaf, M. (1995) 'Inner city redevelopment in China', *Cities*, 12 (3):149–62.

Leask, A. and Fyall, A. (2000) 'World heritage sites: current issues and future implications', paper presented at the conference Tourism 2000, Sheffield Hallam University, 2–7 September.

Lee, Chien-Chiang and Chang, Chun-Ping. (2008) 'Tourism development and economic growth: a closer look at panels', *Tourism Management*, 29 (1):180–92.

Lee, C.K. and Han, S.H. (2002) 'Estimating the use and preservation values of national parks' tourism resources using a contingent valuation method', *Tourism Management*, 23 (5):531–40.

Lehle, K. (1982) *Urlaub auf dem Bauernhof-Analyse und Perspektiven des Nebenbetriebes Tourismus in der Landwirtschaft*, AID-Schriftenreihe 190, Bonn-Bad Godesberg: Bundesministerium Ernahrung, Landwirtschaft und Forsten.

Lehmbruch, G. (1998) 'Zwischn Institutionentransfer und Eigendynamik: sektorale Transformation-spfade und ihre Bestimmungsgrüde', in R. Czarda and G. Lehmbruch (eds) *Transformationspfade in Ostdeutschland: Beiträge zur sektoralen Vereinigungspolitik*, Frankfurt am Main: Campus.

Lever, A. (1987) 'Spanish tourism migrants: the case of Lloret de Mar', *Annals of Tourism Research*, 14 (4): 449–70.

Levine, R.S., Hughes, M., Mather, C.R., Dumreicher, H. and Lu, H. (2005) 'The proto-sustainable Chinese village as generator of the future Chinese City', paper presented at the International Conference on Sustainability, Jinma Conference Center, China Agricultural University, Beijing, 22 April.

Lewin, A.Y. and Minton, J.W. (1986) 'Determining organizational effectiveness: another look, and an agenda for research', *Management Science*, 32 (5): 514–38.

Li Chunman (2000) *Corporate Governance of State-Owned Enterprises in China: the capital management of China's state-owned asset and the evaluation of its result and achievement*, Beijing: Organisation for Economic Co-Operations and Development: Development Research Centre of the State Council of the PRC and Asian Development Bank.

350 Bibliography

Li Lunliang (2006) 'Spatial development strategy for renewal of Confucianism—strategic thinking about the construction of the historic city of Confucianism in Qufu', *Planners*, 22 (4): 85–88.

Li, F.M.S. (2005) 'Chinese common knowledge, tourism and natural landscapes: gazing in "Bie you tian di"—an altogether different world', unpublished thesis, Murdoch University, Perth, Western Australia.

Li, M., Wu, B. and Cai, L. (2008) 'Tourism development of world heritage sites in China: a geographic perspective', *Tourism Management*, 29 (2): 308–19.

Li, Y. (2007) 'Community participation in tourism management of world natural heritage sites: the case of Wulingyuan', unpublished thesis, School of Tourism, Central South University of Forestry and Technology, Hunan, China.

Li, Y., Zhang, J. and Chen, Y. (2006) 'Image of landscapes of ancient water towns: case study on Zhouzhuang and Tongli of Jiangsu Province', *Chinese Geographical Science*, 16 (4): 371–7.

Li. Z. (2000) *Discussion on Enriching Rural Tourism's Cultural Content*, Tian Fu Xin Lun.

Li Lei Lei (2002) 'Counter industrialization and industrial heritage tourism development', *World Geography Research* 5:13–18.

Lin Junhua (2004) 'Comment on the multi-culture in Kangba Region', *Journal of Kangding Nationality Teachers College*: 3.

Lindberg, K., Dellaert, B.G.C. and Rassing, C.R. (1999) 'Resident tradeoffs: A choice modeling approach', *Annals of Tourism Research*, 26 (3): 554–69.

Lindberg, K. and Johnson, R.L. (1997) 'Modeling resident attitudes toward tourism', *Annals of Tourism Research*, 24 (2): 402–24.

Liu Xiaohui (2006) 'Main problems of tourism culture in Ganzi', *Journal of Kangding Nationality Teachers College*: 5.

Liu Zhaoping (1998) 'A review of the social impacts of tourism on destinations: a trail survey on the tourism development of Yesanpo', *Tourism Tribune*, 1: 50–4.

Liu, K.L. (1999) 'From forest farm to national forest park', in X.Y. Li (ed.) *The Growth of Zhangjiajie*, Changsha: Yuelu Press.

Liu, S.H. (1998) *Understanding Confucian Philosophy*, Oxford: Praeger.

Liu, W. (2000) 'Theory considerations on community participations in tourism development', *Tourism Studies*, 1.

Liu, W.H. (2000) 'Some theoretical thoughts about community involved tourism development', *Tourism Tribune*, 1: 47–62.

Liu, X.T. and Sheng, Z.F. (2006) 'Tourism—power industry for structuring harmonious society', *Journal of Changchun University*, 1.

Liu, X. and Yang, Z. (2002) 'Several considerations on "tourism alleviating poverty"', *Economical Geography*, 22 (2): 241–44.

Liu, Y.D. (2005) *Strategies on the Increase of Tourism Tax Revenues in Zhangjiajie*, Online. Available HTTP: <http://www.hnce.cn/news/Article_Print.asp?ArticleID=377> (accessed 17 February 2007).

Liu, Y.W. (2006) *Zhangjiajie Tourism Development with a Promising Future*, Online. Available HTTP: <http://www.zjj-trip.com/news/NewsInfo.aspx?ID=3568and pageUp=5> (accessed 17 February 2007).

Logan, P. (1998) *Wooden Architecture in Ganzi*, Kham Aid Foundation Report, Online. Available HTTP: <http://www.asianart.com/cers/archrep.html> (accessed 31 January 2008).

Long, P.T., Perdue, R.R. and Allen, L. (1990) 'Rural resident tourist perceptions and attitudes by community level of tourism', *Journal of Travel Research*, 28 (3): 3–9.

Lu, M. (2001) 'Evaluation pattern on the sustainability of tourist destination', unpublished dissertation, Wuhan: Postgraduate Program, the Tourist College, Central China Normal University.

Lu Meng, Zhou Xiaoling and Yingjie (2006) 'On the cultural tourism development in Ganzi', *Journal of Kangding Nationality Teachers College*: 5.

Lundtorp, S. and Wanhill, S. (2001) 'Resort life cycle theory: generating processes and estimation', *Annals of Tourism Research*, 28 (4): 947–64.

Luo, Y. (2006) 'Perception of recreational impacts and its relation to visitors' experience: a case study of Zhangjiajie National Forest Park', unpublished dissertation, Central South University of Forestry and Technology, Changsha, China.

Luzhi Ancient Town Protection and Restoration Plan (1999) Shanghai Tongji Urban Planning Research Center.

Luzhi Tourism Development Plan (2001) Edited by Suzou Luzhi Tourism Development Company and the Department of Tourism, Humanities School, Suzhou University.

Ma, Jian (2006) *Stick out your tongue*, London: Chatto and Windus.

Ma, X.L., and Yang, X.J. (2003) 'A study on the 4A tourism area (spots) in China: spatial characteristics and industrial distribution', *Economic Geography*, 23 (5): 713–16.

MacCanell, D. (1999) *The Tourist: a new theory of the leisure class*, Berkeley and Los Angeles: University of California Press.

McGibbon, J. (2000) *The Business of Alpine Tourism in a Globalising World: an anthropological study of international tourism in the village of St. Anton am Arlberg in the Tirolean Alps*, Rosenheim: Vetterling Druck.

McGurn, B. (1999) 'Rebuilding Rome for the big day', *Our Sunday Visitor* (26 December): 6.

Mack, A. (1999) 'Complexity and economy in pilgrimage centres of the Vijayanagara', workshop on the Cosmology and Complexity of Pilgrimage, Indira Gandhi National Centre for the Arts, New Delhi, 5–9 January 1999.

McKay, A.C. (2001) 'Truth, perception and politics: the British construction of an image of Tibet', in Dodin, Raether.

McKercher, B. and du Cros, H. (2005) *Cultural Tourism: the partnership between tourism and cultural heritage management*, Binghampton, NY: The Haworth Press, Inc.

McNeil, P. (1990) *Research Methods*, 2nd edn, London: Routledge.

Maddox, R.N. (1985) 'Measuring satisfaction with tourism', *Journal of Travel Research*, 23 (3): 2–5.

Manente, M. and Pechlaner, H. (2006) 'How to define, identify and monitor the decline of tourist destinations: towards an early warning system', in R.W. Butler (ed.) *The Tourism Area Life Cycle. Vol. 2: conceptual and theoretical issues*, Clevedon: Channel View Publications.

Mao, T. (1942) 'Rectify the party's style of work (1 February 1942)', in *Selected Works, vol. III. From Quotations from Chairman Mao*, Beijing.

Marschalek, I. (2008) 'The concept of participatory local sustainability projects in seven Chinese villages', *Journal of Environmental Management*, 87 (2): 226–35.

Marshall, A. (1920) *Principles of Economics*, London: Macmillan and Co., Ltd.

Martilla, J.A. and James, J.C. (1977) 'Importance-performance analysis', *Journal of Marketing Research*, 41 (January): 77–79.

Martin, B. and Uysal, M. (1990) 'An examination of the relationship between carrying capacity and the tourism lifecycle: management and policy implications', *Journal of Environmental Management*, 31: 327–33.

Marton, A.M. (2000) *China's Spatial Economic Development: restless landscapes in the lower Yangzi Delta*, London: Routledge.

Mason, J. (2002) *Qualitative Researching*, London: Sage.

Mathieson, A. and Wall, G. (1982a) *Tourism: economic, physical and social impacts*, London: Longman.

—— (1982b) *Tourism Impacts*, Harlow: Longmans.

Meller, H. (2001) *European Cities 1890–1930s: history, culture and the built environment*, Chichester: John Wiley.

Meng, P.Y. (2004) Relationship between Man and Nature in Traditional Human Rights Concepts of China and West, Online. Available HTTP: <http://www.humanrights.cn/zt/magazine/200402004811100410.htm>

Meyer-Arendt, K.J. (1985) 'The Grand Isle, Louisiana resort cycle', *Annals of Tourism Research*, 12 (3): 449–65.

Ming Shen Ding and Ping Lei Qian (2004) *The Study of Real Estate and Tourism*, Fudan: Fudan University Press.

Minichiello, V., Aroni, R., Timewell, E. and Alexander, L. (1995) *In-depth Interviewing, Principles, Techniques, Analysis*, 2nd edn, Melbourne: Longman.

Ministry of Agriculture, PRC. (1993, 1994, 1995, 1996, 1997) Zfongguo xiangzhen qiye nianjian.

Ministry of Civil Administration, PRC (1992) *Zhongguo xian qing daquan Vol. 1–6 (Encyclopedia of Chinese Counties)*, Beijing: Shehui chubanshe.

Mitchell, C.J.A. (1998) 'Entrepreneurialism, commodification and creative destruction: a model of post-modern community development', *Journal of Rural Studies*, 14: 273–86.

—— (2003) 'The Heritage Shopping Village: profit, preservation and production', in G. Wall (ed.) *Tourism: people, places and products*, Department of Geography Publication Series, Waterloo: University of Waterloo.

Mitchell, C.J.A. and Coghill, C. (2000) 'The creation of a heritage townscape: Elora, Ontario', *The Great Lakes Geographer*, 7 (2): 88–105.

Mitchell, C.J.A., Nolan, R. and Hohol, F. (1993) 'Tourism and community economic development: a case study of St. Jacobs, Ontario', in D. Brunce and M. Whitla (eds) *Tourism Strategies for Rural Development*, Mount Alison University: Sackville.

Mitchell, C.J.A., Atkinson, R.G. and Clark, A. (2001) 'The creative destruction of Niagara-on-the-lake', *The Canadian Geographer*, 45: 285–99.

Moore, S. and Wen, J. (2004) 'Economic reform and business management in China today', *International Journal of Applied Management*, 5 (2): 66–84.

—— (2007) 'Strategic management in Australia and China: the great leap forward or an illusion', *Journal of Technology Management in China*, 2 (1): 10–21.

Morey, R.C. and Dittman, D.A. (1995) 'Evaluating a hotel GM's performance: a case study in benchmarking', *Cornell Hotel and Restaurant Administration Quarterly*, 36 (5): 30–5.

Moscardo, Gianna (1999) *Making Visitors Mindful: principles for creating sustainable visitor experiences through effective communication*, Champaign, IL: Sagamore Publishing.

Murphy, P.E. (1983) 'Perceptions and attitudes of decision-making groups in tourism centers', *Journal of Travel Research*, 21 (3): 8–12.

—— (1985) *Tourism—A Community Approach*, New York: Methuen.

—— (1988) 'Community driven tourism planning', *Tourism Management*, 9 (2): 96–104.

Naisbitt, J. (1995) *Megatrends Asia*, London: Nicholas Brealey Publishing Limited.

Narayan, P.K. (2004) 'Economic impact of tourism on Fiji's economy: empirical evidence from the computable general equilibrium model', *Tourism Economics*, 10 (4): 419–33.

National Department of Construction (2007) *Urban Construction Statistic Annual Report of China 2007*, Beijing: China Construction Industry Press.

Newsome, S.A., Moore and Dowling, R.K. (2002) *Natural Area Tourism: ecology, impacts and management*, Clevedon: Channel View Publications.

Nolan, M.L. and Nolan, S. (1992) 'Religious sites as tourism attractions in Europe', *Annals of Tourism Research*, 19 (1): 68–78.

Nyiri, P. (2006) *Scenic Spots. Chinese Tourism, the State and Cultural Authority*, Seattle: University of Washington Press.

Oliver, R.L. (1980) 'A cognitive model of the antecedents and consequences of satisfaction decisions', *Journal of Marketing Research*, 17 (4): 460–9.

—— (1993) 'Cognitive, affective, and attribute bases of the satisfaction response', *Journal of Consumer Research*, 20 (3): 418–30.

Olsen, D.H. (2003) 'Heritage, tourism, and the commodification of religion', *Tourism Recreation Research*, 28 (3): 99–104.

—— (2006) 'Managing issues for religious heritage attractions', in D.J. Timothy and D.H. Olsen (eds) *Tourism, Religion and Spiritual Journeys*, London: Routledge.

Olsen, D.H. and Timothy, D.J. (2006) 'Tourism and religious journeys', in D.J. Timothy and D.H. Olsen (eds) *Tourism, Religion and Spiritual Journeys*, London: Routledge.

Oppermann, M. (1996) 'Rural tourism in southern Germany', *Annals of Tourism Research*, 23 (1): 86–102.

Organization of American States (1997) 'Sustaining tourism by managing its natural and heritage resources. Inter-American travel congresses permanent secretariat', paper prepared for XVII Inter-American Travel Congress, San José, Costa Rica, 7–11 April. Online. Available HTTP: <http://www.oas.org/tourism/docnet/iatc3 en.htm> (accessed 15 January 2008).

Orland, B. and Bellafiore, V.J. (1990) 'Development directions for a sacred site in India', *Landscape and Urban Planning*, 19: 181–96.

Overmyer, D.L. (1986) *Religions of China: the world as a living system*, San Francisco: Harper and Row.

Panish, P. (2008) *The Artistry of Chinese Garden Shines*, 14 February, Online. Available HTTP: <http://www.latimes.com/features/home/la-hm-huntington14feb14, 1,1212107.story>

Parasuraman, A., Zeithaml, V.A. and Berry, L.L. (1985) 'A conceptual model of service quality and its implications for future research', *Journal of Marketing Research*, 49 (Fall): 41–50.

Parilla, J.C., Font, A.R. and Nadal, J.R. (2008) 'Tourism and long-term growth: a Spanish perspective', *Annals of Tourism Research*, 34 (3): 709–26.

Park, A. and Minggao Shen (2001) 'Joint liability lending and the rise and fall of China's Township and Village Enterprises', William Davidson Working Paper Number 462.

Pastorelli, J. (2003) *Enrich the Experience: an interpretive approach to tour guiding*, NSW, Australia: Hospitality Press.

Pearce, D.G. (1980) 'Tourism and regional development: a genetic approach', *Annals of Tourism Research*, 7 (1): 69–82.

—— (1989) *Tourist Development*, Harlow: Longmans.

—— (1993) 'Comparative studies in tourism research', in D.G. Pearce and R.W. Butler (eds) *Tourism Research: critiques and challenges*, London: Routledge.

—— (2000) 'Tourism plan reviews: methodological considerations and issues from Samoa', *Tourism Management*, 21 (2): 191–203.

Pearce, P.L., Moscardo, G. and Ross, G. (1996) *Tourism Community Relationships*, Oxford: Elsevier.

Peng, Y. (1999) 'Agricultural and non-agricultural growth and inter-county inequality in China, 1985–1991', *Modern China*, 25 (3): 235–63.

Pennington-Gray, L., Reisinger, Y., Kim, J. and Thapa, B. (2005) 'Do US tour operators' brochures educate the tourist on culturally responsible behaviours? A case study for Kenya', *Journal of Vacation Marketing*, 11 (3): 265–84.

People's Daily (1998) Report on the Economic Working Conference of Central Committee of CCP. 1998-12-06-(1).

Peoples Republic of China (PRC) (1989) *Submission to UNESCO for WHS Listing for Huangshan Scenic Beauty and Historic Interest Reserve*, Huangshan: Huangshan Administrative Committee in Charge of Sites of Scenic Beauty and Historic Interest.

Perdue, R.R., Long, P.T. and Allen, L. (1990) 'Resident support for tourism development', *Annals of Tourism Research*, 17 (4): 586–99.

Peretz, M. (1988) 'Jerusalem diarist: the sealed bottle', *The New Republic*, 199 (19): 42.

Peterson, G.L. (1974) 'Evaluating the quality of the wilderness environment: congruence between perception and aspiration', *Environment and Behaviour*, 6 (2): 169–93.

Petersen, Ying Yang (1995) 'The Chinese landscape as a tourist attraction: image and reality', in Alan A. Lew and Lawrence Yu (eds) *Tourism in China. Geographical, Political and Economic Perspectives*, Boulder, CO: Westview Press.

Petrick, J.F. and Backman, S.J. (2002) 'An examination of the determinants of golf travellers' satisfaction', *Journal of Travel Research*, 40 (3): 252–8.

Philip, J. and Mercer, D. (1999) 'Commodification of Buddhism in contemporary Burma', *Annals of Tourism Research*, 26 (1): 21–54.

Phillinore, J. and Goodson, L. (eds) (2004) *Qualitative Research in Tourism: ontologies, epistemologies and methodologies*, London: Routledge.

Phillips, A. (2002) *Management Guidelines for IUCN Category V Protected Areas: protected landscapes/seascapes*, Cambridge, UK, and Gland, Switzerland: IUCN.

Pike, S. (2002) 'Destination image analysis—a review of 142 papers from 1973 to 2000', *Tourism Management*, 23 (5): 541–9.

Po-Jen Chen and Kerstetter, D.L. (1999) 'International students' image of rural Pennsylvania as a travel destination', *Journal of Travel Research*, 29 (February): 256–66.

Porat, M.U. (1977) *The Information Economy: sources and methods for measuring the primary information sector*, Report No. OT-SP-77-12(2), Washington: National Science Foundation.

Potts, T.D. and Harill, R. 1998. Enhancing Communities for Sustainability: a Travel Ecology Approach. *Tourism Analysis*. 3:133–142.

Prabhu, P.P. (1993) 'Religion's responsibilities in cultural tourism', in W. Nuryati (ed.) *Universal Tourism: enriching or degrading culture?* Yogakarta: Gadjah Mada University Press.

Preda, P. and Watts, T. (2003) 'Improving the efficiency of sporting venues through capacity management: the case of the Sydney (Australia) cricket ground trust', *Event Management*, 8 (2): 83–9.

Prichard, A. and Morgan, N. (1996) 'Selling the Celtic Arc to the USA: a comparative analysis of the destination brochure images used in the marketing of Ireland, Scotland and Wales', *Journal of Vacation Marketing*, 2 (4): 346–65.

Qiu, B.X. (2006) *Report on the Renovation of National Parks*, Online. Available HTTP: <http://www.cin.gov.cn/ldjh/jsld/200701/t20070109_40317.htm> (accessed 5 November 2007).

Quan H. (2003) 'Theories and practices of ecotourism construction: take Zhangjiajie as an example', unpublished doctoral dissertation, The Institute of Geographic Sciences and Natural Resources Research, Chinese Academy of Sciences, Beijing, China.

Rawson, Philip and Legeza, Laslo (1973) *Tao. The Chinese Philosophy of Time and Change*, London: Thames and Hudson.

Ren Zhuge (2000) 'Questionnaire survey on participation of local communities in nature reserve management', in *Study of Sustainable Policy for China's Nature Reserves—Chinese National Committee for Man and the Biosphere Programme*. Beijing: Chinese National Committee for MAB.

Richards, G. and Hall, D. (2000) *Tourism and Sustainable Community Development*, London: Routledge.

Ritzer, G. (2004) *The McDonaldization of Society*, Thousand Oaks, CA: Pine Forge Press.

Ropp, P.S. (ed.) (1992) *Heritage of China: contemporary perspectives on China*, Berkeley: University of California Press.

Ross, D.E.L. and Iso-Ahola, S.E. (1991) 'Sightseeing tourists: motivation and satisfaction', *Annals of Tourism Research*, 18 (2): 226–37.

Rostow, W.W. (1956) 'The take off into self sustained growth', *The Economic Journal*, 66 (261): 25–48.

Ruan, Y. and Shao, Y. (1996) 'Characteristics and protection of water towns in Jiangnan area', *Tongji University Journal (Humanities and Social Science Section)*, 7 (1): 21–8 (in Chinese).

Russell, R. (2006) 'The contribution of entrepreneurship theory to the TALC model', in R.W. Butler (ed.) *The Tourism Area Life Cycle (Vol. 2): conceptual and theoretical issues*, Clevedon: Channel View Publications.

Russell, R., and Faulkner, B. (1999) 'Movers and shakers: chaos makers in tourism development', *Tourism Management*, 20: 411–23.

Ryan, C. (1995a) 'Conversations in Majorca—the over 55s on holiday', *Tourism Management*, 16 (3): 207–17.

—— (1995b) *Researching Tourist Satisfaction Issues, Concepts, Problems*, London: Routledge.

—— (2002) 'Tourism and cultural proximity: examples from New Zealand', *Annals of Tourism Research*, 29 (4): 952–67.

—— (2003) *Recreational Tourism: demand and impacts*, Clevedon: Channel View Publications.

Ryan, C. and Aicken, M. (2005) *Indigenous Tourism: the commodification and management of culture*, Oxford: Pergamon.

Ryan, C. and Cooper, C. (2004) 'Residents' perceptions of tourism development: the case of Raglan, New Zealand', *Tourism Review International*, 8 (1): 1–17.

Ryan, C. and Gu, H. (2007a) 'Perceptions of Chinese hotels', *Cornell Hotel and Restaurant Quarterly*, 48 (4): 380–91.

—— (2007b) 'The social impacts of tourism in a Beijing hutong—a case of environmental change', *China Tourism Research*, 3 (2): 235–71.

—— (2009) Reflexivity and Culture in tourism research: understanding the 4th Buddhist Festival, Wutaishan, China forthcoming Tourism Management.

Ryan, C. and Stewart, M. (2009) 'The paradoxical social construction of ecotourism—the case of a luxury desert lodge at Al Maha, Dubai', *Journal of Sustainable Tourism*.

Ryan, C., Scotland, A. and Montgomery, D. (1998) 'Resident attitudes to tourism development—a comparative study between the Rangitikei, New Zealand and Bakewell, United Kingdom', *Progress in Tourism and Hospitality Research*, 4 (2): 115–30.

Said, E. (1978) *Orientalism. Western Conceptions of the Orient*, Harmondsworth: Penguin.

Sarkis, J. and Talluri, S. (2004) 'Performance based clustering for benchmarking of US airports', *Transportation Research Part A*, 38 (5): 329–46.

Say, J. (1963) *Conspectus of Political Economics (1803)*, Beijing: Commercial Press.

Schoppner, A. (1988) 'Urlaub auf dem Bauemhof. Eine fremdenverkehrsgeographische Untersuchung', *Bochumer Materialien zur Raumordnung*, 36.

Schumpeter, J. (1943) *Capitalism, Socialism and Democracy*, New York: Harper and Row.

Selmer, J. (1998) *International Management in China: cross-cultural issues*, London: Routledge.

Shackley, M. (2001) *Managing Sacred Sites: service provision and visitor experience*, London: Continuum.

—— (2006) 'Empty bottles at sacred sites: religious retailing at Ireland's National Shire', in D.J. Timothy and D.H. Olsen (eds) *Tourism, Religion and Spiritual Journeys*, London: Routledge.

Shanghai Statistical Bureau (2005) *Shanghai Statistical Yearbook 2005*, Beijing: China Statistical Publishing House.

Shanxi Tourism Bureau (2002) 'The 13th Wutai Mountain International Tourism Month ended today', *Shanxi Daily*. Online. Available HTTP: <http://www.sxta.com.cn/news/wts13jbm.htm> (accessed 15 March 2005).

Shao, Q.W. (2006) *Yearbook of Tourism Statistics*, Beijing: Chinese Tourism Press.

Sharpley, R. (1994) *Tourism, Tourists and Society*, Huntingdon, Cambridgeshire: ELM.

—— (2002) 'The consumption of tourism', in R. Sharpley and D. Telfer (eds) *Tourism and Development: concepts and issues*, Clevedon: Channel View Press.

Shen, F. (2004) 'Agritourism sustainability in mountain rural areas in China: Chongdugou happy-in-farmhouse case study', unpublished thesis, Wageningen University.

Shi, F. (2007) 'Commoditised religious souvenirs and visitor experience at Chinese Buddhist sites', unpublished thesis, Nottingham Business School, Nottingham Trent University.

Shi, Q. (2005) *Environmental Impact Assessments of Forest Parks*, Beijing: China Science Press.

Simpson, B. (1993) 'Tourism and tradition from healing to heritage', *Annals of Tourism Research*, 20 (1): 164–81.

Sirakaya, E. (1997) 'Attitudinal compliance with ecotourism guidelines', *Annals of Tourism Research*, 24 (4): 919–50.

Smith, R.A. (1992) 'Beach resort evolution: implications for planning', *Annals of Tourism Research*, 9 (2): 304–22.

Smith, S.L.J. (1983) *Recreational Geography*, Harlow: Longman.

—— (1989) *Tourism Analysis*, Harlow: Longman.

Sofield, T.H.B. (2003) *Empowerment of Sustainable Tourism Development*, London: Pergamon.

Sofield, T.H.B. and Li, F.M.S. (1998a) 'China: tourism development and cultural policies', *Annals of Tourism Research*, 25 (2): 323–53.

—— (1998b) 'Tourism development and cultural policies in China', *Annals of Tourism Research*, 25 (2): 362–92.

—— (2003) 'Processes in formulating an ecotourism policy for nature reserves in Yunnan Province, China', in D. Fennell and R. Dowling (eds) *Ecotourism: policy and strategy issues*, London: CAB International Academic.

—— (2007) 'China: ecotourism and cultural tourism, harmony or dissonance', in J. Higham (ed.) *Critical Issues in Ecotourism: understanding a complex tourism phenomenon*, Oxford: Butterworth Heinemann.

Stansfield, C. (1978) 'Atlantic City and the resort cycle', *Annals of Tourism Research*, 5 (2): 238–51.

State Department (2006) Rule of National Parks.

State Statistical Bureau, PRC (1989) *Zhongguo fenxian nongcun jingji tongji gaiyao 1980–1987*.

State Statistical Bureau, PRC (1991) *Zhongguo fenxian nongcun jingji tongji gaiyao 1989 (Summary Statistics of the Rural Economy of Chinese Counties 1989)*, Beijing: China Statistical Publishing House.

State Statistical Bureau, PRC. (1992a) *Statistical Yearbook of Chinese Cities, Vol. 1992*, Beijing: China Statistical Publishing House.

State Statistical Bureau, PRC (1992b) *Zhongguo xiaokang biaozhun (The Criteria of Affluence in China)*, Beijing: China Statistical Publishing House.

State Statistical Bureau, PRC (1993) *Zhongguo fenxian nongcun jingji tongji gaiyao 1991 (Summary Statistics of the Rural Economy of Chinese Counties 1991)*, Beijing: China Statistical Publishing House.

State Statistical Bureau, PRC (1996) *China Statistical Yearbook 1996*, Beijing: China Statistical Publishing House.

State Statistical Bureau, PRC (1997a) *China Labour Statistical Yearbook 1997, Compiled by State Statistical Bureau and the Ministry of Agriculture (PRC)*, Beijing: China Statistical Publishing House.

State Statistical Bureau, PRC (1997b) *Rural Statistical Yearbook of China 1997*, Beijing: China Statistical Publishing House.

Stefan, G., Paul, P., Jean, P.C., Ghislain, D., Trista, P. and Robert, B.R. (2005) 'The eco-efficiency of tourism', *Ecological Economics*, 54 (15): 417–34.

Stephenson, J. (2008) 'The cultural values model: an integrated approach to values in landscapes', *Landscape and Urban Planning*, 84: 127–39.

Stevens, T. (1988) 'The ministry of welcome: tourism and religious sites', *Leisure Management*, 8: 41–4.

Stokes, R. (2007) 'Tourism strategy making: insights to the events tourism domain', *Tourism Management*, 29 (2): 252–62.

Sun, J.X. (2005) 'Positive effects of community participation in tourism upon the protection of ethnic traditional culture', *Journal of Guangxi University for Nationalities (Philosophy and Social Science Edition)*, 27 (4): 35–9.

Sun, J.X. and Bao, J.G. (2005) *On Tourism Anthropology Involving Community Participation: a*

Suzhou City Luzhi Town (2004) Maps.

Tang Shun-ying (2004) 'Qufu: Confucius' hometown's cultural tourism', *Social Scientist*, 2 (1)14–19.

Tang, B.S., Wong, S.W. and Lau, M.C.H. (2008) 'Social impact assessment and public participation in China: a case study of land requisition in Guangzhou', *Environmental Impact Assessment Review*, 28 (1): 57–72.

Tang, D.J. and Sui, L.N. (2006) 'Tourism and harmonious society', *Economic and Social Development*, 2006 (4).

Tang, S.T. (1998) 'Communitification of tourist destination and community tourism', *Geographical Research*, 2: 145–9.

Tao Wei (2004) 'Comparative study on tourism development mode in historic towns: Venice and Lijiang', *Urban Planning Forum*, 1: 8–15.

Telfer, D. (1996) 'Development through economic linkages: tourism and agriculture in Indonesia', unpublished thesis, University of Waterloo.

—— (2002) 'The evolution of tourism and development theory', in R. Sharpley and D. Telfer (eds) *Tourism and Development: concepts and issues*, Clevedon: Channel View Press.

Tellenbach, Hubertus and Bin, Kimura (1989) 'The Japanese concept of "nature"', in J. Baird Callicott and Roger T. Ames (eds) *Nature in Asian Traditions of Thought: essays in environmental philosophy*, New York: State University of New York Press.

Teye, V., Sonmez, S.F. and Sirakaya, E. (2002) 'Residents' attitudes towards tourism development', *Annals of Tourism Research*, 29 (3): 668–88.

Tilden, F. (1977) *Interpreting our Heritage*, 3rd edn, Chapel Hill: University of North Carolina Press.

Tilson, D.J. (2001) 'Religious tourism, public relations and church–state partnership', *Public Relations Quarterly*, 46 (3): 35–9.

Tonts, M. and Greive, S. (2002) 'Commodification and creative destruction in the Australian rural landscape the case of Bridgetown, Western Australia', *Australian Geographical Studies*, 40 (1): 58–70.

Tooman, L.A. (1997) 'Application of the lifecycle model in tourism', *Annals of Tourism Research*, 24 (1): 214–34.

Trandis, H.C. (1990) 'Cross-cultural studies of individualism and collectivism', in J.J. Berman (ed.) *Cross-Cultural Perspectives*, Lincoln: University of Nebraska Press.

Trauer, B. and Ryan, C. (2005) 'Destination image, romance and place experience—an application of intimacy theory in tourism', *Tourism Management*, 26 (4): 481–92.

Tribe, J. and Snaith, B. (1998) 'From SERVQUAL to HOLSAT: holiday satisfaction in Varadero, Cuba', *Tourism Management*, 19 (1): 25–34.

Trilling, L. (1972) *Sincerity and Authenticity*, London: Oxford University Press.

Tsaur, S.H., Chiang, C.I. and Chang, T.Y. (1999) 'Evaluating the operating efficiency of international tourist hotels using the modified DEA model', *Asia Pacific Journal of Tourism Research*, 14 (1): 73–8.

Tschang, C. (2007) 'In China, dharma confronts the dollar', *Business Week Online*, 16 August: 20. Online. Available HTTP: <http://search.ebscohost.com/login.aspx?direct=trueanddb=buhandAN=26278585andsite=ehost-live> (accessed 1 October 2007).

Tu, Wei-Ming (1989) 'The continuity of being: Chinese visions of nature', in J. Baird Callicott and Roger T. Ames (eds) *Nature in Asian Traditions of Thought: essays in environmental philosophy*, New York: State University of New York Press.

Twigger-Ross, C.L. and Guzzle, D.L. (1996) 'Place and identity processes', *Journal of Environmental Psychology*, 16: 205–20.

Ullian, R. (1998) *Frommer's Israel*, New York: MacMillan.

UNESCO (1990) *Convention Concerning the Protection of the World Cultural and Natural Heritage. Report of the World Heritage Committee, Fourteenth Session, Banff, Alberta, Canada, 7–12 December 1990*. CLT-90/CONF.004/13, 12 December 1990.

UNESCO World Heritage Centre (2005) *Mount Wutai Administrative Bureau*, Online. Available HTTP: <http://whc.unesco.org/en/tentativelists/1621/> (accessed 10 October 2005).

UNESCO World Heritage Centre (2006) *World Heritage List*, Online. Available HTTP: <http://whc.unesco.org/> (accessed 1 June 2006).

United Nations Environment Programme (1996) *World Conservation Monitoring Centre. Mount Emei and Leshan Giant Buddha*, Online. Available HTTP: <http://www.unep-wcmc.org/sites/wh/emei.htm> (accessed 16 January 2008).

United Nations World Tourism Organization (1992) *An Integrated Approach to Resort Development*, Madrid: UNWTO.

United Nations World Tourism Organization (1998) *Asian Experiences in Tourism Development*, Madrid: UNWTO.

United Nations World Tourism Organization (1999) *Silk Road Tourism—Current Issues*, Madrid: UNWTO.

United Nations World Tourism Organization (2005) *Cultural Tourism and Poverty Alleviation—The Asia-Pacific Perspective*, Madrid: UNWTO.

United Nations World Tourism Organization (2006) *Poverty Alleviation through Tourism—A Compilation of Good Practices*, Madrid: UNWTO.

Urry, J. (1990) *The Tourist Gaze: leisure and travel in contemporary societies*, London: Sage.

—— (2002) *The Tourist Gaze: leisure and travel in contemporary societies*, London: Sage.

Uysal, M. and Jurowski, C. (1993) 'Testing the push and pull factors', *Annals of Tourism Research*, 21 (4): 844–6.

Uzzell, D.L. (1995) 'Conferring a sense of place identity: the role of museums', *The International Journal of Heritage Studies*, 1: 4.

Vassallo, M. (1979) *From Lordship to Stewardship: religion and social change in Malta*, The Hague: Mouton.

Veverka, J. (2007) *What is Interpretation? An Overview of Interpretive Philosophy and Principles*, Online. Available HTTP: <http://www.heritageinterp.com/library.html> (accessed 18 March 2007).

Vukonic, B. (1996). *Tourism and Religion*, New York: Pergamon.

—— (1998) 'Religious tourism: economic value or an empty box?' *Zagreb International Review of Economics and Business*, 1 (1): 83–93.

—— (2002) 'Religion, tourism and economics: a convenient symbiosis', *Tourism Recreation Research*, 27 (2): 59–64.

Wagner, J.E. (1997) 'Estimating the economic impacts of tourism', *Annals of Tourism Research*, 24 (3): 592–608.

Waitt, G. (2003) 'Social impacts of the Sydney Olympics', *Annals of Tourism Research*, 30 (1): 194–215.

Wakefield, K.L. and Blodgett, J.G. (1994) 'The importance of servicescapes in leisure service settings', *Journal of Services Marketing*, 8 (2): 66–76.

Wall, G. (2001) 'Tourism in Hainan: prospects and problems', *Journal of Guilin Institute of Tourism*, 12 (4): 5–7.

Wall, G. and Stone, M. (2005) 'Ecotourism and community development: the case of Jian Feng Ling National Forest Park, Hainan, China', *China Tourism Research*, 1 (1): 78–100.

Wan, Y., Du, G.F. and Tian, J.R. (2004) 'Visitors from South Korea: an emerging market to Zhangjiajie', *Zhangjiajie Daily*, 10 October.

Wang, H. (2003) *China's New Order: society, politics, and economy in transition*, ed. T. Huters, Cambridge: Harvard University Press.

Wang Jiajun (1997) 'Visitors' choice of destinations: research on tourist decision making behaviours', *Travel Geographic Tribune*.

Wang, J.M. (2003) 'Discussion on "Why are the World Heritage Sites in China so difficult to manage?"' *Chinese Business Herald*, 20 November.

Wang, M.X. (2004) 'A study on rural tourism's community involvement mechanism', unpublished dissertation, Zhejiang University.

Wang, N. (1997) 'Vernacular house as an attraction: illustration from hutong tourism in Beijing', *Tourism Management*, 18 (8): 573–80.

—— (1999) 'Rethinking authenticity in tourism experience', *Annals of Tourism Research*, 26 (2): 349–70.

Wang, Y. and Wall, G. (2009) Tourism Development in China in (eds.) W. Gartoure C. Hsu Tourism in the Contemporary World: A Handbook of Research, New York, Routledge.

Wang, Li and Wang, Ning (2000) 'Geothermal development in Xiong County, Hebei Province, China', in *Proceedings World Geothermal Congress 2000, Kyushu—Tohoku, Japan, May 28–June 10, 2000*.

Wang, Y.L. and Yang, X.J. (1999) 'Study on tourism sustainable development of national parks general planning—take Shuanglong national park as an example, in Jinhua, Zhejiang Province', *Resources Science*, 21 (1): 37–43.

Wang, Y. and Zhang, X. (1996) 'Evaluation on social benefits of Zhangjiajie National Forest Park in Hunan Province', *Forestry Economics*, 5: 44–54.

Wang, Z., Zhou, W. and Li, S. (2001) 'An analysis for market area of Chinese national park based on railway corridor', *Acta Geographica Sinica*, 56 (2): 206–13.

Wang, D.G., Lu, L. and Liu, C.X. (2003) 'A study on improving the present management system of scenic areas in China: taking Tianzhushan national park as an example', *Tourism Tribune*, 18 (3): 67–72.

Wang, T., Xu, Y. and Chen, M.B. (2003) 'Why are the World Heritage Sites in China so difficult to manage?' *China Business Herald*, 20 November.

Wang, Y., Li, G. and Bai, X.M. (2005) 'A residential survey on urban tourism impacts in Harbin', *China Tourism Research*, 1 (1): 116–28.

Wang, F.C., Huang, W.T. and Shang, J.K. (2006) 'Measuring pure managerial efficiency of inter national tourist hotels in Taiwan', *The Service Industries Journal*, 26 (1): 59–71.

Ward, L.M. and Russell, J.A. (1981) 'Cognitive set and the perception of place', *Environment and Behaviour*, 13 (5): 610–32.

Warner, M. (2003) *The Future of Chinese Management*, London: Frank Cass.

Weaver, D.B. (2000) 'A broad context model of destination development scenarios', *Tourism Management*, 21 (3): 217–24.

Weber, A. (1929) *The Theory of the Location of Industries*, Chicago: Chicago University Press.

Wei, Q.L. (1998) *Efficiency Evaluating Method: DEA—a new field of operational research*, Beijing: China Renmin University Press.

Weizenegger, S. (2006) 'The TALC model and protected natural areas: African examples', in R.W. Butler (ed.) *The Tourism Area Life Cycle, Vol. 2: conceptual and theoretical issues*, Clevedon: Channel View Publications.

Wicks, B. and Schuett, M. (1991) 'Examining the role of tourism promotion through the use of brochures', *Tourism Management*, 12 (4): 301–12.

Wilkinson, P. (1987) 'Tourism in small island nations: a fragile dependence', *Leisure Studies*, 26 (2): 127–46.

Williams, J. and Lawson, R. (2001) 'Community issues and resident opinions of tourism', *Annals of Tourism Research*, 28 (2): 269–90.

Willis, K.G. (1994) 'Paying for heritage: what price for Durham Cathedral?' *Journal of Environmental Planning and Management*, 37 (3): 267–78.

Winchester, S. (2007) 'China's ancient skyline', *The New York Times*, 15 July. Online. Available HTTP: <http://travel.nytimes.com/2007/07/15/travel/15wuling.html> (accessed 5 August 2007).

Winter, M. and Gasson, R. (1996) 'Pilgrimage and tourism: cathedral visiting in contemporary England', *International Journal of Heritage Studies*, 2 (3): 172–82.

Witt, C.A. and Mühlemann, A.P. (1994) 'The implementation of total quality management in tourism: some guidelines', *Tourism Management*, 15 (6): 416–24.

Wöber, K.W. (2007) 'Data envelopment analysis', *Journal of Travel and Tourism Marketing*, 21 (4): 91–108.

Wöber, K.W. and Fesenmaier, D.R. (2004) 'A multi-criteria approach to destination benchmarking: a case study of state tourism advertising programs in the United States', *Journal of Travel and Tourism Marketing*, 16 (2/3): 1–18.

Wood, D.A., and Erskine, J.A. (1976) 'Strategies in canonical correlation with application to behavioral data', *Educational and Psychological Measurement*, 36 (4): 861–78.

Worden, R.L., Savada, A.M. and Dolan, R.E. (1987) *China: A Country Study*, Washington: GPO for the Library of Congress. Online. Available HTTP: <http://countrystudies.us/china/112.htm> (accessed 25 June 2007).

World Heritage Centre (1990) Online. Available HTTP: <http://whc.unesco.org/archive/advisory_body_evaluation/547> (accessed 2 Feb 2005).

World Tourism Organization (2003) *Global Code of Ethics for Tourism*, Online. Available HTTP: <www.world-tourism.org> (accessed 12 December 2005).

—— (2006) *Tourism Industry Performance in 2005*, Online. Available HTTP: <www.world-tourism.org> (accessed 20 April 2006).

Wu Bihu (2003) 'Review on the master plan of tourism development for Shandong Province', *Human Geography*, 8.

Wu County Year Book (1999, 2000, 2001) Edited by Jiangsu Province Archives Bureau.

Wu, W. (2003) 'Implementing community participation strategy and foster traditional culture: an issue of traditional culture protection and development in folk area's tourism development', *Guilin Tourism College Post*, 14 (4).

Wu, Z. and Liu, M. (2003) 'Tourism impacts on local residents: an analysis of Wulingyuan Scenic Area, Zhangjiajie', in Z. Wu (ed.) *Assessment of Environmental Resources in Forest Tourism Areas*, Beijing: China Environmental Science Press.

Wutai Mountain International Cultural Festival of Buddhism Organising Committee (2004) *Bulletin of the Organising Committee of Wutai Mountain International Cultural Festival of Buddhism*, Online. Available HTTP: <http://www.wutai-shan.com/wtb/xwsd/content.asp?id=97> (accessed 22 August 2005).

Wutais, (2005) *Star-Rated Hotels*, Wutaishan Tourism Bureau. Online. Available HTTP: <http://www.wutais.com/xjbg.asp> (accessed 7 July 2005).

Wutaishan Tourism Bureau (2006) *General Introduction to Wutai Mountain*, Wutaishan Tourism Bureau. Online. Available HTTP: <http://www.wutais.cn/wutais.asp> (accessed 10 September 2006).

www.bjsch.net (2006) Accessed 13 October 2006.

Xia Z. (2004) 'A historical examination of the modern tourism development in Zhangjiajie', unpublished doctoral dissertation, Hunan Normal University.

Xiao, H.G. (2003) 'Leisure in China', in A. Lew, L. Yu, J. Ap and G. Zhang (eds) *Tourism in China*, Haworth Hospitality Press.

Xie, N. (2000) 'Protect natural and cultural heritage and revive the traditional civilization of conservation', *Chinese Landscape Architecture*, 2: 36–8.

Xie, P.F. (2001) 'Authenticating cultural tourism: folk villages in Hainan, China', unpublished thesis, Department of Planning, University of Waterloo.

Xie, P.F. and Lane, B. (2006) 'A life cycle model for aboriginal arts performance in tourism peripheries: perspectives on authenticity', *Journal of Sustainable Tourism*, 14 (6): 545–61.

Xin S. (2004) 'Inferring wetland hydrological conditions from remote sensing: a case study of Lake Baiying, China', unpublished thesis, International Institute for Geo-Information Science and Earth Observation, Enschede.

Xiong, X.X., Zhang, S. and Zhou, J. (2002) 'Problems and strategies of the canal towns' tourism in Jiangnan: research on tourism in Zhouzhuan, Tongli and Luzhi', *Urban Planning Forum*, 6: 61–3 (in Chinese).

Xiu Jun Pan (2005) 'The study of tourism and real estate development for OCT', *Corporate Research* (May): 39.

Xu, G.G. (2007) *Sinascape: contemporary Chinese cinema*. Lanham, MD: Rowman & Littlefield Publishing Company, Inc.

Xu, H. (2003) 'Managing side effects of cultural tourism development: the case of Zhouzhuang', *Systems Analysis Modelling Simulation*, 41 (2): 175–8.

Xu W. (2004) 'Reed land change and its relationship to water level change in Baiyang Lake', unpublished thesis, International Institute for Geo-Information Science and Earth Observation, Enschede.

Xu, F.F., Liu, P.L. and Bai, X.C. (2004) 'Applying "entropy technical based on AHP" to national parks planning evaluation', *Geographical Research*, 23 (3): 395–402.

Xuzhihui, Dingdengshan and Xiangdong L. (2006) 'Discussing the model of cultural tourism development and the keystone integration in Nanjing of China', *Human Geography* 20 (1): 37–45.

Yamamoto, D. and Gill, A. (1999) 'Emerging trends in Japanese package tourism', *Journal of Travel Research*, 38 (2): 134–43.

Yamamura, T. (2005) 'Dongba art in Lijiang, China: indigenous culture, local community and tourism', in C. Ryan and M. Aicken (eds) *Indigenous Tourism: the commodification and management of culture*, Oxford: Pergamon.

Yan Lili (2006) *The Principle and Model of the World Heritage Tourism Development in China. The Modernization of Business* Nanjing: Nanjing University.

Yan Liu (2004) 'The thinking for tourism and real estate development' *Tian Fu Review*, 12: 23–4.

Yang, G.H. (2001) 'A study on the positive environmental impacts of community residents involvement in tourism', *Inquiry into Economic Problems*, 11: 124–6.

Yang Hongying (2006) *Application of Story Matrix in Tour Guide Interpretation*, Singapore: International Forum on Tour Guide Education and Hotel Management.

Yang M. and Zhou, G. (2005) 'Study on effective measures of environmental protection for sustainable development of tourism industry in the world natural heritage Wulingyuan', *Ecological Economy of China*, 1 (2): 84–8.

Yang, Y.C. (2003) *Investigating the Problems of Self-Management in Chinese Villages*, Shanghai: Fu Dan University Press.

Ying, T. and Zhou, Y. (2007) 'Community, governments and external capitals in China's rural culture tourism: a comparative study of two adjacent villages', *Tourism Management*, 28: 96–107.

Ying, T. and Zhou, Y. (2007) 'Community, governments and external capital in China's rural cultural tourism: a comparative study of two adjacent villages', *Tourism Management*, 29 (1): 96–107.

Yu, D.S. and Gu, Y. (1988) 'The planning of national park Wutai Mountain', *Urban Planning*, 1: 42–3.

Yu, X. (2007) ' "Development of Buddhism" and "Buddhism in development": future of China's Buddhism seen from phenomenon of Shaolin Temple', *Henan Social Sciences*, 3: 7–12.

Zahra, A. and Ryan, C. (2005) 'National tourism organisations—politics, functions and form: a New Zealand case study', *Anatolia: An International Journal of Tourism and Hospitality Research*, 16 (1): 5–26.

Zaidman, N. (2003) 'Commercialisation of religious objects: a comparison between traditional and new age religions', *Social Compass*, 50 (3): 345–60.

Zeng, B. (2006) 'The role of tourism in eliminating poverty in China: a review of literature', *Tourism Tribune*, 21 (2): 89–94.

Zhang Guangrui (2006) 'China's outbound tourism: an overview', paper presented at the China: The Future of Travel, World Travel Market—China Contact Conference, London, 6 November.

Zhang, J. (2002) The Sightseeing Life: a monster in the cliff, Online. Available HTTP: <http://www.cctv.com/geography/news/20020906/15.html> (accessed 14 July 2007).

Zhang Lei (2007) 'Alliances instead of opponents: a new perspective towards TVEI environmental management in Chinese small towns', in C. Frescata (ed.) *Ecological Rural Development: green China—an information bridge China-Europe*, Portugal:

Zhang, W. (2003) 'Measuring stakeholder preparedness for tourism planning in Leshan, China', UMP-Asia Occasional Paper, No. 57. Online. Available HTTP: <http://www.serd.ait.ac.th/ump> (accessed 18 January 2008).

Zhang, X. (1999) *Tibetan Buddhism in Wutai Mountain*, Anistoriton. Online. Available HTTP: <http://www.anistor.co.hol.gr/english/enback/v993.htm> (accessed 2 July 2005).

Zhang, H.Q. and Chow, I. (2004) 'Application of importance—performance model in tour guides' performance: evidence from mainland Chinese outbound visitors in Hong Kong', *Tourism Management*, 25 (1): 81–91.

Zhang, Y. and Fang, K. (2004) 'Is history repeating itself? From urban renewal in the United States to inner city redevelopment in China', *Journal of Planning Education and Research*, 23 (3): 286–98.

Zhang, Z. and Ouyang, H. (2004) Tourism Circle of Southern Hunan, China, Online. Available HTTP: <http://www.hnphoenix.com/fhcweb/fhyj/fhyjnr.asp?id=337> (accessed 29 May 2006).

Zhang, P. and Wang, B. (2003) 'Inspiration for our country from overseas community involvement tourism development: a case on South Pembroke Shire of England', *Fujian Geography*, 18 (4): 38–41.

Zhang, C. and Yang, B.G. (1991) *Computation Geography*, Beijing: High Education Press.

Zhang, X.Q., Li, H. and Dong, X.W. (2003) 'Study on the theory of tourist resistance side', *Scientia Geographica Sinica*, 23 (2): 240–4.

Zhang, H.Q., Pine, R. and Lam, T. (2005) *Tourism and Hotel Development in China: from political to economic success*, Binghamton, NY: The Haworth Press.

Zhang, W., An, Y.Y. and Sun, H.L. (2007) 'The impacts of rural tourism on the social and economic development in rural areas—a case study of the suburbs of Beijing', *China Tourism Research*, 2 (4): 546–62.

—— (2008) 'Community involvement in rural tourism development—evidence from Pinggu, Yanqing, and Miyun districts, Beijing Province', in H. Gu and C. Ryan (eds) *Chinese Tourism Destination Management—issues and examples*, New York: Routledge.

—— (forthcoming) 'Community involvement in rural tourism development—evidence from Pinggu, Yanqing, and Miyun districts, the municipality of Beijing', in C. Ryan and H.M. Hu (eds) *Chinese Destination Planning*, London: Routledge.

Zhangjiajie National Forest Park Administration (2006) *Tourism Development in Zhangjiajie*, Zhangjiajie: Zhangjiajie National Forest Park Administration.

Zhao, C.G. (1982) 'The tourism planning assumption on national park of "three mountains and one town" in Dazu, Sichuan Province', *Urban Planning*, 3: 47–50.

Zhao, Y.J. (2001) 'Consideration on the governmental system of national parks', *Planner*, 17 (1): 91–5.

Zhe, X. (2000) *Community Practice: Supper Village's development*, Hangzhou: Zhejiang People Publication.

Zheng Benfa (1999) 'The drawback of tourism industry and some solutions' *Gansu Social Science*, 5: 48–50.

Zheng, L. (1999) *Tourism Development of Zhangjiajie*, Beijing: China Tourism Press.

Zhong, G. (2004) 'Authenticity rules of tourism experience and tourism attraction management', *Guilin Tourism College Post*, 15 (4).

Zhou Changchun (2003) 'Classification and evaluation of tourism resources in Qufu, Shandong Province', *Fujian Geography*, 18 (2): 141–147.

Zhou, X. (2002) 'Pay attention to the core essence of "tourism alleviating poverty"', *Tourism Studies*, 17 (1): 17–21.

Zhou Xiao (2003) 'Anthropologic visual angle: research on the essence and impacts on social culture of tourism', *Journal of Hubei University*, 5: 114–16.

Zhou, N. and Yu, K. (2004) 'The urbanization of national park and its countermeasures. Urban Planning Forum, 1, 57–61.

Zhou, Y.G. and Ma, E. (2008) 'Maintaining the authenticity of rural tourism experiences through community participation—the case of two Baiyang Lake Island villages', in H. Gu and C. Ryan (eds) *Chinese Tourism Destination Management—issues and examples*, Oxford: Elsevier.

—— (2008) 'Maintaining the authenticity of rural tourism experiences through community participation—the case of two Baiyang Lake Island villages', in H. Gu and C. Ryan (eds) *Chinese Destination Management—issues and case studies*, Oxford: Pergamon.

Zhu Xiao Di (2005) 'The perception of social and physical change, especially residential environment, relating to urban renewal—a case in Suzhou, China', paper presented at the Conference Doing, thinking, feeling home: the mental geography of residential environments, 14 and 15 October. http://citg.tudelff.nl/live/binaries/2e2a5607_3f77-4d71-61d1-33aa897e794aa/doc/conference_paper_di.pdf accessed 2nd February 2008.

中国国家旅游局 (CNTA). (1997). 走遍中国—中国优秀导游词精选 (综合篇). 北京: 中国旅游出版社.

中国国家旅游局 (CNTA). (1999年5月14日). 导游人员管理暂行规定. 2007年3月18日摘自 http://www.cnta.com/news_detail/newsshow.asp?id=A2006627153455604539.

中国国家旅游局 (CNTA). (2003年). 全国导游人员、旅行社经理人员人力资源状况调查报告. 2007年3月10日摘自 http://www.cnta.com/news_detail/newsshow.asp?id=A2006622106453640548.

付岗. (2002). 人文景观导游讲解与旅游文化审美整合. 燕山大学学报 (哲学社会科学版), 3(2): 77–80.

吴必虎, 等. (1999). 旅游解说系统的规划和管理. 旅游学刊, 1: 44–46.

杨红英. (2005). 英语导游解说教程. 西安: 陕西人民出版社.

焦国标. (2000). 导游也该换脑了. 时代潮, 6: 46–47.

王绪昂. (2000). 自然解说在赏鲸事业的必要性与解说员在赏鲸过程中应扮演的角色. 第四届台湾海洋环境大会暨第八届鲸类生态与保育研讨会.

王连义. (2002). 导游技巧与艺术. 北京: 旅游教育出版社.

邵琪伟. (2006年10月27日). 邵琪伟在全国导游大会上的讲话. 2007年3月18日摘自 http://www.cnta.com/news_detail/newsshow.asp?id=A2006122611283541315.

邵琪伟. (2007年1月18日). 邵琪伟在2007年全国旅游工作会议上的讲话. 2007年3月18日摘自 http://www.cnta.com/news_detail/newsshow.asp?id=A2007124901553241563.

黄婕, 林云. (2003). 浅谈导游实践讲解中的艺术. 江西广播大学学报, 4: 73–74.

Contributors

Note on Sequencing of Names

It is generally recognized that in Chinese culture the family name precedes the individual name, but often when writing in English, Chinese scholars will adopt the western rubric and reverse the sequence of their names. For the purposes of this book, names are listed in the manner preferred by the author, with the last name in the sequence being used to decide the alphabetical sequence.

Ma Aiping is Professor of Tourism Marketing at Beijing International Studies University, and is a member of the China International Management Research Academy. Among her publications is *Tourist Product Marketing Strategy and Tactics in 21st Century*, published in June 2007 by China Water Publishing Company. Her research interests include the dissemination of information to visitors, the role of tourism information centres and the impacts of tourism.

Wolfgang Georg Arlt has an MA (Sinology) and PhD (Political Sciences) from FU Berlin, Germany, and has undertaken further studies in Taiwan and Hong Kong. He was formerly the owner of an outbound and inbound tour operator with offices in Germany and China. He is currently Professor and Study Program Director for International Tourism Management at West Coast University of Applied Sciences in Heide, Germany. He holds the position of Visiting Professor at several Chinese universities and at the University of Sunderland in the United Kingdom, and is Research Fellow of JSPS Japanese Society for the Promotion of Science. A part-time lecturer at Sun Yat-sen University Guangzhou, China, Professor Arlt is also the Director of the China Outbound Tourism Research Institute (COTRI).

Ji-gang Bao attained his BA and PhD from Sun Yat-sen University and master's degree from Peking University, P. R. China. His research interests include tourism geography, theme parks, tourism planning, urban tourism, tourism impacts and community tourism. He is the Professor and Dean of the School of Tourism and the Dean of the School of Geography and Planning, Sun Yat-sen University. He is the Chair of the Commission on the Geography of Tourism, the Geographical Society of China. Dr Bao is the author of the text-book, *Geography of Tourism* (in Mandarin) published by High Education Press, Beijing, in 1999, which has sold over 100,000 copies. He has undertaken many tourism planning assignments throughout China. These projects include tourism planning for the City of Guilin, the City of Suzhou, the City of Huangshan and the Province of Hubei.

Xiang Baohui earned her BS in Forest Recreation from the Central South Forestry College (now renamed as Central South University of Forestry and Technology), China (1998), and an MS in Ecotourism from Southern West Forestry College, China (2004). She has been an assistant lecturer and lecturer with the Southern West Forestry College from 1998 to 2005. Since 2005, she has held the post of lecturer at China Woman's University. Her current research at China Woman's University focuses on tourism management, ecotourism planning and hospitality management.

Wang Bihan teaches Tourism Economics at the Tourism College, Tianjin University of Finance and Economics. Born in Tianjin, China, she received both her undergraduate and master's degrees in marketing from the Marketing Department of Nan Kai University, Tianjin, China.

Chennan (Nancy) Fan completed her undergraduate studies at Nankai University and then completed postgraduate studies under the Ecoplan China scholarships at the University of Waterloo, Canada, under the supervision of Professor Geoff Wall. Her research interests are related to the Chinese water towns of Luzhi, Tongli and Mudu in the municipality of Suzhou. She is currently Liaison Officer at Columbia International College, Hamilton, Ontario, where she has primary responsibility for international students.

Li Hong is currently Professor and Head of the Department of Tourism Management at Hebei University of Economics and Business. She is also Manager of the Tourism Research Institute at that university. By the end of 2007, she had published more than 30 academic articles at the level of both provincial and state publications in Mandarin, and has also undertaken more than 20 scientific studies, publically published 11 academic works and teaching materials and has twice won prizes for outstanding achievement in literary society branch research above the provincial level. Li has also undertaken a number of consultancy projects for hotels of different star classifications.

Zhang Hongfei is currently undertaking a master's degree in Tourism Management at Beijing International Studies University and is undertaking research with Professor Ma Aiping.

Xu Honggang gained her master's and doctoral degrees from the Asian Institute of Technology, having obtained her first degree from Beijing University. Her major research interests lie in sustainable development, system dynamics, urban tourism and resource management. She currently holds the post of Professor at the Research Centre of Tourism Planning at Sun Yat-Sen University, Guangzhou, PRC, while also holding visiting Professorships at the Universities of Rikkyo (Japan) and Angers (France).

Yang Hongying is Associate Professor at the School of Tourism, Xi'an International Studies University, China. Her first degree was in English Language and Literature and she subsequently gained her master's degree in Tourism Studies at James Cook University, Australia. She has over 20 years professional experience in training foreign language speaking tour guides in China and in offering training and lectures to China National Tourism Administration (CNTA) on tour guide training. Her areas of expertise and research interest include tour guide interpretation, international tourism and culture, visitor satisfaction and tourist attractions in Xi'an. Among her publications in the last two years are *Cross-cultural Interaction in Tour Guiding Practice* (2007) and *International Tourism and Culture* (2005), both published in Mandarin.

Rong Huang gained her doctoral degree from the University of Derby, UK, and is a Lecturer in Tourism Marketing and an International Student Tutor at the University of Plymouth (UK). Her research interests focus on aspects of the tourism phenomenon, including student travel, conference and incentive tourism and tea tourism. Among her publications are items on tea tourism in *China Tourism Research*, while she has also published on issues relating to tourism education in *The Journal of Hospitality, Leisure, Sport and Tourism Education*. Among her publications in Mandarin, she has considered cross-cultural issues, notably in the *Journal of Changsha University of Science and Technology*. Rong is also a member of the Small Business and Services Research Unit (SBSRU) at the University of Plymouth.

Chen Hui is a lecturer at the School of Tourism, Xi'an International Studies University, China. She holds her first degree, a BA from Xi'an International Studies University and is currently undertaking a postgraduate programme in Tourism Management at that university. Hui has 13 years professional experience in teaching tourism English and seven years in training English-speaking tour guides in Xi'an. She is an Assessor and Judge for the oral test of the National Tour Guide Certificate Examination in Shaanxi Province. Her main areas of research interest include international tourism and culture and tour guide interpretation. Among her recent publications in Mandarin are *On Training Interpretive Tour Guides in China* (2005) and *International Tourism and Culture* (2005).

Gu Huimin obtained her doctoral degree from Renmin University of China, Beijing, in economics. She is Deputy Dean of the School of Tourism Management at Beijing International Studies University and Vice President of the China Hotel Institute. She has held Visiting Scholar positions at the Conrad Hilton College, University of Houston and the School of Hospitality and Tourism Management at Hong Kong Polytechnic University. She has won several awards for both teaching and for services to the Chinese hospitality industry, and been involved in developing regional tourism plans in China. She has published widely in both Mandarin and English, has completed books on Crisis Management and the Chinese Hotel industry and published in English language journals including *Tourism Management*, *Cornell Quarterly* and *The International Journal of Hospitality Management*. Her current research projects relate to best practice in the hotel industry and tolerances of pollution by tourists.

Deng Jingyang was born and raised in a rural village in central China. He earned a BS in Forestry from the Central South Forestry College (now renamed as Central South University of Forestry and Technology), China (1987), and an MS in Forest Recreation from the same university (1990), and a PhD in Recreation and Leisure Studies from the University of Alberta, Canada (2004). He is currently Assistant Professor at West Virginia University. From 1997 to 1998, he was Visiting Scholar in the Department of Hospitality, Tourism, and Marketing, Victoria University, Melbourne, Australia.

Fung Mei Sarah Li has lectured in Tourism at the Universities of Hong Kong Polytechnic, James Cook, Queensland and Tasmania. She has been undertaking research in China for more than a decade and has published research in both English and Mandarin, the former including articles in *Annals of Tourism Research*. Her research interests lie in culture, heritage, natural areas and the Chinese perspectives of these issues.

Si Lina is from Shihezi City, Xinjiang Uygur Autonomous Region, China, and received her Master of Tourism Marketing degree from Beijing International Studies University in June 2008, having previously gained a Bachelor of Tourism Management from Wuhan Polytechnic University. She has worked as a receptionist in Wuhan Changjiang Hotel (4-star) and on the Three Gorges cruise on board the *Princess Jenny*. Her other industry experiences include marketing the Chengteh Mountain Resort in Hebei Province while also being responsible for network marketing and designing activities with car clubs in Beijing. Lina has had many articles and papers published in *China Tourism News* and has presented at The Symposium on the Management and Innovation of Heritage Tourism. Most of her research focuses on tourist behavior and perceived destination image.

Zhong Linsheng earned his BS in Forestry from Jiangxi Agriculture University, China (1994), and an MS in Forest Recreation from the Central South Forestry College (now renamed as Central South University of Forestry and Technology), China (1997), and a PhD in Ecotourism at the Institute of Applied Ecology, Chinese Academy of Sciences (CAS) (2000). Dr Zhong joined the Institute of Geographical Sciences and Natural Resources Research in 2000 as a Post-Doctoral Research Fellow. Since 2002, he has held the post of Associate Professor of Institute of Geographical Sciences and Natural Resources Research, CAS. He has been Visiting Scholar at the International Centre for Ecotourism Research at Griffith University. His research field included ecotourism, recreation ecology, tourism planning and protected area management. His English language publications include work published in *Tourism Management* and *Annals of Tourism Research*.

Danging Liu is an Associate Professor of Tourism and Hospitality Management at Tianjin University of Commerce, P.R. China. She gained her master's degree in Education from Hunan Normal University, P.R. China, and gained a second master's degree in Tourism and Hospitality Management from Florida International University, US. She has taught and researched in the field of tourism and hospitality management for over 10 years and her publications focus on cultural issues in the tourism and hospitality industry, and tourism and hospitality education.

Jumei Liu completed her master's degree in Tourism Management at Beijing International Studies University, Beijing, China, in 2007. She now works as an instructor in the School of Tourism Management of North College of Beijing University of Chemical Technology. Her main research interests include community involvement in rural areas and other tourism destinations in China, and the social and cultural impacts of tourism on the famous places of interest in China, especially on the world heritage destinations. Her current research focus is on consumer behaviour in rural areas, and the motivations of tourists visiting rural China world heritage sites.

Emily Ma is a PhD student at the School of Hotel and Restaurant Administration, Oklahoma State University. She received her bachelor's degree in Tourism Management from Zhejiang University and master's degree from the School of Hotel and Tourism Management, the Hong Kong Polytechnic University. Her research interests are tourism and hospitality education, rural community tourism and convention service quality. Studying in different places of the world has, she feels, provided her with both a broader view of and a wish to know tourism and hospitality education systems. After completion of her doctoral degree she intends to continue her career as an educator and researcher in tourism and hospitality.

Xiao-Long Ma gained his doctoral degree from Sun Yat-Sen University in Guangdong under the supervision of Professor Jigang Bao. During his studying for his doctoral degree, he gained '985 project' scholarship from Sun Yat-Sen University for Study Abroad and completed further research while a Visiting Scholar at the Department of Tourism and Hospitality Management, University of Waikato Management School. His research interests lie in rural and natural areas as tourism resources with specific reference to China's National Parks and reserves.

Fang Meng, PhD, is Assistant Professor in the School of Human and Consumer Sciences, College of Health and Human Services, Ohio University, US. She received her BA and MA degrees from Beijing International Studies University in China, and her PhD from Virginia Polytechnic Institute and State University, US. Her research interests include destination marketing, consumer behaviour and international tourism. Dr Meng has published articles in *Journal of Travel Research*, *Tourism Management*, *Tourism Analysis*, *Journal of Sustainable Tourism* and *Journal of Vacation Marketing*. Dr Meng is a member of the Travel and Tourism Research Association (TTRA), the International Council of Hotel, Restaurant and Institutional Education (I-CHRIE) and the Asia Pacific Tourism Association (APTA). She is a member of the editorial board of *Journal of Vacation Marketing* and serves as a reviewer for several academic journals.

Clare Mitchell completed her undergraduate studies at the University of Guelph before completing her master's and doctoral degrees at the University of Waterloo where she has now lectured in the Faculty of Environmental Studies since 1988. Her research interests lie in population change and economic consequences, including retail locations and small towns with special reference to heritage centres. With reference to China she has published work on Zhu Jia Jiao and Luzhi in *Annals of Tourism Research* and *Tourism Management*.

Zhang Ning obtained a master's degree from the School of Tourism Management, Beijing International Studies University. Her research areas are in tourism management and tourism impacts. She has carried out three research projects under Professor Zhang Wen, including a study of Tourism Human Resource Development in Beijing, World Heritage Sustainable Development in China, and observation of tourism impacts in China.

Chris Ryan obtained his doctoral degree from the Aston University Management School, UK, and is editor of *Tourism Management* and Professor of Tourism at the University of Waikato Management School. An elected Fellow of the International Academy for the Study of Tourism, he is also Honorary Professor of the University of Wales at the University of Wales Institute Cardiff, and Visiting Professor at Beijing International Studies University, China, and the Emirates Academy, Dubai. His work has been published in all the leading tourism journals, while he has advised companies and organizations that include APEC Tourism Ministers, UNWTO and small adventure tourism operators in New Zealand.

Fangfang Shi gained a BA in English from Xi'an University of Architecture and Technology (China). She developed an interest in tourism studies during her own travel experience and her part-time work as a voluntary tour guide. Because of this, she went to the University of Surrey (UK) and completed a MSc in Tourism Management in 2002, followed by a PhD at Nottingham Trent University (UK) in 2008. She is now a Lecturer in Tourism and International Business at Nottingham Trent University. Her research interests include religious tourism, visitor attraction management, tourism in China, tourist behaviours and impacts of tourism development.

Trevor H. B. Sofield has held Professorial positions in Tourism at the Universities of Tasmania and Queensland and is past co-ordinator of the Australian National Research Centre for Sustainable Tourism for Western Australia. He was head of a task force for the World Tourism Organization on Sustainable Tourism as a tool for Eliminating Poverty (STEP), Madrid, Spain; and an expert for the Australian Minister for the Environment's National Task Force on Heritage Tourism. He has been Team Leader for the Mekong Tourism Development Program, Cambodia and Vietnam and has had a long interest in China, its culture, resources and tourism. His past career includes work in Australia's diplomatic corps and he is an Honorary Lifetime Member of EcoClub.

Geoff Wall is Professor at the University of Waterloo's Faculty of Environmental Studies and obtained his doctoral degree from Hull University in the UK. He has acted as a consultant for such agencies as the Asia Development Bank, Canadian International Development Agency, the International Joint Commission, Office of Technology Assessment (US), Environment Canada, the Federal Department of Communications, the Department of Canadian Heritage, the Ontario Ministries of Culture and Recreation, Citizenship and Culture, Treasury and Economics, and Natural Resources. He has had a long association with China and is Honorary Professor of Nanjing University and Dalian University of Technology and in 2000 was given a Friendship Award by the Province of Hainan, China. He is a founding member and past President of the International Academy for the Study of Tourism.

Zhang Wei hopes to obtain her doctoral degree from the University of Waikato, New Zealand, in 2009 and lectures in tourism and hospitality management at Beijing City University. Her publications include book chapters in English, and she has had research published in Mandarin in *Tourism Tribune*, *Tourism Science* and *China Tourist Hotels*. Her doctoral studies relate to outbound Chinese tourism.

Zhao Xin lectures in Tourism and Business Management at Hebei University of Economics and Business. He also undertakes research under the auspices of the Tourism Research Institute at that university and has worked on various projects with Professor L. Hong.

An Yanyan gained her Master of Management degree from Beijing International Studies University. Yanyan's research interests lie in the relationships between rural tourism development and local residents' economic, social and cultural environments and the impacts of tourism development. Among her publications are articles on community-based tourism published (in Mandarin) in the *Journal of Beijing International Studies University*, and a study of rural tourism in the suburbs of Beijing in both English and Mandarin in *China Tourism Research* with Wen Zhang and Hongli Sun. She currently holds a position at Beijing Wuzi University.

Wen Zhang is now Dean of the College of Translation at Beijing International Studies University having previously been Dean of the School of Tourism Management. She has held the position of President of Asia Pacific CHRIE and has published widely in English and Mandarin including in *Tourism Management* and *Tourism Tribune*. Her past research has included studies of community tourism and the impacts of tourism.

Liang Zhi has been Chief Professor and Deputy Dean of the Tourism College, Tianjin University of Finance and Economics since 2003. Liang Zhi worked at Tianjin International Travel Service between 1974 and 1987. He studied at the graduate school of Johnson and Wales University, Providence, Rhode Island, US, between 1988 and 1990 and achieved a degree of Master of Science of Hospitality Administration in 1990. He worked at Royal Princess Hotel at Ocean City, Maryland, and Practicum Properties of Johnson and Wales University between 1990 and 1992. He was Associate Professor and Dean of Academic Studies of China Tourism Management Institute, Tianjin between 1992 and 2003. In 2002, he obtained a doctoral degree in economics at Nankai University, Tianjin.

Yong-guang Zhou is an Associate Professor of the College of Tourism, Zhejiang University. He received his bachelor degree on city planning from Hangzhou University in China and his master's and doctoral degrees from Fukui University in Japan. He has a broad research interest in tourism planning, rural tourism, ecotourism and heritage tourism. Dr Zhou has published in leading refereed journals in China and Japan. Besides his academic achievements, Dr. Zhou has also participated in more than 30 tourism planning projects both in China and Japan and is widely consulted by the tourism industry in those countries.

Index

animal species, 15
annual average revenue, 60
annual budget of tourism development
    and marketing, 31
annual family income, 46
annual personal income, 58
A-rating tourist sites, 27
archaeological exploration, 191
architectural heritage, 299
architectural heritage of the hutong,
    325
architecture, 169
artistic activities, 185
artistic and religious performances, 190
art of communication of place, history,
    heritage and culture, 227
arts, 144
Asian nations, 5
assertive action, 21
assess community interests, 306
assessment of resource efficiency, 72
asset formation in the Parks, 85
associated environmental, 60
A to AAAAA, 28
attachment to tourist attractions, 34
attitude toward the general public, 17
attractions, 11, 15, 18, 26, 34, 35, 36,
    42, 45, 68, 88, 98, 111, 170,
    184, 199, 201, 229, 230, 231,
    296, 303
Attractiveness of tourist resources, 26
authenticity, 296, 304
authenticity experience, 304
authenticity of the area, 13
Availability, quality and uniqueness of
    tourism facilities and services,
    30
average per capita urban disposable, 2

B
Ba Da Ling Great Wall, 69
Bai, 4, 44
Baiyang Lake, 2, 8, 151, 258, 294, 298,
    299, 301, 302, 305
Baiyang Lake Hot Spring Town, 299
Baiyang Lake Island Villages Tourism
    Development, 298
Baiyang Lake island village tourism
    development, 305
Baiyang Lakes, 185
Baiyang Lake's island tourism develop-
    ment, 306
Baiyang Lake's island villages, 297
banking practices, 2

banned practices, 71
Baosheng Temple, 106, 112
basic principles of interpretation, 233
beach front and high-rise hotels, 19
beach locations, 19
bed and breakfast accommodation, 3
behave in ethical ways, 14
behaviour of tourists, 218
Beidaihe Resort, 32
Beiddaihe, 17
Beijing, 2, 17, 253
Beijing airport, 4
Beijing Badaling Great Wall, 32
Beijing economy, 325
Beijing hutong, 8
Beijing Olympics, 1
Beijing suburban tourism community,
    281
Beijing Tourism Administration, 248
beneficial for host communities, 14
benefits to the community, 13
Best Tourist City, 30
bidirectional causality relationships, 7
biodiversity has been lost in their devel-
    opment, 20
birthplace of Confucius, 194
branding and image, 31
brochures, guides and other tourism
    services, 224
Buddha's nose, 15
Buddhism, 4
Buddhism and the Cultural Revolution,
    202
Buddhist and Taoist faiths, 4
Buddhist faith, 156
Buddhist sacred site, 202
Buddhist stone pillar, 106
building and operating attractions, 306
building boom, 253
'Building China's Outstanding Tourist
    City', 27
business and official business, 5
business and residential properties, 93
business ethics, 29

C
cable cars, 163
cable chair, 59
cable trams, 62
cableway, 51
cableway system, 61
calligraphy, 158, 166
calligraphy and stele tour, 190
camping sites, 34

Chinese state, 4, 143
Chinese State Bodies, 16
Chinese tour guides, 226
Chinese tourism, 1, 5, 8, 16, 19, 36, 150, 151, 185, 276
Chinese tourism industry, 147, 149
Chinese tourism resources, 229
Chinese tourists travelled overseas, 5
Chinese visitors, 5, 158, 162, 164, 165
Chi Qiao in Shanxi Province, 249
Chongdugou Village in Henan Province, 255
cities are major economic contributors, 27
City's ecological environment, 28
classifications and benchmarking used in tourism planning, 35
classified at national and provincial levels, 31
classifying tourism resources, 187
Clause 2 in the Provisional Act of Scenic Area Management, 64
cleanliness and sanitation, 25
Cleanliness and sanitation, 25
cleanliness of the overall city environment, 28
cleanliness (public areas, restrooms, waste management, food sanitation), 34
clinic service, 34
coastal/island destinations, 39
Codes for Scenic Area Planning (GB50 298 ~ 1999), 19
Collaboration and cooperation among governments, tourism business, and local community, 31
collaborative approaches, 14
collective farming, 2
Collective operations, 246
collective village entrepreneurship, 2
combined functions of sightseeing and vacation, 34
commerce, 203
commercial, 304
commercial development, 64
commercial, education, health care and cultural institutions, 92
commercialisation, 209, 210
commercialisation at sacred sites, 200
commercialization of the streets, 106
commercialized tourism zone, 248
Commercial realities and time frames involved, 11
commissioned reports, 14

communication barriers, 228
communication with local community residents, 34
communicative interpretation, 232, 234
Communicative interpretation, 236
community, 13
community and heritage based tourism, 150
community approach in tourism development, 282
community in tourism, 8
community involvement, 13, 269, 281
community involvement in tourism planning, 21
Community management system, 34
community participation, 303
Community Participation and Perspectives, 8
community responsibility, 274
community's economy, culture and environment, 282
community support and tourism development, 270
community sustainable development, 281
community tourism, 244, 252, 268
complaint processing and recording, 29
complete regulations and policies, 34
comprehensive and detailed research, 98
comprehensive facilities, 25
concepts of cultural tourism, 184
Conference on Tourism Planning and Management in Developing Conference, 22
Confucian culture, 187, 192
Confucian family cemetery, 187
Confucian family traditions, 192
Confucian mansion, 187, 189
Confucian Temple gourmet tours, 190
Confucian traditions, 193
Confucian traditions of respect, 149
Confucius Cultural Festival, 189, 193
Confucius' Dream, 190, 192
Confucius family cemetery, 183
Confucius Home study tour, 190
Confucius Institute, 187, 188
Confucius International Tourism Co. Ltd., 192
Confucius Research Institute of China, 183
Confucius Six Arts city, 194
Confucius Six Arts City, 187
Confucius Study Tour Festival, 192

An environmentally friendly book printed and bound in England by www.printondemand-worldwide.com

PEFC Certified

This product is
from sustainably
managed forests
and controlled
sources

www.pefc.org

PEFC/16-33-415

This book is made entirely of sustainable materials; FSC paper for the cover and PEFC paper for the text pages.

#0057 - 130613 - C0 - 234/156/22 [24] - CB